**PART OF THE MIT-PAPPALARDO SERIES IN MECHANICAL ENGINEERING**

Series Editor: Mary C. Boyce
*Gail E. Kendall Professor of Mechanical Engineering*
*Head of Department, Massachusetts Institute of Technology*

SECOND
EDITION

# ENERGY AND THE ENVIRONMENT

## SCIENTIFIC AND TECHNOLOGICAL PRINCIPLES

James A. Fay
Department of Mechanical Engineering
Massachusetts Institute of Technology

Dan S. Golomb
Department of Environmental, Earth, and Atmospheric Sciences
University of Massachusetts Lowell

New York     Oxford
OXFORD UNIVERSITY PRESS

Oxford University Press, Inc., publishes works that further Oxford University's objective of excellence in research, scholarship, and education.

Oxford   New York
Auckland   Cape Town   Dar es Salaam   Hong Kong   Karachi
Kuala Lumpur   Madrid   Melbourne   Mexico City   Nairobi
New Delhi   Shanghai   Taipei   Toronto

with offices in
Argentina   Austria   Brazil   Chile   Czech Republic   France   Greece
Guatemala   Hungary   Italy   Japan   Poland   Portugal   Singapore
South Korea   Switzerland   Thailand   Turkey   Ukraine   Vietnam

For titles covered by Section 112 of the US Higher Education Opportunity Act, please visit www.oup.com/us/he for the latest information about pricing and alternate formats.

Published by Oxford University Press, Inc.
198 Madison Avenue, New York, New York 10016
http://www.oup.com

**Library of Congress Cataloging-in-Publication Data**

Fay, James A.
Energy and the environment: scientific and technological principles / James A. Fay,
    Dan S. Golomb. 2nd ed.
        p. cm.
ISBN 978-0-19-976513-3
1. Power resources—Environmental aspects.   I. Golomb, D.   II. Title.
TD195.E49F39 2002
333.79'14—dc21                                    2010044579

# CONTENTS

*Tables*   xiii
*Preface*   xv

1   **ENERGY AND THE ENVIRONMENT**   1

1.1   Introduction   1
    1.1.1   An Overview of This Text   2
1.2   Energy   7
    1.2.1   Electric Power   9
    1.2.2   Transportation Energy   10
    1.2.3   Energy as a Commodity   10
1.3   The Environment   11
    1.3.1   Managing Industrial Pollution   13
Bibliography   13

2   **GLOBAL ENERGY USE AND SUPPLY**   14

2.1   Introduction   14
2.2   Global Energy Consumption   15
2.3   Global Carbon Emissions   17
2.4   Global Energy Sources   18
2.5   Global Electricity Consumption   20
2.6   End-Use Energy Consumption in the United States   21
    2.6.1   Industrial Sector   22
    2.6.2   Residential Sector   24

2.6.3    Commercial Sector    24

2.6.4    Transportation Sector    25

2.7    Global Energy Supply    26

2.7.1    Coal Reserves    26

2.7.2    Petroleum Reserves    27

2.7.3    Unconventional Petroleum Resources    29

2.7.4    Natural Gas Reserves    29

2.7.5    Unconventional Gas Resources    30

2.7.6    Summary of Fossil Reserves    31

2.8    Conclusion    31

Problems    32

Bibliography    33

3    THERMODYNAMIC PRINCIPLES OF ENERGY CONVERSION    34

3.1    Introduction    34

3.2    The Forms of Energy    35

3.2.1    The Mechanical Energy of Macroscopic Bodies    35

3.2.2    The Energy of Atoms and Molecules    36

3.2.3    Chemical and Nuclear Energy    36

3.2.4    Electric and Magnetic Energy    37

3.2.5    Total Energy    37

3.3    Work and Heat Interactions    37

3.3.1    Work Interaction    38

3.3.2    Heat Interaction    39

3.4    The First Law of Thermodynamics    39

3.5    The Second Law of Thermodynamics    40

3.6    Thermodynamic Properties    41

3.7    Steady Flow    43

3.8    Heat Transfer and Heat Exchange    43

3.9    Ideal Heat Engine Cycles    44

3.9.1    The Carnot Cycle    45

3.9.2    The Rankine Cycle    47

3.9.3    The Otto Cycle    50

3.9.4    The Brayton Cycle    52

3.9.5    Combined Brayton and Rankine Cycles    54

3.10    The Vapor Compression Cycle: Refrigeration and Heat Pumps    55

3.11    Energy Processing: First and Second Law Constraints    57

3.11.1   Fuel Heating Value    58

3.11.2   Free Energy Change    59

3.11.3   Separating Gases    60

3.12   Fuel (Thermal) Efficiency    60

3.13   Conclusion    61

Problems    62

Bibliography    63

4    THERMODYNAMICS OF FOSSIL, BIOMASS, AND
     SYNTHETIC FUELS    64

4.1   Introduction    64

4.2   Fossil Fuels    64

4.3   Combustion of Fossil Fuel    67

4.3.1   Fuel Heating Value    68

4.4   Biomass Fuels    69

4.5   Synthetic Fuels    70

4.5.1   Examples of Fossil Fuel Synthesis    71

4.5.1.1   Coal to Gas    72

4.5.2   Examples of Biochemical Synthesis    73

4.6   Biochemical Production of Ethanol from Biomass    74

4.7   Electrochemical Reactions    74

4.7.1   Fuel Cells    75

4.7.2   Practical Fuel Cell Systems    79

4.8   The Hydrogen Economy    80

4.8.1   Hydrogen Fuel for Vehicle Propulsion    81

4.8.2   Synthetic Hydrogen from Fossil Fuels with
        Carbon Capture and Storage    81

4.8.3   Hydrogen as Energy Storage for Intermittent
        Electric Power Plants    82

4.8.4   Hydrogen as a Substitute for Pipeline Natural Gas    83

4.9   Conclusion    83

Problems    83

Bibliography    84

5    ELECTRICAL ENERGY GENERATION, TRANSMISSION, AND
     STORAGE    85

5.1   Introduction    85

5.2   Electromechanical Power Transformation    88

5.3   Electric Power Transmission    92

5.3.1   AC/DC Conversion    93

5.4    Energy Storage    94
    5.4.1    Electrostatic Energy Storage    94
    5.4.2    Magnetic Energy Storage    95
    5.4.3    Electrochemical Energy Storage    96
        5.4.3.1    Lead-Acid Storage Battery    97
        5.4.3.2    Lithium-Ion Storage Battery    98
        5.4.3.3    Other Storage Batteries    99
    5.4.4    Mechanical Energy Storage    99
        5.4.4.1    Pumped Hydropower    99
        5.4.4.2    Flywheel Energy Storage    100
    5.4.5    Properties of Energy Storage Systems    101
5.5    Conclusion    102
Problems    102
Bibliography    103

6    FOSSIL-FUELED POWER PLANTS    104
6.1    Introduction    104
6.2    Fossil-Fueled Power Plant Components    105
    6.2.1    Fuel Storage and Preparation    106
    6.2.2    Burner    106
    6.2.3    Boiler    108
    6.2.4    Steam Turbine    109
        6.2.4.1    Impulse Turbine    110
        6.2.4.2    Reaction Turbine    110
    6.2.5    Gas Turbine    111
    6.2.6    Condenser    112
    6.2.7    Cooling Tower    113
        6.2.7.1    Wet Cooling Tower    113
        6.2.7.2    Dry Cooling Tower    114
    6.2.8    Generator    114
    6.2.9    Combustion Stoichiometry    115
    6.2.10    Emission Control    116
        6.2.10.1 Control of Products of Incomplete Combustion
             and Carbon Monoxide    117
        6.2.10.2 Control of Particles    117
        6.2.10.3 Sulfur Control    121
        6.2.10.4 Nitrogen Oxide Control    125
        6.2.10.5 Mercury Control    128
        6.2.10.6 Toxic Metals    129

6.2.11   Waste Disposal    129
6.3   Advanced Cycles    129
     6.3.1   Combined Cycle    129
     6.3.2   Coal Gasification Combined Cycle    130
     6.3.3   Cogeneration    132
     6.3.4   Fuel Cell    132
6.4   Conclusion    133
Problems    134
Bibliography    136


7   **NUCLEAR-FUELED POWER PLANTS**    137
7.1   Introduction    137
7.2   Nuclear Energy    138
     7.2.1   Nuclear Energy from Fission    138
7.3   Radioactivity    141
     7.3.1   Decay Rates and Half-Lives    142
     7.3.2   Units and Dosage    143
             7.3.2.1   Health Effects of Radiation    144
             7.3.2.2   Radiation Protection Standards    144
7.4   Nuclear Reactors    145
     7.4.1   Boiling Water Reactor    147
     7.4.2   Pressurized Water Reactor    149
     7.4.3   Gas-Cooled Reactor    151
     7.4.4   Breeder Reactor    151
7.5   Nuclear Fuel Cycle    153
     7.5.1   Mining and Refining    153
     7.5.2   Gasification and Enrichment    154
     7.5.3   Spent Fuel Reprocessing    155
     7.5.4   Temporary Waste Storage    156
     7.5.5   Permanent Waste Storage    156
7.6   Fusion    156
     7.6.1   Magnetic Confinement    157
     7.6.2   Laser Fusion    158
7.7   Energy Evolvement in Nuclear Fission and
       Fusion Reactions    158
7.8   Conclusions    159
Problems    159
Bibliography    161

8    **RENEWABLE ENERGY**    162

8.1    Introduction    162

8.2    Hydropower    164

    8.2.1    Environmental Effects    167

8.3    Biomass Energy    168

    8.3.1    Photosynthesis    169

    8.3.2    Biofuels    169

        8.3.2.1    Bioethanol    171

        8.3.2.2    Biodiesel    171

    8.3.3    Wood as Biofuel    171

    8.3.4    Environmental Effects    172

8.4    Geothermal Energy    173

    8.4.1    Environmental Effects    176

8.5    Solar Energy    176

    8.5.1    Flat Plate Collector    181

    8.5.2    Focusing Collectors    183

        8.5.2.1    Solar Thermal Farms    185

    8.5.3    Photovoltaic Cells    185

        8.5.3.1    Photovoltaic Farms    190

    8.5.4    Environmental Effects    192

8.6    Wind Power    192

    8.6.1    Aerodynamics of Wind Turbine Operation    193

    8.6.2    Mechanical and Electrical Components    196

    8.6.3    Wind Resources    198

        8.6.3.1    Capacity Factor    200

        8.6.3.2    Effectiveness    202

        8.6.3.3    Wind Variability and Predictability    202

    8.6.4    Economical Turbine Designs    204

    8.6.5    Wind Farms    205

    8.6.6    Integrating Wind Farms into the Electric Power Network    208

        8.6.6.1    Averaging an Array of Wind Farms    209

    8.6.7    Environmental Effects    210

8.7    Tidal Power    210

    8.7.1    Tidal Current Power    214

    8.7.2    Environmental Effects    216

8.8    Ocean Wave Power    216

    8.8.1    Ocean Wave Energy and Power    217

    8.8.2    Ocean Wave Power Systems    220

8.8.3    Wave Power Farms    224

8.8.4    Environmental Impacts    224

8.9    Ocean Thermal Power    225

8.10    Capital Cost of Renewable Electric Power    226

8.11    Conclusion    228

Problems    228

Bibliography    231


9    AUTOMOTIVE TRANSPORTATION    233

9.1    Introduction    233

9.2    Internal Combustion Engines for Highway Vehicles    236

9.2.1    Combustion in SI and CI engines    238

9.3    Engine Power and Performance    240

9.3.1    Engine Efficiency    242

9.4    Vehicle Power and Performance    244

9.4.1    Connecting the Engine to the Wheels    246

9.5    Vehicle Fuel Efficiency    248

9.5.1    U.S. Vehicle Fuel Efficiency Regulations and Test Cycles    248

9.5.2    Improving Vehicle Fuel Economy    251

9.5.2.1    Improving Vehicle Performance    251

9.5.2.2    Improving Engine Performance    252

9.6    Electric-Drive Vehicles    254

9.6.1    Vehicles Powered by Storage Batteries    254

9.6.2    Hybrid Vehicles    255

9.6.3    Fuel Cell Vehicles    256

9.7    Vehicle Emissions    259

9.7.1    U.S. Vehicle Emission Standards    259

9.7.2    Reducing Vehicle Emissions    261

9.7.2.1    Reducing Engine-Out Emissions    263

9.7.2.2    Catalytic Converters for Exhaust Gas Treatment    263

9.7.2.3    Evaporative Emissions    265

9.7.2.4    Reducing CI Engine Emissions    265

9.7.2.5    Fuel Quality and Its Regulation    266

9.8    Conclusion    267

Problems    268

Bibliography    269

**10   ENVIRONMENTAL EFFECTS OF FOSSIL FUEL USE   271**

10.1   Introduction   271

10.2   Air Pollution   272

    10.2.1   U.S. Emission Standards   273

    10.2.2   U.S. Ambient Standards   275

    10.2.3   Health and Environmental Effects of Fossil-Fuel-Related Air Pollutants   277

    10.2.4   Air Pollution Meteorology   279

    10.2.5   Air Quality Modeling   281

        10.2.5.1   Modeling of Steady-State Point Source   281

        10.2.5.2   Plume Rise   283

        10.2.5.3   Steady-State Line Source   284

        10.2.5.4   Steady-State Area Source   284

    10.2.6   Photo-Oxidants   285

        10.2.6.1   Photo-Oxidant Modeling   287

    10.2.7   Acid Deposition   290

        10.2.7.1   Acid Deposition Modeling   293

    10.2.8   Regional Haze and Visibility Impairment   295

10.3   Water Pollution   297

    10.3.1   Acid Mine Drainage and Coal Washing   298

    10.3.2   Solid Waste from Power Plants   299

    10.3.3   Water Use and Thermal Pollution from Power Plants   299

    10.3.4   Atmospheric Deposition of Toxic Pollutants onto Surface Waters   300

        10.3.4.1   Toxic Metals   300

        10.3.4.2   Polycyclic Aromatic Hydrocarbons   300

10.4   Land Pollution   301

10.5   Conclusion   302

Problems   303

Bibliography   304

**11   GLOBAL WARMING AND CLIMATE CHANGE   305**

11.1   Introduction   305

11.2   What Is the Greenhouse Effect?   306

    11.2.1   Solar and Terrestrial Radiation   307

    11.2.2   Sun–Earth–Space Radiative Equilibrium   309

    11.2.3   Modeling Global Warming   311

    11.2.4   Global Warming Potential   312

    11.2.5   Radiative Forcing   313

11.2.6  Results of Global Warming Modeling    316

11.2.7  Observed Trend of Global Warming    316

11.3  Associated Effects of Global Warming    317

11.3.1  Sea Level Rise    317

11.3.2  Water Vapor and Precipitation Changes    318

11.3.3  Hurricanes and Typhoons    318

11.3.4  Climate Changes    318

11.4  Greenhouse Gas Emissions    319

11.4.1  Carbon Dioxide Emissions and the Carbon Cycle    319

11.4.2  Methane    320

11.4.3  Nitrous Oxide    322

11.4.4  Chlorofluorocarbons    322

11.4.5  Ozone    323

11.5  Conclusion    323

Problems    324

Bibliography    324


12  MITIGATING GLOBAL WARMING    326

12.1  Introduction    326

12.2  Controlling Halocarbon Emissions    326

12.3  Controlling Nitrous Oxide Emissions    327

12.4  Controlling Methane Emissions    327

12.4.1  Controlling Methane Generated by Coal Mining    328

12.4.2  Controlling Methane from Petroleum and Natural Gas Operations    328

12.4.3  Controlling Landfill Methane    328

12.5  Controlling Carbon Dioxide Emissions    329

12.5.1  Controlling $CO_2$ Emissions from Fossil-Fueled Electric Power Plants    330

12.5.1.1  Shift from Coal or Oil to Natural Gas Fuel    330

12.5.1.2  Natural Gas-Fired Combined Cycle Plants    330

12.5.1.3  Capturing $CO_2$ from the Flue Gas by Chemical Absorption    330

12.5.1.4  Oxyfuel Combustion with $CO_2$ Capture    332

12.5.1.5  Integrated Coal Gasification Combined Cycle Plants with $CO_2$ Capture    333

12.5.1.6  Capturing $CO_2$ after Gasification by Physical Absorption    334

12.5.1.7 Capturing $CO_2$ after Gasification by Membrane
Separation    335

12.6  Thermal Efficiency and Cost of Controlling $CO_2$ Emissions from
Power Plants    336

12.7  $CO_2$ Sequestration    337

12.7.1  Sequestration in Oil and Gas Reservoirs    337

12.7.2  Sequestration in Coal Seams    338

12.7.3  Sequestration in Deep Sedimentary Basins    339

12.7.4  Sequestration in the Deep Ocean    341

12.7.5  $CO_2$ Removal from the Atmosphere    344

12.7.5.1  Afforestation    344

12.7.5.2  Ocean Fertilization    344

12.7.5.3  Mineral Sequestration    344

12.7.5.4  $CO_2$ Utilization    344

12.8  Conclusion    346

Problems    346

Bibliography    348

13  CONCLUDING REMARKS    349

13.1  Energy Resources    349

13.2  Regulating the Environmental Effects of
Energy Use    350

13.3  Global Climate Change    352

13.3.1  Coping with Climate Change    353

13.4  Conclusion    355

*Appendix A: Measuring Energy*    356
*Index*    361

# TABLES

| | | |
|---|---|---|
| **2.1** | Population, Energy Use, Gross Domestic Product (GDP), Energy Use per Capita, and Energy Use per $GDP in Several Countries, 2008 | 16 |
| **2.2** | Carbon Emissions, Carbon Emissions per Capita, and Emissions per $GDP of the Largest Emitting Countries, 2006 | 18 |
| **2.3** | Composition and Characteristics of Coal | 26 |
| **2.4** | World's Proven Fossil Fuel Reserves | 31 |
| **3.1** | Fuel (Thermal) Efficiencies of Current Power Technologies | 61 |
| **4.1** | Thermodynamic Properties of Fuel Combustion in Air at 25°C and One Atmosphere Pressure | 66 |
| **4.2** | Thermal Efficiencies of Synthetic Fuel Production | 71 |
| **5.1** | Typical Properties of Storage Battery Systems | 100 |
| **5.2** | Properties of Energy Storage Systems | 101 |
| **7.1** | Some Isotopes in the Nuclear Fuel Cycle, with Half-Lives and Radiation | 143 |
| **8.1** | Annual Average Energy Flux in Renewable-Energy Systems | 164 |
| **8.2** | Hydropower Development in the United States in 1980 | 165 |
| **8.3** | Installed Electrical and Thermal Power of Geothermal Systems in 1993 | 174 |
| **8.4** | Clear-Sky Irradiance at 40° N Latitude | 179 |
| **8.5** | Solar Thermal Farms | 186 |
| **8.6** | Photovoltaic Farms | 191 |
| **8.7** | Properties of Wind Energy Turbine Systems | 198 |
| **8.8** | Tidal Power Plant Characteristics | 212 |
| **8.9** | Capital Cost of Renewable Electric Power | 227 |

**9.1**   U.S. Transportation Vehicle Use, 1995                                      235

**9.2**   2010 Model Year Light-Duty Passenger Vehicle
          Characteristics (SI Engines)                                               243

**9.3**   2010 Model Year Electric Vehicle Characteristics                           254

**9.4**   Characteristics of 2010 Fuel Cell Vehicles                                 258

**9.5**   U.S. Vehicle Exhaust Emission Standards                                    261

**10.1**  U.S. NSPS Emission Standards for Fossil Fuel Steam Generators with Heat
          Input > 73 MW (250 MBtu/hr)                                                274

**10.2**  U.S. 2000 National Ambient Air Quality Standards                           277

**10.3**  Effects of Criteria Air Pollutants on Human Health, Fauna and Flora, and
          Structures and Materials                                                   278

**10.4**  Pasquill–Gifford Stability Categories                                      280

**11.1**  Greenhouse Gas (GHG) Concentrations, Global Warming Potential and
          Radiative Forcing of Several Anthropogenic GHG, and Other Radiative
          Forcing                                                                    313

**12.1**  Large Stationary $CO_2$ Emission Sources                                   329

**12.2**  Thermal Efficiency, Normalized Incremental Cost of Electricity Production,
          and $CO_2$ Emission Avoidance of Power Plants with $CO_2$ Control
          Technologies                                                               336

**A.1**   SI Units                                                                   357

**A.2**   U.S. Commercial Units                                                       358

**A.3**   Measured Quantities                                                         359

**A.4**   SI Unit Prefixes                                                            359

# PREFACE

This book is the successor to an earlier textbook[1] based upon the authors' many years of teaching and research on the environmental consequences of energy use in industrialized nations. Since that earlier publication, the urgency of preventing further damage to the natural environment and human populations caused by the growth of energy consumption has undergone a global metamorphosis. Scientists and engineers in many nations are developing new energy systems that have less severe environmental consequences, yet produce sufficient energy to provide better economic opportunities for all humans. This textbook incorporates what has been learned to date that allows a newcomer to understand potential developments in this important undertaking of fellow scientists and engineers the world over.

## CHANGES TO THE 2nd EDITION

The more significant changes since the 2002 edition include:

- Updating data, references, and bibliography to the 2008–2010 time frame in all chapters
- Adding a new chapter (Chapter 4: Thermodynamics of Fossil, Biomass, and Synthetic Fuels) that details the transformability of these fuels into various forms
- Additional developments in renewable energy systems, giving recent examples of capital cost, land (or water) area requirements, and capacity factors
- Updating the automotive technology, emissions, and fuel economy in Chapter 9 (Automotive Transportation)
- Expanding the climate change discussion to two chapters, Chapters 11 (Global Warming and Climate Change) and 12 (Mitigating Global Warming), to include the latest research and analysis of international scientific panels

---

[1] Fay, J. A., and D. S. Golomb. *Energy and the Environment*. New York: Oxford University Press, 2002.

- Including a brief discussion of prospective programs for controlling global climate change in Chapter 13 (Concluding Remarks).


## ORGANIZATION OF THE TEXT

Our book is intended for upper-level undergraduate and graduate students and for informed readers who have had a solid dose of science and mathematics. While we do try to refresh the student's and reader's memory on some fundamental aspects of physics, chemistry, engineering, and geophysical sciences, we are not bashful about using some advanced concepts, the appropriate mathematical language, and chemical equations. Each chapter is accompanied by a set of numerical and conceptual problems designed to stimulate creative thinking and problem solving.

Chapter 1 (Energy and the Environment) is a general introduction to the subject of energy, its use, and its environmental effects. It is a preview of the subsequent chapters and sets the context of their development.

In Chapter 2 (Global Energy Use and Supply) we survey the world's energy reserves and resources. We review historic trends of energy usage and estimates of future supply and demand. This is done globally, by continent and country, by energy use sector, and by proportion to population and gross domestic product. The inequalities of global energy supply and consumption are discussed.

Chapter 3 (Thermodynamic Principles of Energy Conversion) is a refresher of the thermodynamic principles of energy use. It reviews the laws that govern the conversion of energy from one form to another—that is, the first and second laws of thermodynamics and the concepts of work, heat, internal energy, free energy, and entropy. Special attention is given to the combustion of fossil fuels. Various ideal thermodynamic cycles that involve heat or combustion engines are discussed in detail—for example, the Carnot, Rankine, Brayton, and Otto cycles. Also, advanced and combined cycles are described.

Chapter 4 (Thermodynamics of Fossil, Biomass, and Synthetic Fuels) is a continuation of Chapter 3 that explains in detail the combustion of fuels with air and the role such combustion plays in the power-producing cycles previously mentioned in Chapter 3. Chapter 4 further delves into the production of synthetic fuels based upon fossil fuels or biomass resources, emphasizing the role of the limitations imposed by the third law of thermodynamics.

The generation and transmission of electrical power and the storage of mechanical and electrical energy are covered in Chapter 5 (Electric Energy Generation, Transmission, and Storage). Electrostatic, magnetic, and electrochemical storage of electrical energy are treated, as well as various mechanical energy storage systems.

The generation of electricity in fossil-fueled power plants is thoroughly discussed in Chapter 6 (Fossil-Fueled Power Plants). The complete workings of a fossil-fueled power plant are described, including fuel storage and preparation, burners, boilers, turbines, condensers, and generators. Special emphasis is given to emission control techniques, such as particulate matter control with electrostatic precipitators, sulfur oxide control with scrubbers, and nitric oxide control with low-NOx burners and flue gas denitrification. Alternative fossil-fueled power plants are discussed, such as natural gas combined cycle, integrated coal gasification combined cycle, fuel cell power plant, and cogeneration.

In Chapter 7 (Nuclear-Fueled Power Plants) we describe electricity generation in nuclear-fueled power plants. Here we review the fundamentals of nuclear energy: atoms, isotopes, the nucleus and electrons, protons and neutrons, radioactivity, nuclear stability, fission, and fusion. The nuclear fuel cycle is described, including mining, purification, enrichment, fuel rod preparation, and spent fuel (radioactive waste) disposal. The workings of nuclear reactors are discussed, including control rods, moderators, neutron economy, and the different reactor types—boiling water, pressurized water, and breeder reactors.

The principles of renewable energy utilization are explained in Chapter 8 (Renewable Energy). This includes hydropower, biomass, geothermal, solar thermal and photovoltaic, wind, tidal, ocean wave, and ocean thermal power production. Special attention is given to the capacity factor and capital cost of these systems.

Chapter 9 (Automotive Transportation) is devoted principally to the automobile, by far the major component of road vehicles that consume approximately one third of all primary energy, and also because the transportation of people and goods is so dependent upon them. The characteristics of the internal combustion engine are described for both gasoline and diesel engines. The importance of vehicle characteristics for vehicle fuel efficiency is stressed. Electric drive vehicles are described, including battery-powered and hybrid vehicles. Vehicle emissions are explained, and the technology for reducing them is described.

A survey of the environmental effects of fossil fuel usage begins in Chapter 10 (Environmental Effects of Fossil Fuel Use). In this chapter we discuss urban and regional air pollution, the transport and dispersion of particulate matter, sulfur oxides, nitrogen oxides, carbon monoxide, and other toxic pollutants from fossil fuel combustion, and the effects of these pollutants on human health, biota, materials, and aesthetics. The phenomena of photochemical smog, acid deposition, and regional haze are also described. Also treated are the impacts of energy usage on water and land.

Chapter 11 (Global Warming and Climate Change) treats the fundamentals of global warming and climate change. The perturbation is described of the Sun–Earth–Space radiative equilibrium due to the increased addition to the Earth's atmosphere of infrared-absorbing, polyatomic gases—the so-called greenhouse gases, or GHGs—such as carbon dioxide, methane, nitrous oxide, and halocarbons. We discuss radiative forcing, global warming potential, and the modeling of expected global warming if more GHGs are added to the atmosphere. The associated effects of global warming are described, such as climate change, ice cap melting, and sea level rise. The GHG emission inventory is given as well as the global carbon cycle.

In Chapter 12 (Mitigating Global Warming) we describe the various technologies that could be applied to reduce emissions of anthropogenic GHG. We start with the minor contributors to radiative forcing—halocarbons, nitrous oxide, and methane—followed by tackling the more difficult problem of controlling anthropogenic emissions of carbon dioxide, which is a consequence of fossil fuel combustion. Special attention is given to a new generation of coal-fueled power plants, such as oxyfuel combustion, chemical absorption of flue gas, and integrated coal gasification combined cycle with carbon capture. Underground and deep ocean sequestration methods of captured carbon dioxide are described.

We conclude with Chapter 13 (Concluding Remarks), a reemphasis of the important relationships among the science, technology, and economics of energy usage and its environmental effects. We note the limited success of regulation of urban and regional air pollution in industrialized nations and the great challenge that lies ahead in dealing with global climate change.

Finally, we include Appendix A, an explanation of the scientific and engineering units that are commonly used in energy studies, easing the translation from one set to another.

## ACKNOWLEDGMENTS

The authors express their appreciation to colleagues who aided them in discussions during the manuscript preparation. We also thank Alex Brown for preparing many of the tables and figures and aiding with software preparation.

Special thanks are due to Claire Sullivan, Rachel Zimmermann, and Shelby Peak, our editors at Oxford University Press, and their reviewers: John M. Mativo, Chunbao Xu, Ralph G. Lightner, Christine Ehlig-Economides, Russ Houldin, Frank R. Leslie, Kendrick Aung, Darin W. Nutter, Milivoje Kostic, and Michael M. Ohadi.

The authors dedicate this book to Gay Fay and Claire Golomb and in memory of Maya Golomb.

James A. Fay
Massachusetts Institute of Technology

Dan S. Golomb
University of Massachusetts Lowell

# Energy and the Environment

## 1.1 INTRODUCTION

Modern societies are characterized by a substantial consumption of fossil and nuclear fuels needed to provide for the operation of the physical infrastructure upon which these societies depend: the production of food and water, clothing, shelter, transportation, communication, and other essential human services. The amount of this energy use and its concentration in the urban areas of industrialized nations have caused the environmental degradation of air-, water-, and land-dependent ecosystems on a local and regional scale, as well as adverse health effects in human populations. Recent scientific studies have forecast potentially adverse global climate changes that would result from the accumulation of gaseous emissions to the atmosphere, principally carbon dioxide from energy-related sources. This accumulation is aggravated by an expected expanding consumption of energy, both by industrialized nations and by developing nations seeking to improve the living standards of their growing populations. The nations of the world, individually and collectively, are attempting to limit the damage to human health and natural ecosystems that accompanies these problems and to forestall the development of even more severe issues in the future. But because the source of the problem, energy usage, is so intimately involved in nations' and the world's economies, it will be difficult to ameliorate this environmental degradation without some adverse effects on the social and economic circumstances of national populations.

To comprehend the magnitude of intensity of human use of energy in current nations, we might compare it with the minimum energy needed to sustain an individual human life, that of the caloric value of food needed for a healthy diet. In the United States, which ranks among the most intensive users of energy, the average daily fossil fuel use per capita amounts to 56 times the necessary daily food energy intake. On the other hand, in India, a developing nation, the energy used is only 3 times the daily food calorie intake. U.S. nationals expend 20 times the energy used by Indian nationals, and their per capita share of the national gross domestic product is 50 times greater. Evidently, the economic well-being of populations is closely tied to their energy consumption.

When agricultural technology began to displace that of the hunter–gatherer societies about 10,000 years ago, other activities than acquiring food became possible. Eventually other sources of mechanical energy, that of animals, wind, and water streams, were developed, augmenting human labor and further enhancing both agricultural and nonagricultural pursuits. As the world population increased, the amount of crop and pasture land increased in proportion, permanently replacing natural forest and grassland ecosystems by less diverse ones. Until the beginning of the industrial revolution several centuries ago, this was the major environmental impact of human activities. Today, we are approaching the limit of available land for agricultural purposes, and only more intensive use of it can provide food for future increases of the world population.

The industrial revolution drastically changed the conditions of human societies by making available large amounts of energy from coal (and later oil, gas, and nuclear fuel) far exceeding that available from the biofuel, wood. Some of this energy was directed to increasing the productivity of agriculture, freeing up a large segment of the population for other beneficial activities. Urban populations grew rapidly as energy-using activities, such as manufacturing and commerce, concentrated themselves in urban areas. Urban population and population density increased, while those of rural areas decreased.

By the middle of the 20th century, nearly all major cities of the industrialized world experienced health-threatening episodes of air pollution, and today this type of degradation has spread to the urban areas of developing countries as a consequence of the growing industrialization of their economies. Predominantly, urban air pollution is a consequence of the burning of fossil fuels within and beyond the urban region itself. This pollution can extend in significant concentrations to rural areas at some distance from the pollutant sources so that polluted regions of continental dimensions even include locations where there is an absence of local energy use.

Despite the severity of urban pollution, it is technically possible to reduce it to harmless levels by limiting the emission of those chemical species that cause the atmospheric degradation. The principal pollutants comprise only a very small fraction of the materials processed and can be made even smaller, albeit at some economic cost. In industrialized countries, the cost of abating urban air pollution is but a minor slice of a nation's economic pie.

While the industrialized nations grapple with urban and regional air pollution, with some success, and developing nations lose ground to the intensifying levels of harmful urban air contamination, the global atmosphere experiences an untempered increase in greenhouse gases, those pollutants that are thought to cause the average surface air temperature to rise and climate to be modified. Unlike the urban pollutants, most of which are precipitated from the atmosphere within a few days of their emission, greenhouse gases accumulate in the atmosphere for years, even centuries. The most common greenhouse gas is carbon dioxide, which is released when fossil fuels are burned. Because it is not possible to utilize the full energy of fossil fuels without forming carbon dioxide, it will be very difficult to reduce the global emissions of carbon dioxide while still providing enough energy to the world's nations for the improvement of their economies. While there is technology available or being developed that would make possible substantial reductions in global carbon dioxide emissions, the cost of implementation of such control programs will be much larger than that for curbing urban air pollution.

### 1.1.1  An Overview of This Text

This book describes the technology and scientific understanding by which the world's nations could ameliorate the growing urban, regional, and global environmental problems associated with energy use while still providing sufficient energy to meet the needs of populations for a humane existence. The book focuses on the technology and science, the base on which any effective environmental control program must be built. It does not prescribe control programs, because they must include social, economic, and political factors that lie outside the scope of this book. We describe the scientific and technological principles that determine how energy is used and how it can be converted from one form to another. We describe the environmental consequences of energy use and how they might be mitigated. We present a bibliography in each chapter for the reader who wants to pursue some aspects of this exposition in greater depth.

The major sources of energy for modern nations are fossil fuels, nuclear fuels, and hydropower. Nonhydro renewable energy sources, such as biomass, wind, geothermal, solar thermal, and photovoltaic power, account for only a small portion of current energy production. Like other mineral deposits, fossil fuels are not distributed uniformly around the globe, but are found mostly on continents and their margins that were once locations of great biomass production. They need to be discovered and removed, and often processed, before they can be available for energy production. Current and expected deposits would appear to last for a few centuries at current consumption rates. Within the time horizon of most national planning, there is no impending shortage of fossil fuel despite the continual depletion of what is a finite resource. In contrast, renewable energy sources are not depletable, being supplied ultimately by the flux of solar insolation that impinges on the earth.

Like food, energy needs to be stored and transported from the time and place where it becomes available to that where it is to be used. Fossil and nuclear fuels, which store their energy in chemical or nuclear form indefinitely, are overwhelmingly the preferred form for storing and transporting energy. Electrical energy is easily transmitted from source to user, but there is no electrical storage capability in this system. Hydropower systems store energy for periods of days to years in their reservoirs. For most renewable energy sources, there is no inherent storage capability, so they must be integrated into the electrical network. Many forms of mechanical and electrical energy storage are being developed to provide for special applications where storage in chemical form is not suitable. Efficient transformation of energy from mechanical to electrical form is an essential factor of modern energy systems.

Although fossil fuels may be readily burned to provide heat for space heating, cooking, or industrial and commercial use, producing mechanical or electrical power from burning fuels required the invention of power-producing machines, beginning with the steam engine and subsequently expanding to the gasoline engine, diesel engine, gas turbine, and fuel cell. The science of thermodynamics prescribes the physicochemical rules that govern how much of a fuel's energy can be transformed to mechanical power. While perfect machines can convert much of the fuel's energy to work, practical and economic ones only return between one quarter and one half of the fuel energy. Nevertheless, the technology is rich and capable of being improved through further research and development, but large increases in fuel efficiency are not likely to be reached without a considerable cost penalty.

Initially, steam engines were used to pump water from mines, to power knitting mills, and to propel trains and ships. Starting in the late 19th century, electrical power produced by steam engines became the preferred method for distributing machine power to distant end users. By the time electricity distribution had become universal, supplying mechanical power, light, and communication signals, the generation of electrical power in steam power plants had become the largest segment of energy use. Currently, 55% of world fossil fuel is consumed in electric power plants.

The modern fossil-fueled steam-electric power plant is quite complex (see Figure 1.1). Its principal components—the boiler, the turbine, and the condenser—are designed to achieve maximum thermal efficiency. But combustion of the fuel produces gaseous and solid pollutants, among which are oxides of carbon, sulfur, and nitrogen; soot; toxic metal vapors; and ash. Removing these

**FIGURE 1.1**  A large coal-fired steam-electric power plant whose electrical power output is nearly 3,000 megawatts. In the center are the power house and tall stacks that disperse the flue gas. To the left, a cooling tower provides cool water for condensing the steam from the turbines. To the right, high-voltage transmission lines send the electric power to consumers. (Copyright Brian Hayes.)

pollutants from the flue gases requires complex machinery, such as scrubbers and electrostatic pre-cipitators, which increases the operating and capital costs of the power plant and consumes a small percentage of its electrical output. The removed material must be disposed of safely in a landfill. But because of the size and technical sophistication of these plants, they provide a more certain avenue of improvement in control than would many thousands of small power plants of equal total power.

Nuclear power plants utilize a steam cycle to produce mechanical power, but steam for the turbine is generated by heat transfer from a hot fluid that passes through the nuclear reactor or by direct contact with the reactor fuel elements. The main disadvantage of a nuclear power plant, which does not release any ordinary pollutants to the air, is the difficulty of assuring that the immense radioactivity of its fuel is never allowed to escape by accident. Nuclear power plant technology is technically quite complex and expensive. The environmental problems associated with preparing the nuclear fuel and sequestering the spent fuel have become very difficult and expensive to manage. In the U.S., all of these problems make new nuclear power plants more expensive than new fossil fuel power plants.

Renewable energy sources are of several kinds. Wind turbines and ocean wave energy systems convert the energy of the wind and ocean waves that stream past the power plant to electrical power. Hydropower and ocean tidal power plants convert the gravitational energy of dammed up water to electrical power (see Figure 1.2). Geothermal and ocean thermal power plants make use of streams of hot and cold fluid to generate electric power in a steam power plant. A solar thermal power plant absorbs sunlight to heat steam in a power cycle. Photovoltaic systems create electricity by direct absorption of solar radiation on a semiconductor surface. Biomass-fueled power plants directly burn biomass in a steam boiler or utilize a synthetic fuel made from biomass. Most of these energy systems experience low energy flux intensity, so that large structures are required per unit of power output compared with fossil-fueled plants. On the other hand, they emit no or few pollutants, while contributing no net carbon dioxide emissions to the atmosphere. Their capital cost per unit of power output is higher than that of fossil plants, so that renewable plants may not become economical sources of power until fossil fuel prices rise.

Solar electric power requires much greater land area than is needed for conventional power plants. Figure 1.3 is an aerial view of a small solar thermal addition to a natural gas-fired power plant. Although its power is only 2% of that of the conventional plant, the land area needed for the 2% is several times as great. Renewable energy facilities are usually located farther away from the market than fossil-fueled plants, where land costs are very low.

Transportation energy is a major sector of the energy market in both industrialized and developing nations. Automobiles are a major consumer of transportation energy and emitter of urban air

**FIGURE 1.2**   A run-of-the-river hydropower plant on the Androscoggin River in Brunswick, Maine (United States). In the center is the power house, on the right is the dam/spillway, and on the left is a fish ladder to allow anadromous species to move upriver around the dam. Except when occasional springtime excessive flows are diverted to the spillway, the entire river flow passes through the power house.

**FIGURE 1.3**   An add-on solar thermal power plant (in background) augments the output of a conventional gas-fired combined cycle power plant (in foreground) by 2%. (Copyright John Van Beekum/*The New York Times*/Redux Pictures.)

pollutants. The technology of automobiles has advanced considerably in the past several decades under regulation by governments to reduce pollutant emissions and improve energy efficiency. Current automobiles emit much smaller amounts of pollutants than their uncontrolled predecessors as a consequence of complex control systems. Considerable gains in energy efficiency seem possible by the introduction of lightweight body designs and electric drive systems powered by electric storage systems, onboard-engine-driven electric generators, or combinations of these technologies.

Air pollutants emitted into the urban atmosphere by fossil fuel users and other sources can reach levels harmful to public health. Some of these pollutants can react in the atmosphere by absorbing sunlight so as to form even more harmful toxic products. This soup of direct and indirect pollutants is termed photochemical smog. One component of smog is the toxic oxidant ozone, which is not directly emitted by any source. Because of the chemical complexity of these photochemical atmospheric reactions, great effort is required to limit all the precursors of photochemical smog if it is to be reduced to low levels.

Carbon dioxide and other greenhouse gases warm the lower atmosphere by impeding the radiative transfer of heat from the earth to outer space. Limiting the growth rate of atmospheric carbon dioxide requires either (a) reducing the amount of fossil fuel burned or (b) capturing and storing the carbon dioxide below the earth's or ocean's surface. To maintain or increase the availability of energy while fossil fuel consumption is lowered, renewable or nuclear energy must be used. Of course, improving the efficiency of energy use can result in the lowering of fossil fuel use while not reducing the social utility of energy availability. By the combination of all these methods, the rate of rise of atmospheric carbon dioxide can be ameliorated or reversed at an economic and social cost that may be acceptable.

The amelioration of environmental degradation caused by energy use is a responsibility of national governments. By regulation and by providing economic incentives, governments induce energy users to reduce pollutant emissions by changes in technology or use practices. Bilateral or global treaties can bring about coordinated multinational actions to reduce regional or global environmental problems, such as acid deposition, ozone destruction, and climate change. The role of technology is to provide the necessary reduction in emissions while still making available energy at the minimum increase in cost needed to attain that end.

## 1.2  ENERGY

There is a minimum amount of energy needed to sustain human life. The energy value of food is the major component, but fuel energy is needed for cooking and, in some climates, for heating human shelter. In an agricultural society, additional energy is expended in growing, reaping, and storing food; making clothing; and constructing shelters. In modern industrial societies, much more energy than this minimum is consumed in providing food, clothing, shelter, transportation, communication, lighting, materials, and numerous services for the entire population.

It is a basic principle of physics that energy cannot be destroyed, but can be transformed from one form to another. When a fuel is burned in air, the chemical energy released by the rearrangement of fuel and oxygen atoms to form combustion products is transformed to the random energy of the hot combustion product molecules. When food is digested in the human digestive tract, some of the food energy is converted to energy of nutrient molecules and some warms the body. When human societies "consume" energy, they transform it from one useful form to a less useful form, in the process providing a good or service that is needed to maintain human life and societies.

A quantitative measure of the ongoing good that energy "consumption" provides to society is the time rate of transformation of the useful energy content of energy-rich materials, such as fossil and nuclear fuels. In 2008, the world-wide consumption rate amounted to 500 exajoules per year (EJ/y)[1], about 74.3 gigajoules per year (GJ/y) per capita.[2] Of this world total, the United States consumed 105 EJ/y, or 21%, and had a per capita energy consumption rate 4.6 times the global average, the largest of any nation.

Figure 1.4 displays graphically the imbalance among different geographical regions of the earth, as measured by the average energy consumption per capita within them (ordinate), as a function of their individual fractions of global population (abscissa). Seven regions (A through G) comprise the geographic distribution. Approximately 80% of the global population (regions A through D) consumes energy per capita at one-half the global average, while 20% (regions E through G) consume three times the global average. The first group includes the developing regions and the latter the developed ones. In terms of total energy consumption, three regions (B, G, and E, in order of total consumption) account for 77% of the global total. These distinctions are related to

---

[1] One exajoule (EJ) = 1E(18) joule (J); 1 exajoule/year (EJ/y) = 0.317 E(12) watt (W). See Appendix A for a specification of scientific notation for physical units.

[2] This rate, 2.36 kW, is 20 times the per capita food energy consumption rate (metabolic rate) of 120 W. However unevenly distributed among the world's population, the world energy consumption rate far exceeds the minimum food energy required to sustain human life.

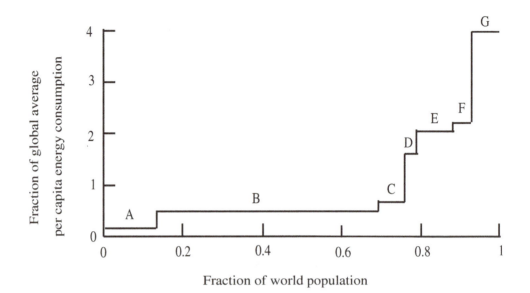

**FIGURE 1.4**   The distribution of energy consumption per capita among populations of geographic regions of the world: (A) Africa, (B) Asia and Oceana, (C) Central and South America, (D) Middle East, (E) Europe, (F) Eurasia, and (G) North America. (Data from U.S. Department of Energy, Energy Information Administration, *International Energy Annual 2004,* 2008.)

the respective economic outputs of these regions and their levels of industrialization. Of course, the nations within each group will differ in their per capita energy consumption, but a nation-by-nation distribution will look quite similar to Figure 1.4.

The capacity to consume energy at this rate is a consequence of the technology developed in industrialized nations to permit the efficient extraction and utilization of these fuels by only a small fraction of the population.[3] But the earth's fossil and nuclear fuel resources are being depleted at a rate that will render them extremely scarce in future centuries, even if they are used more efficiently than in the past. The current cost of these fuels, however, has remained low for decades because recovery technology has improved enough to offset the distant threat of scarcity.

There are other less energy-rich sources of energy that are not depletable. These are the so-called renewable energies, such as solar insolation, wind, flowing river currents, tidal flows, and biomass fuels. In fact, these are the energies that were developed on a small scale in preindustrial societies, providing for ocean transportation, cooking, sawing of lumber, and milling of grain. Industrial age technologies have made it possible to develop these sources today on a much larger scale, yet in aggregate they constitute less than 8% of current world energy consumption. Renewable energy is currently more costly than fossil energy, but not greatly so, and may yet become more economical when pollution abatement costs of fossil and nuclear energy are factored in.

---

[3]This situation is analogous to the industrialization of agriculture in advanced economies, whereby a small percentage of the population provides food for all.

How is energy used? It is customary to divide energy usage among four sectors of economic activity: industrial (manufacturing, material production, agriculture, and resource recovery), transportation (cars, trucks, trains, airplanes, pipelines, and ships), commercial (services), and residential (homes). In the United States in 2006, these categories consumed, respectively, 32, 29, 18, and 21% of the total energy. Considered all together, energy is consumed in a myriad of individual ways, each of which is an important contributor to the functioning of these sectors of the national economy.

One prominent use of energy, principally within the industrial and commercial sectors, is the generation of electric power. This use of energy now constitutes 36% of the total energy use worldwide, but 44% in the United States. Combined with the transportation sector, these two uses comprise 70% of the total U.S. energy use. For this reason, electric power production and transportation form the core energy uses discussed in this text.

How is energy supplied? Except for renewable energy sources (including hydropower), the main sources of energy are fossil and nuclear fuels, depletable minerals that must be extracted from the earth, refined as necessary, and transported to the end user in amounts needed for the particular uses. Given the structure of modern industrialized economies, supplying energy is a year-round activity in which the energy is consumed within months of being extracted from its source.[4] While there are reserves of fossil and nuclear fuels that will last decades to centuries at current consumption rates, these are not extracted until they are needed for current consumption.[5] Since fossil and nuclear fuel reserves are not uniformly distributed within or among the continents, some nations are fuel poor and others fuel rich. The quantities of fuel traded among nations comprise a significant fraction of overall energy production.

## 1.2.1  Electric Power

One hallmark of industrialization in the 20th century has been the growth of the electric power sector, which today consumes about 36% of the world's energy in the production of an annual average of 1.4 TW of electric power. In the United States, 44% of total energy is used to generate an annual average of 0.4 TW of electrical power. Nearly all of this electric power is produced in large utility plants, each generating in the range of 100 to 1,000 megawatts (MW). Fossil and nuclear fuels supply 63 and 17%, respectively, of the total electric power, with the remainder being generated in renewable energy plants, of which hydropower (19%) is the overwhelming contributor.[6] The generation and distribution of electric power to numerous industrial, commercial, and residential consumers is considered a requirement for both advanced and developing economies.

The electric energy produced in power plants is very quickly transmitted to the customer, where it is instantaneously consumed for a multitude of purposes: providing light, generating mechanical power in electric motors, heating space and materials, powering communication equipment, and so on. There is practically no accumulation of energy in this system, in contrast with the storage of fuels (or water in hydrosystems) at power plants, so that electric energy is produced and consumed

---

[4]The United States has established a crude oil reserve for emergency use to replace a sudden cut off of foreign oil supplies. The reserve contains only several months' supply of imported oil.

[5]In contrast, food crops are produced mostly on an annual basis, requiring storage of food available for marketing for the better part of a year.

[6]When comparing the amount of hydropower energy with that of fossil and nuclear energy, the former is evaluated on the basis of fuel energy needed to generate the hydroelectric power output of these plants.

nearly simultaneously.[7] Electric power plants must be operated so as to maintain the flow of electric power in response to the instantaneous aggregate demand of consumers. This is accomplished by networking together the electric power produced by many plants so that a sudden interruption in the output of one plant can be replaced by the others.

### 1.2.2   Transportation Energy

Transportation of goods and people among homes, factories, offices, and stores is a staple component of industrialized economies. Ground, air, and marine vehicles powered by fossil-fueled combustion engines are the principal means for providing this transportation function.[8] Transportation systems require both vehicle and infrastructure: car, truck and highway, train and railway, airplane and airport, ship and marine terminals. Ownership, financing, and construction of the infrastructure are often distinct from that of the vehicle, with public ownership of the infrastructure and private ownership of the vehicle being most common.

In economic terms, the largest transportation sector is that of highways and highway vehicles. Worldwide, highway vehicles now number about 750 million, with 250 million of them in the United States. In the United States, 96% of the road vehicles are passenger automobiles and light-duty trucks. The world and U.S. vehicle populations are growing at annual rates of 2.2 and 1.7%, respectively. On average, U.S. vehicles are replaced every 13 years or so, providing an opportunity to implement improvements in vehicle technology relatively quickly.[9]

Transportation fuels are nearly all petroleum derived. In the United States, transportation consumes 70% of the petroleum supply, or 32% of all fossil fuel energy. Highway vehicles account for 46% of petroleum consumption, or 21% of all fossil fuel energy. Transportation systems are especially vulnerable to interruptions in the supply of imported oil, which now exceeds the supply from domestic production. Unlike some stationary users of oil, transportation vehicles cannot substitute coal or natural gas for oil in times of scarcity.

While substantial reduction in highway vehicle air pollutant emissions has been achieved in the United States since 1970 and more reductions are scheduled for the first decades of the 21st century, the focus of vehicle technology has shifted to improving vehicle fuel economy. Doubling current fuel efficiencies without penalizing vehicle performance is technically possible, at a vehicle manufacturing cost penalty that will be offset in part by fuel cost savings. Automobiles promise to be one of the more cost-effective ways for reducing oil consumption and carbon emissions.

### 1.2.3   Energy as a Commodity

Because of the ubiquitous need for energy and the ability to store and utilize it in many forms, energy is marketed as a commodity and traded internationally at more or less well-established prices. For example, in recent decades the world crude oil price has ranged from about 15 to 100$/barrel, or about 2 to 13$/GJ.[10] Coal is generally cheaper than oil, while natural gas is more

---

[7]In some renewable energy electric power systems, such as wind and photovoltaic power systems, there is usually no energy storage; these systems can comprise only part of a reliable electric energy system.

[8]In developing countries, human-powered bicycles may be important components of ground transportation.

[9]In contrast, fossil and nuclear power plants have a useful life of 40 years or more.

[10]A barrel of crude oil contains about 6 GJ (6 MBtu) of fuel heating value.

expensive. The difference in price reflects the different costs of recovery, storage, and transport. Nuclear fuel refined for use in nuclear electric power plants is less expensive than fossil fuels per unit of heating value.

Coal is the cheapest fuel to extract, especially when mined near the earth's surface. It is also inexpensive to store and transport, both within and between continents. But it is difficult to use efficiently and cleanly, and in the United States it is used mainly as an electric utility fuel. Oil is more expensive than coal to recover, being more dispersed within geologic structures, but is more easily transported by pipeline and intercontinentally by supertanker. Oil is almost exclusively the fuel of transportation vehicles and also the fuel of choice for industrial, commercial, and residential use in place of coal. Like oil, natural gas is recovered from wells, but is not easily stored or shipped across oceans. It commands the highest price because of the greater cost of recovery, but is widely used in industry, commerce, and residences because of its ease, efficiency, and cleanliness of combustion.

In contrast with fossil and nuclear fuels, renewable energy is not transportable (except in the form of electric power) or storable (except in hydropower and biomass systems). Renewable hydropower electricity is a significant part of the world electric power supply and is sold as a commodity intra- and internationally.

Synthetic fuels, such as hydrogen, ethanol, and producer gas, are commonly manufactured from other fuels, principally fossil or biomass. By transforming the molecular structure of a natural fossil fuel to a synthetic form while preserving most of the heating value, the secondary fuel may be stored or utilized more easily or provide superior combustion characteristics, but is inevitably more expensive than its parent fuel.[11]

On the time scale of centuries, the supplies of fossil and nuclear fuels will be severely depleted, leaving only deposits that are difficult and expensive to extract. The only sources that could supply energy indefinitely beyond that time horizon are nuclear fusion and renewable energy. These are both capital-intensive technologies. Their energy costs will inevitably be competitive with that of fossil and fission fuels when the latter become scarce enough.[12]

## 1.3   THE ENVIRONMENT

The 20th century, during which industrialization proceeded even faster than population growth, marked the beginning of an understanding, both popular and scientific, that human activity was having deleterious effects upon the natural world, including human health and welfare. These effects included increasing pollution of air, water, and land by the by-products of industrial activity, permanent loss of natural species of plants and animals by changes in land and water usage and human predation, and, more recently, growing indications that the global climate was changing because of the anthropogenic emissions of so-called greenhouse gases.

---

[11]Plutonium-239, a fissionable nuclear fuel, is formed from uranium-238, a nonfissionable natural mineral, in nuclear reactors. In this sense it is a synthetic nuclear fuel, which can produce more energy than is consumed in its formation, unlike fossil fuel-based synthetic fuels.

[12]If fusion power plants will be no more expensive than current fission plants, at about $1 per thermal watt of heat input, then the capital cost of supplying the current U.S. energy consumption of about 3 TW would be 3 trillion dollars. The cost of this energy would be several times current costs.

At first, attention was focused on recurring episodes of high levels of air pollution in areas surrounding industrial facilities, such as coal-burning power plants, steel mills, and mineral refineries. These pollution episodes were accompanied by acute human sickness and the exacerbation of chronic illnesses. After midcentury, when industrialized nations' economies recovered rapidly from World War II and expanded greatly above their prewar levels, many urban regions without heavy industrial facilities began to experience persistent, chronic, and harmful levels of photochemical smog, a secondary pollutant created in the atmosphere from invisible volatile organic compounds and nitrogen oxides produced by burning fuels and the widespread use of manufactured organic materials. Concurrently, the overloading of rivers, lakes, and estuaries with industrial and municipal wastes threatened both human health and the ecological integrity of these natural systems. The careless disposal on land of mining, industrial, and municipal solid wastes despoiled the purity of surface and subsurface water supplies.

As the level of environmental damage grew in proportion to the rate of emission of air and water pollutants, which themselves reflected the increasing level of industrial activity, national governments undertook to limit the rate of these emissions by requiring technological improvements to pollutant sources. As a consequence, by the century's end ambient air and water pollution levels were decreasing gradually in the most advanced industrialized nations, even though energy and material consumption was increasing. Nevertheless, troubling evidence of the cumulative effects of industrial waste disposal became evident, such as acidification of forest and agricultural soils, surface waters (including the oceans), contamination of marine sediments with municipal waste sludge, and poisoning of aquifers with drainage from toxic waste dumps. Not the least of the impending cumulative waste problems is the disposal of used nuclear power plant fuel and its reprocessing wastes.

Environmental degradation is not confined to urban regions. In preindustrial times large areas of forest and grassland ecosystems were replaced by much less diverse cropland. Subsequently, industrialized agriculture has expanded the predominance of monocultured crops and intensified production by copious applications of pesticides, herbicides, and inorganic fertilizers. Valuable topsoil has eroded at rates above replacement levels. Forests managed for pulp and lumber production are less diverse than their natural predecessors, with the tree crop being optimized by use of herbicides and pesticides. In the United States, factory production of poultry and pork has created severe local animal waste control problems.

The most threatened, and most diverse, natural ecosystems on earth are the tropical rain forests. Tropical forest destruction for agricultural or silvicultural uses destroys ecosystems of great complexity and diversity, extinguishing irreversibly an evolutionary natural treasure. It also adds to the burden of atmospheric carbon dioxide in excess of what can be recovered by reforestation.

The most sobering environmental changes are global ones. The recent appearance of stratospheric ozone depletion in polar regions, which could increase harmful ultraviolet radiation at the earth's surface in midlatitudes should it increase in intensity, was clearly shown by scientific research to be a consequence of the industrial production of chlorofluorocarbons. (By international treaty, these chemicals are being replaced by less harmful ones, and the stratospheric ozone destruction will eventually be reduced.) But the more ominous global pollutants are infrared-absorbing molecules, principally carbon dioxide, but including nitrous oxide and methane, that are inexorably accumulating in the atmosphere and promising to disturb the earth's thermal radiation equilibrium with the sun and outer space. It is currently believed by most scientists that this disequilibrium will cause the average atmospheric surface temperature to rise, with probable adverse climatic consequences. Since carbon dioxide is formed ineluctibly in the combustion of fossil fuels that produce much of the current and expected future energy use and is known to accumulate in the atmosphere for centuries, its continued emission into the atmosphere presents a problem that cannot be managed except on a global scale. Controlling this problem would greatly affect the future course of energy use for centuries to come.

### 1.3.1   Managing Industrial Pollution

To address the problem of a deteriorating environment, industrialized nations have attempted to regulate the emission of pollutants into the natural environment, whether it be air, water, or land. The concept that underlies governmental control is that the concentration of pollutants in the environment must be kept below a level that will assure no harmful effects in humans or ecological systems. This can be achieved by limiting the mass rate of pollutant emissions from a particular source so that, when mixed with surrounding clean air or water, the concentration is sufficiently low to meet the criterion of harmlessness.[13]

In the case of multiple sources located near each other, such as automobiles on a highway or many factories crowded together in an urban area, the additive effects require greater reduction per source than would be needed if only one isolated source existed. In industrialized countries and regions, the cumulative effects of emissions into limited volumes of air or water result in widespread contamination, with both local and distant sources contributing to local levels.

The ultimate example of cumulative effects is the gradual increase in the global annual average atmospheric carbon dioxide concentration caused by the worldwide emissions from burning of fossil fuels and forests. Because the residence time of carbon dioxide in the atmosphere is of the order of a century, this rise in atmospheric concentration reflects the cumulative emissions over many prior decades. Unlike urban or regional air pollutant emission reduction, reducing carbon dioxide emissions will not immediately reduce the ambient carbon dioxide level; rather, it will only slow its inexorable rise.

The scientific and technological basis for national and international management of environmental pollution is the cumulative understanding of the natural environment, the technology of industrial processes that release harmful agents into the environment, and the deleterious effects upon humans and ecological systems from exposure to them. By itself, this knowledge cannot secure a solution to environmental degradation, but it is a requisite to fashioning governmental programs for attaining that purpose.

# BIBLIOGRAPHY

Evans, R. L., *Fueling Our Future: An Introduction to Sustainable Energy*. Cambridge: Cambridge University Press, 2007.

---

[13]In regulatory procedures, it is usually not necessary to prove absolute harmlessness, but only the absence of detectable harm.

# Global Energy Use and Supply

## 2.1  INTRODUCTION

The industrial revolution has been characterized by very large increases in the amount of energy available to human societies compared with their predecessors. In preindustrial economies only very limited amounts of nonhuman mechanical power were available, such as that of domesticated animals, the use of wind power to propel boats and pump water, and the use of water power to grind grain. Wood and charcoal were the principal fuels to cook food, to heat dwellings, and to smelt and refine metals. Today, in industrial nations, or in the urban–industrial areas of developing nations, the availability of fossil and nuclear fuels has vastly increased the amount of energy that can be expended on economic production and personal consumption, helping to make possible a standard of living that greatly exceeds the subsistence level of preindustrial times. Furthermore, the population of the world increased several fold since the preindustrial era, thus requiring the recovery of ever-increasing amounts of energy resources. However, these resources are not evenly distributed among the countries of the world, and they are finite.

The principal sources of energy in present societies are fossil energy (coal, petroleum, and natural gas), nuclear, and hydroenergy. Other energy sources, the so-called renewables, are currently supplying a small fraction of the total energy consumption of the world. The renewables include solar, wind, geothermal, biomass, ocean–thermal, and ocean–mechanical energy. Indeed, hydroenergy may also be called a renewable energy source, although usually it is not classified as such. Increased use of renewable energy sources is desirable because they are deemed to cause less environmental damage, and their use would extend the available resources of fossil and nuclear energy.

In this chapter we describe the supply and consumption patterns of energy in the world today and the historical trends, with emphasis on available resources and their rate of depletion. In recent years the increase of atmospheric concentration of $CO_2$ resulting from increased consumption of fossil fuels has become an international concern. In examining global energy use, it is useful to include in our accounting the concomitant $CO_2$ emissions to provide a perspective on the problem of managing the potential threat of global climate change resulting from these emissions.

## 2.2  GLOBAL ENERGY CONSUMPTION

The trend of world annual energy consumption from 1980 to 2008 and projections to 2030 is depicted in Figure 2.1.[1] The worldwide energy consumption in 2008 was 498.1 EJ/y.[2] In Figure 2.2 is plotted the energy consumption trend for the Organization for Economic Co-operation and Development (OECD) countries and the non-OECD countries. In 2008, the OECD and non-OECD countries consumed roughly one half each of the world's energy. Energy demand in the OECD countries is expected to grow at an annual rate of 0.6%, whereas energy consumption in the non-OECD countries is expected to grow by an average of 2.3% per year. Thus, it is expected that in 2030 the non-OECD countries will consume roughly 59% of the world's energy while the OECD countries will consume 41%. In the United States, energy consumption increased 1.7%/y over the decade 1996–2006; China 9.7%/y; India 5.3%/y. Most of the growth is the result of increased fossil fuel consumption.

Table 2.1 lists the 2008 population, total energy use, gross domestic product (GDP), energy use per capita, and energy use per GDP of several OECD and non-OECD countries. The United States is the largest consumer of energy (105.4 EJ annually), followed by China (77.9 EJ), Russia (31.9 EJ), Japan (24.1 EJ), and India (18.4 EJ). The United States consumes 25.6% of the world's energy with 4.6% of the world's population; Western Europe consumes 18.3% of the world's energy

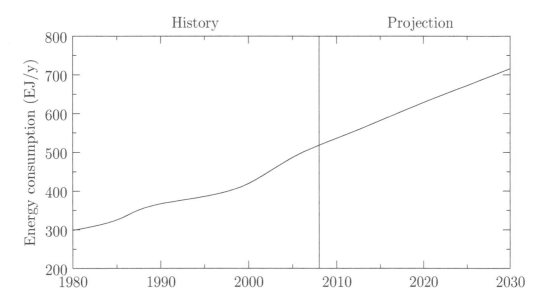

**FIGURE 2.1**   Trend of world's energy consumption for 1980–2008 and a projection to 2030.[1]

---

[1]U.S. Department of Energy, Energy Information Agency, 2009. *International Energy Outlook*. Data given are for calendar year 2008.

[2]One exajoule (EJ) = 1 E(18) joule (J) = 0.948 quadrillion Btu (British thermal units) or "quads" (Q). 1Q = 1E(15) Btu. See Tables A.1, A.2, and A.4 in Appendix A.

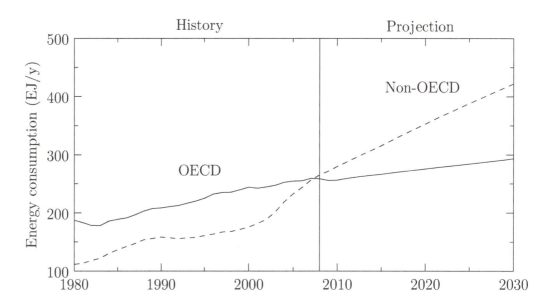

**FIGURE 2.2**  Energy consumption trend for the Organization for Economic Co-operation and Development (OECD) countries and the non-OECD countries.[1]

**TABLE 2.1**  Population, Energy Use, Gross Domestic Product (GDP), Energy Use per Capita, and Energy Use per $GDP in Several Countries, 2008

| Country | Population (million) | Energy use (EJ/y) | GDP ($T/y) | Per capita (GJ/cap/y) | Per $GDP (MJ/$GDP) |
|---|---|---|---|---|---|
| **Developed** | | | | | |
| United States | 307.21 | 105.4 | 14.29 | 343.09 | 7.38 |
| Canada | 33.49 | 14.7 | 1.31 | 438.97 | 11.22 |
| Russia | 140.04 | 31.9 | 2.23 | 227.79 | 14.3 |
| Japan | 127.08 | 24.1 | 4.35 | 189.65 | 5.54 |
| Germany | 82.33 | 15.5 | 2.86 | 188.27 | 5.42 |
| France | 64.06 | 12.0 | 2.10 | 187.33 | 5.71 |
| Italy | 58.13 | 8.5 | 1.82 | 146.23 | 4.67 |
| United Kingdom | 61.11 | 10.4 | 2.23 | 170.18 | 4.66 |
| Norway | 4.66 | 1.9 | 0.26 | 407.68 | 7.31 |
| **Less developed** | | | | | |
| China | 1338.61 | 77.9 | 3.9 | 58.19 | 19.97 |
| India | 1166.08 | 18.4 | 1.2 | 15.78 | 15.33 |
| Indonesia | 240.27 | 4.6 | 0.51 | 19.15 | 9.02 |
| Brazil | 198.74 | 10.2 | 1.6 | 51.32 | 6.38 |
| Mexico | 111.21 | 7.7 | 1.1 | 69.24 | 7.0 |
| **World total** | 6707 | 498.1 | 60.1 | 74.27 | 8.29 |

with 6.5% of the world's population. China consumes about 15.6% of the world's energy with 20% of the world's population; India consumes 3.7% of the energy with 17.4% of the population.

Among the listed countries, Canada, Norway, and the United States are the world's highest users of energy per capita, with 439, 408, and 343 GJ per capita per year, respectively. The Russian Federation consumes 228, Japan 190, the UK 170, Germany 188, and France 187 GJ/cap y. The less-developed countries consume much less energy per capita. For example, Mexico consumes 69, China 58, Brazil 51, Indonesia 19, and India 16 GJ/cap y. The world average consumption is 74 GJ/cap y.

If we compare the energy consumption per GDP, a different picture emerges. Among developed countries, Canada uses 11.22, Norway 7.31, the United States 7.38, the UK 4.66, Germany 5.42, France 5.71, Italy 4.67, and Japan 5.54 MJ/$GDP. Canada, Norway, and the United States use more energy per $GDP than the other Western European countries and Japan, in part because of the colder climate, larger living spaces, longer driving distances, and larger automobiles. On the other hand, Russia, China, and India spend a higher rate of energy per $GDP than the United States, the European countries, and Japan. This is an indication that much of the population in these countries does not (yet) contribute significantly to the GDP. Furthermore, their industrial facilities, power generation, and heating (or cooling) systems apparently are less efficient or in other ways more wasteful of energy than in the United States, Western Europe, and Japan.

## 2.3   GLOBAL CARBON EMISSIONS

Currently, worldwide carbon emissions amount to approximately 7.8 Gt C/y (to obtain $CO_2$ emissions, multiply by 44/12, the molecular weight ratio of $CO_2$ and C). Table 2.2 lists the carbon emissions, carbon emissions per capita, and emissions per GDP of the largest emitting countries.

In terms of absolute quantities, the United States and China are the largest emitters of carbon, with 1.57 and 1.66 Gt C/y, respectively, followed by Russia with 0.43 Gt C/y. In terms of per capita emissions, the United States and Canada are the largest emitters, with 5.1 and 4.4 t C/cap y, followed by Russia, with 3.05 t C/cap y. In countries where nonfossil energy is used for electricity generation and other purposes, the per capita carbon emissions are lower. Thus, while the energy consumption per capita in Germany and France is similar, at 188 and 187 GJ/cap y, respectively, the carbon emissions are quite different, 2.7 and 1.6 t C/cap y, respectively. This reflects the greater use of nuclear energy for electricity generation in France compared with Germany. The world average is 1.1 t C/cap y. The United States emits nearly five times as much carbon per capita as the world's average.

In terms of carbon emissions per $GDP, an interesting picture emerges. The ratio in the United States and Canada is 0.1 kg carbon per $GDP (reckoned in 2006 $U.S.), respectively, whereas in Japan, Germany, France, Italy, and the UK it ranges from 0.04 to 0.1 kg/$. In part this stems from the higher consumption of energy per unit of GDP in the United States and Canada, but also from the fact that the United States and Canada use more fossil fuel per capita for space heating, cooling, and transportation than the European countries and Japan. In Russia and China, the ratio of carbon emission per $GDP is about 0.2, India and Indonesia 0.1, and Mexico 0.1 kg/$. In these countries fossil fuel is not used as efficiently in the production of GDP. The exception is Brazil, where the ratio is 0.05 kg/$, probably on account of Brazil's greater use of hydro and biomass energy. (Note that emissions from forest burning are not included in these estimates.)

**TABLE 2.2**  Carbon Emissions, Carbon Emissions per Capita, and Emissions per $GDP of the Largest Emitting Countries, 2006

| Country | Carbon emissions | | |
| --- | --- | --- | --- |
| | Mt/y | kg/cap y | kg/$GDP |
| United States | 1568.81 | 5106.62 | 0.11 |
| Canada | 148.55 | 4435.63 | 0.11 |
| Russia | 426.73 | 3047.19 | 0.19 |
| Japan | 352.75 | 2775.79 | 0.08 |
| Germany | 219.57 | 2666.95 | 0.08 |
| France | 104.49 | 1631.2 | 0.05 |
| Italy | 129.31 | 2224.55 | 0.07 |
| United Kingdom | 155.05 | 2537.24 | 0.07 |
| Norway | 10.97 | 2353.88 | 0.04 |
| Switzerland | 11.41 | 1500.05 | 0.04 |
| New Zealand | 8.31 | 1973.44 | 0.07 |
| China | 1664.59 | 1243.52 | 0.21 |
| India | 411.91 | 353.25 | 0.13 |
| Indonesia | 90.95 | 378.53 | 0.1 |
| Brazil | 96.14 | 483.76 | 0.05 |
| Mexico | 118.95 | 1069.6 | 0.08 |
| World total | 7754.11 | 1156.12 | 0.13 |

*Source:* Oak Ridge National Laboratory, Carbon Dioxide Information Analysis Center (CDIAC), 2006.

## 2.4  GLOBAL ENERGY SOURCES

The primary[3] energy sources supplying the world's energy consumption in 2006 were petroleum and other liquids (35.8%), coal (26.7%), natural gas (23.5%), nuclear–electric (6.4%), hydroelectric (6.5%), and other renewables (geothermal, wind, solar, and wood) (1%).[4] This is depicted in the pie chart of Figure 2.3.

The trend of the growth of annual primary energy consumption from 1980 to 2008 and the prediction to 2030 is given in Figure 2.4.

For 2030, EIA predicts the following share of energy sources: petroleum (33.5%), coal (29%), natural gas (24%), nuclear (6%), and hydro and other renewables (7.5%). Taking the 1980–2008 period, the worldwide coal consumption grew by 0.8%/y, natural gas by 2.45%/y, petroleum by

---

[3]Primary energy is energy produced from energy resources such as fossil, nuclear, or renewable energy. It is distinguished from secondary energy, such as electric power or synthetic fuel, which is derived from primary energy sources.

[4]In converting nuclear and hydroenergy to primary energy in the same units, the U.S. Energy Information Agency uses the thermal energy that would be used in an equivalent steam power plant with a thermal efficiency of 31%.

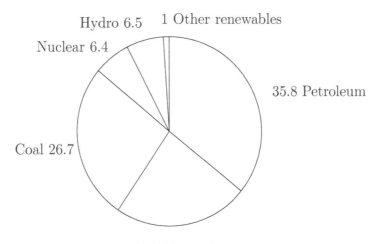

Hydro 6.5    1 Other renewables

Nuclear 6.4

35.8 Petroleum

Coal 26.7

23.5 Natural gas

**FIGURE 2.3** Proportions (%) of the world's energy consumption supplied by primary energy sources, 2008.[1]

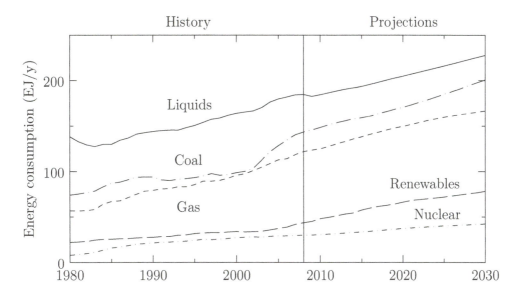

**FIGURE 2.4** Trend of growth of annual primary energy consumption from 1980 to 2008 and prediction to 2030.[1]

1.1%/y, nuclear–electric by 2.2%/y, hydroelectric by 2.1%/y, and other renewables (mainly wind) by 13%/y. Thus, despite the rapid growth of wind and some other renewable energies (e.g., ethanol from corn and sugar cane), their relative contribution to total energy consumption is not expected to change significantly over the next few decades. Currently more than 85% of all primary energy comes from fossil fuels (petroleum, coal, and natural gas). In the next few decades the relative ratio of fossil fuels to total energy consumption is not likely to change very much. Furthermore, in absolute terms, the consumption of fossil fuels is on the increase.

## 2.5   GLOBAL ELECTRICITY CONSUMPTION

Electricity is a secondary form of energy, because primary energy (fossil, nuclear, hydro, geothermal, and other renewable sources of energy) is converted to electricity. The trend of the world's electricity generation from 1990 to 2008 and the prediction to 2030 is depicted in Figure 2.5. In 2006, the world's total electricity generation was 17.3 TWh (1 TWh = E(12) watt hours). By 2030, the generation is predicted to increase to over 30 TWh. Since 1990, growth in electricity generation increased by 2.9%/y. It is predicted that in the near future the electricity generation in non-OECD countries will exceed that of OECD countries.

Of the 2006 electricity generation, 42% was from coal, 19.5% from natural gas, 6% from petroleum, 15% from nuclear, and 18.5% from renewable; the latter includes hydroelectricity (the majority), geothermal, wind, solar, and biomass (mainly wood). This is depicted in the pie chart of Figure 2.6.

In past decades, natural gas became a preferred fuel for electricity generation, and many new power plants were built that employ the technology of Gas Turbine Combined Cycle (GTCC), which is described in Section 6.3.1. Hydropower is a significant contributor to electricity generation in many countries. For example, in Norway, practically all electricity is generated by hydropower, in Brazil 93.5%, New Zealand 74%, Austria 70%, and Switzerland 61%. China and India produce about 19% of their electricity from hydropower; the United States produces 10.7%.

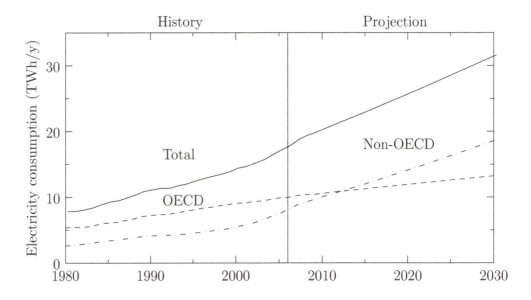

**FIGURE 2.5**   Trend of the world's annual electricity consumption (TWh/y) from 1980 to 2006 and projection to 2030.[1]

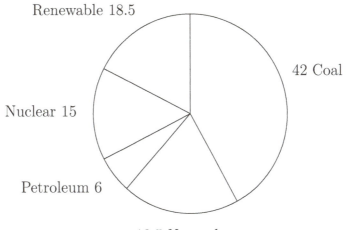

**FIGURE 2.6**  Proportions (%) of the world's electricity generation supplied by primary energy sources, 2006.[1]

While hydroelectricity is a relatively clean source of energy and there is still potential for its greater use worldwide, most of the accessible high "head" hydrostatic dams are already in place. Building dams in remote, inhospitable areas will be expensive and hazardous. Furthermore, there is a growing public opposition to damming up more rivers and streams for environmental reasons and because of the population dislocation that is often involved in creating upstream reservoirs.

Geothermal energy supplies a significant portion of electricity in the following countries: El Salvador 28.5%, Nicaragua 18.5%, Costa Rica 10.3%, New Zealand 5.7%, Iceland 5.3%, Mexico 3.8%, Brazil 2.6%, Indonesia 1.8%, Italy 1.6%, and the United States 0.25%. Geothermal energy has great potential for supplying heat and electricity to many areas of the world. However, at present, geothermal energy is only competitive with fossil energy where the geothermal sources are on, or near, the surface of the earth.

In contrast to fossil-fueled power plants, nuclear power plants do not emit any $CO_2$ into the atmosphere or some other fossil energy-related pollutants ($SO_2$, $NO_x$, particulate matter). However, the fear of nuclear accidents and the unresolved problem of nuclear waste disposal have brought the construction of additional nuclear power plants to a halt in many countries. On the other hand, in some countries, new nuclear power plants are being constructed, and nuclear energy does provide a significant portion of the total electricity generation. For example, in France 76% of the electricity generation is nuclear–electric, South Korea 36%, Germany 29%, Taiwan 27%, Japan 26%, and the United States 22%.

## 2.6  END-USE ENERGY CONSUMPTION IN THE UNITED STATES

To gain some insight as to where the greatest potentials in energy savings lie, it is useful to consider the consumption of energy in each end-use sector. The major sectors are residential, commercial, industrial, and transportation. We shall use the United States as an example. In other countries

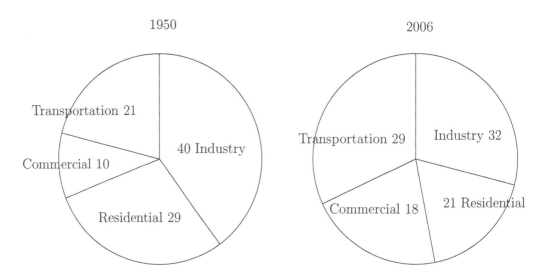

**FIGURE 2.7**  Proportions (%) of U.S. energy consumption provided by primary energy sources by sector, 1950 and 2006.[1]

the end-use pattern may differ, depending on the industrial output of the country (heavy vs. light industry), climate (heating vs. air conditioning or neither), and automobile usage (personal vs. freight; distances traveled). In 2006, the U.S. total primary energy consumption was over 105 EJ. Of this, the industrial sector consumed 32%, residential 21%, commercial 18%, and transportation 29%. In 1950, the shares were industry 40%, residential 29%, commercial 10%, and transportation 21% (Figure 2.7). This reflects the trend of (a) population growth and (b) a shift from an industrial to a service-oriented economy and, within the industrial sector, a shift from heavy ("low-tech") to light ("high-tech") industry. In the United States, over the years 1950–2006 the energy consumption per $GDP also declined. In 1950, it was 18 MJ per constant 2000 dollar; in 2006, it was 9.5 MJ per 2000 dollar. This shift also reflects the increasing share of the service industry to the total economy.

The trend of annual consumption of energy to the four sectors over the years 1980–2008 and projections to 2030 is shown in Figure 2.8. Over the past decade, and for the foreseeable future, the share of industrial energy consumption to total energy consumption is relatively constant; the residential and commercial sectors show a slight but steady increase, and the transportation sector shows a big increase. Let us consider the pattern of energy consumption within each sector.

## 2.6.1  Industrial Sector

Of the total energy used by the industrial sector, about 35% is used for boiler fuel. The major part of the boiler steam is used for direct industrial processes, including electricity generation; a smaller part is used for space heating. Direct process heat consumes 37%, machine drive 16%, nonprocess uses 11%, electrochemical processes 2.4%, and process cooling and other uses 2.4% of the total energy consumed by the industrial sector (Figure 2.9). Even though the industrial sector has become more energy efficient over the past years, there is still room for improvement. Industry could save energy by process modification, better heat exchangers, more efficient drive mechanisms, and

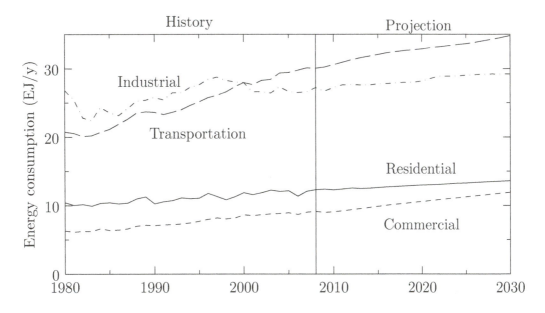

**FIGURE 2.8** Trend of annual U.S. energy consumption for industrial, residential, commercial, and transportation sectors in 1980–2008 and projection to 2030.[1]

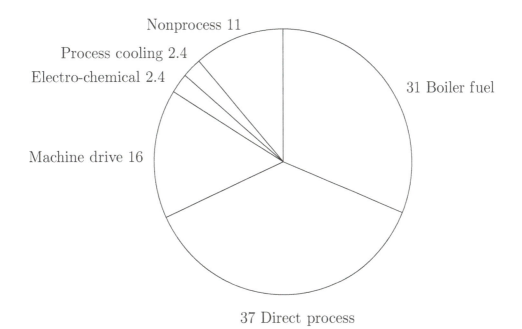

**FIGURE 2.9** Proportions (%) of primary energy use in the U.S. industrial sector. (Data from U.S. Department of Energy, Energy Information Agency, *Manufacturing Energy Consumption Survey*, 2006.)

load-matched, variable speed motors. In some cases, cogeneration can save energy. In cogeneration, heat (steam) and electricity required by the industrial process and the other energy needs for the facility (e.g., space heating) are supplied from the same power plant. The fuel savings are especially pronounced when low-quality heat is adequate for the industrial process or space heating, such as the heat rejected by steam to a condenser after driving a turbine.

### 2.6.2    Residential Sector

Of the total energy used by the residential sector, about 40% is used for appliances and lighting (mainly electricity), 34% for space heating (mainly fossil fuel as petroleum and natural gas), 16% for water heating (mainly electricity and natural gas), and 10% for air conditioning (mainly electricity). This is depicted in Figure 2.10. Significant savings in space heating could be realized by conservation (e.g., lowering the thermostat in winter and raising it in summer), and by better insulation. Solar heating could be more widely utilized for both space and water heating. Appliances can be made more energy efficient and smaller. Lighting could be converted from incandescent to compact fluorescent bulbs (CFL) or light-emitting diodes (LED). Incandescent bulbs convert only 5% of the electricity that heats the filament (usually tungsten) to radiative energy, whereas fluorescent bulbs can convert 20% and LED as much as 30% of the electricity to radiation. However, at present, LED lamps do not deliver the intensity of light output required for domestic uses at a reasonable cost. Because lighting accounts for approximately 9% of household electricity usage in the United States, widespread use of CFL could save as much as 7% of the total U.S. household energy usage.

### 2.6.3    Commercial Sector

Of the total energy used in the commercial sector, about 26% is used for lighting, 19% for space heating, 9% for office equipment, 8.5% for water heat, 7.5% for air conditioning, 7% for ventilation, 5% for refrigeration, 3.5% for cooking, and 14.5% for other uses (Figure 2.11). As in the residential sector, this sector could also realize large energy savings, especially in lighting, space heating, cooling, and ventilation.

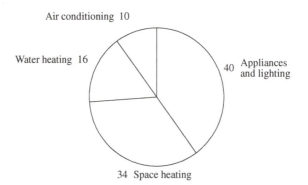

**FIGURE 2.10**   Proportions (%) of primary energy use in the U.S. residential sector. (Data from U.S. Department of Energy, Energy Information Agency, *Residential End Use Energy Consumption*, 1990).

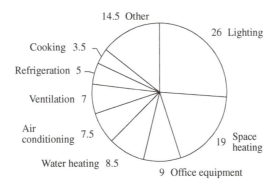

**FIGURE 2.11** Proportions (%) of primary energy use in the U.S. commercial sector. (Data from U.S. Department of Energy, Energy Information Agency, *Energy Markets and End Use*, 1990).

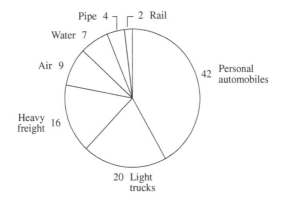

**FIGURE 2.12** Proportions (%) of primary energy use in the U.S. transportation sector. (Data from U.S. Department of Transportation, *Transportation Energy Data Book*, 1993).

### 2.6.4  Transportation Sector

Of the total energy used by the transportation sector, about 42% is used for personal automobile transport, 20% for light trucks (often just used for personal transport), 16% for heavy freight, 9% for air, 7% for water, 4% for pipe, and 2% for rail transport (Figure 2.12). Considering that in the United States almost all transportation fuel is derived from petroleum, that over 60% of petroleum is imported, that automobiles are responsible for about 50% of ground level air pollution, and that the transportation sector contributes about one third of United States $CO_2$ emissions, the transportation sector could realize significant savings in energy consumption and reduction in air pollutant and carbon emissions. This could be accomplished by (a) increased engine fuel efficiency, (b) using fuel cell and battery-powered electric drive cars, (c) using hybrid internal combustion engine–electric powered cars, and (d) increased use of public transportation. Some of these alternatives are further discussed in Chapter 9.

## 2.7    GLOBAL ENERGY SUPPLY

In this section we shall address only the supply of fossil energy, that is, coal, petroleum, natural gas, and unconventional sources of fossil energy, such as oil shale, oil sand, geopressurized methane, and coal seam methane. The supply of nuclear electricity and renewable energy, including hydroelectricity, will be discussed in Chapters 7 and 8, respectively.

### 2.7.1    Coal Reserves

Coal is found on practically every continent and subcontinent. It is found buried deeply in the ground or under the seabed or close to the surface. Coal characteristics vary widely according to its biological origin (forests, low-growing vegetation, swamps) and geological history (age, overburden, temperature, pressure). Thus, the chemical and physical characteristics of coal are also highly variable, such as the content of moisture, minerals (ash), sulfur, nitrogen, and oxygen, as well as heat value, hardness, porosity, etc. Table 2.3 lists the characteristics and composition of several coals. The variability from coal to coal is clearly evident. For example, the fixed carbon content varies from 31.4% for lignite (a relatively young coal) to 80.5–85.7% for anthracite (a relatively old coal). The sulfur content varies from 0.2 to 4% by weight, the higher heating value (HHV) from 15 MJ/kg for lignite to 32 MJ/kg for anthracite.[5]

The world total coal reserves are estimated at 1.037E(12) metric tons.[6] About one half is bituminous and anthracite coal; the other half is subbituminous and lignite coal. Assuming that the average HHV of bituminous and anthracite coal is 29 MJ/kg and that of subbituminous and lignite coal is 19 MJ/kg, the world's coal reserves have a total heating value of about 25,000 EJ. The 2008 world coal consumption amounted to about 140 EJ/y. If the present consumption level were to continue into the future, the world coal reserves would last about 170–190 years. However, worldwide coal consumption keeps increasing at a rate of 0.8%/y. Thus, the lifetime of the world's coal reserves may last only about 100–120 years.[7]

**TABLE 2.3**    Composition and Characteristics of Coal

| Type of coal | Sulfur content (weight %) | Moisture (weight %) | Fixed carbon (weight %) | Ash content (weight %) | Heating value (MJ/kg) |
|---|---|---|---|---|---|
| Anthracite | 0.6–0.8 | 2.8–16.3 | 80.5–85.7 | 9.7–20.5 | 32 |
| Bituminous | 0.7–4 | 2.2–15.9 | 44.2–78.9 | 3.3–11.7 | 29.3 |
| Subbituminous | 0.2–1.4 | 20–30 | 35–45 | 10–15 | 19–26 |
| Lignite | 0.4 | 39 | 31.4 | 4.2 | 15 |

*Source:* International Energy Agency, Clean Coal Centre, *Coal Information*, London, UK, 2009.

---

[5]The HHV includes the latent heat of condensation of the moisture content of the coal and the water vapor formed in combustion, whereas the lower heating value (LHV) excludes it.

[6]U.S. Department of Energy, Energy Information Agency, 2006. *International Energy Outlook*.

[7]The lifetime $T$ (in years) of a reserve under exponential growth is calculated from

$$T = \ln[r(Q_T/Q_0) + 1]/r,$$

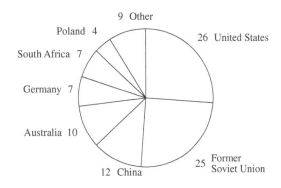

**FIGURE 2.13**   World's coal reserves, in percentages.[6]

The countries where the world's major coal reserves are found are (in percentages of the total) the United States (26), Russia and former Soviet Union countries (25), China (12), Australia (10), Germany (7), South Africa (7), Poland (4), and other countries (9) (Figure 2.13).

In addition to the above reserves, coal may be found in yet unproven reservoirs. Unproven reservoirs are called resources.[8] Some estimates place the coal resources at about 148,000 EJ.[9] The resources could supply the present consumption rate for about 1,000 years. The resources are mainly located in China, the former Soviet Union countries, the United States, and Australia. However, the resources may be located at great depth underground or under the continental shelves. The cost of exploiting these resources will certainly be much greater than that of the reserves.

### 2.7.2   Petroleum Reserves

The terms petroleum and mineral oil, or oil for short, are synonymous. The crude oil found in various parts of the world differs in quality and composition, depending on the biological origin and geological history. Crude oil is found in geological reservoirs underground or under the seabed at depths up to several thousand meters.

A survey of the U.S. Geological Survey estimated the world's oil reserves as 1.6 E(12) barrels (bbl).[10,11] The distribution of the oil reserves among the major oil reservoirs of the world is as follows (in percentages): Middle East (42), North America, including the United States, Canada, and Mexico (15), Russia, including Siberia (14), North and West Africa (7.2), South and Central

---

where $r$ is the rate of growth (per year), $Q_T$ is the total reserve (EJ), and $Q_0$ is present consumption per year (EJ/y).

[8]Reserves of a given fossil fuel are those quantities that geological and engineering information indicate with reasonable certainty to be extractable under existing economic and operating conditions. Resources are those quantities that from geological and engineering information may exist, but their extraction will require different economic and operating conditions.

[9]Anon, World Energy Conference. Guildford, UK: IPC Science and Technology Press, 1978.

[10]U.S. Geological Survey. *Ranking of the World's Oil and Gas Reserves*. USGS Report 97–463, 1997.

[11]1 barrel (bbl) = 42 U.S. gallons = 159 liters. See Table A.2.

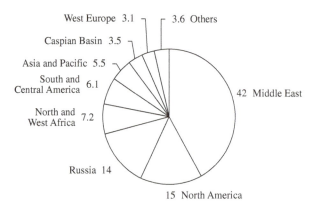

**FIGURE 2.14**   World's oil reserves, in percentages.[10]

America (6.1), Asia and Pacific (5.5), Caspian Basin (3.5), Western Europe, including the North Sea (3.1), and others (3.6) (see Figure 2.14).

Taking an average heating value of crude oil as 6.12 GJ/bbl, the world's oil reserves amount to close to 10,000 EJ. The world's oil consumption in 2008 amounted to about 180 EJ/y. If that consumption rate were to continue into the future, the world's oil reserves would last only about 50–60 years. If oil consumption keeps increasing at a rate of 1.1%/y, the lifetime of the world's oil reserves would be approximately 40–45 years.

The estimates of oil reserves and the lifetime must be taken with great caveat. First, the "proven" reserves are based on estimates from drilling in and around the perimeters of known oil reservoirs and on indirect measurements, such as reflections of sound waves induced by explosives or heavy truck-mounted vibrators. New reservoirs are constantly being discovered on- and offshore. Second, ordinary pumping of oil from reservoirs, called *primary production*, typically produces only 25% of the original oil in place (OIP).[12] Oil is pumped to the surface because the pressure in the production well is less than the lithostatic pressure on the sedimentary rock (mostly sandstone, limestone, or dolomite), the pores of which contain the oil. Another 25% of the OIP can be recovered by injecting highly pressurized water into the rock. The water fractures the rock, increases its permeability, and also dislocates the oil from the rock surface to which it adheres. This is called *secondary recovery*. A further 10–20% of the OIP can be recovered by injecting various gases, solvents, or surfactants into the reservoir. This is called *tertiary recovery* or enhanced oil recovery (EOR). Increasingly popular is the injection of liquid or supercritical carbon dioxide. Liquid $CO_2$ acts as a solvent for high-molecular-weight hydrocarbons, reducing their viscosity and increasing their mobility. In West Texas a significant amount of oil is produced by injecting liquid $CO_2$ that is piped from natural sources in southwestern Colorado. In the future, coal-fueled power plants may be equipped with $CO_2$ capture. The captured $CO_2$ may be piped to semidepleted oil fields for EOR (see Chapter 12).

Considering that new reservoirs may still be discovered and that secondary and tertiary recovery may yield much more oil than what is termed "proven reserves," it is possible that the above-quoted reserves and lifetimes may be much longer. Of course, the cost of new drillings in mostly offshore and inhospitable areas, as well as the cost of enhanced recovery, will be far greater than that of ordinarily pumped oil.

---

[12]Deffeyes, K.S. *Beyond Oil: The View from Hubbert's Peak*, New York. Hill and Wang, 2005.

### 2.7.3   Unconventional Petroleum Resources

In addition to conventional oil reserves, vast amounts of hydrocarbons are distributed in various geological formations, such as oil shales and oil sands. Oil shale deposits are known to exist in the United States in Colorado, Utah, Wyoming, South Dakota, Pennsylvania, Virginia, and West Virginia, as well as in Europe (e.g., Estonia). Oil sands are found in the Canadian Province of Alberta, in Venezuela, and in Colombia. In the United States alone, it is estimated that deposits of oil shale contain perhaps close to 2000 EJ of extractable petroleum. This is about 10 times as much as the proven oil reserves in the United States. The exploitation of these unconventional petroleum resources may require greater financial and technological resources than those for the discovery and extraction of oil reserves. Furthermore, the extraction of petroleum from oil shale may impact the environment to a greater degree than that of pumping oil from on- or offshore wells. On the average, oil shales contain between 60 and 120 liters of petroleum per ton of shale rock. The rock must be excavated and heated in retorts to drive out the liquid petroleum. Thus, a significant fraction of the derived petroleum must be burned to heat the rock for further extraction of petroleum. The process will require complex and expensive control technology for the prevention of air emissions and disposal of liquid and solid waste. After the OPEC oil embargoes in the 1970s, a consortium of oil companies started to produce pilot-scale quantities of petroleum products from oil shale deposits in Colorado. However, after the prices of crude oil fell from a high of $35 per barrel in 1981 to the teens in the late 1980s, all oil shale activities in the United States ceased.

The oil sands in Alberta, Canada, contain vast amounts of hydrocarbons, called bitumen. The Alberta oil sands have the potential to produce 1.7 trillion barrels of oil, equivalent to 10,400 EJ, more than the world's proven oil reserves. In 2005, Canada produced about 1.2 million bbl/day from oil sands. Planned production may increase to 3.5 million bbl/day. Most of the oil is produced *ex situ*. The soil and vegetation overlaying the oil sand deposits is removed. The exposed oil sand is excavated (strip mined) and brought to huge retorts. An enormous auger shovel scoops up the oil sand and delivers it by truck to the treatment facility. The oil sand is treated with hot caustic water, which produces a froth of oily water floating on the top of the retort. The froth is decanted and transferred to a centrifuge. During spinning, the bitumen froth remains in the middle, while the aqueous slurry of clay and sand is thrown to the periphery of the centrifuge. The separated bitumen is sent to a refinery, whereas the aqueous slurry of clay and sand is sent to tailings ponds.

The extraction of bitumen from the oil sands is an expensive process. Oil companies must decide whether the price of oil extracted from oil sands is competitive with regular crude oil. Also, the *ex situ* process wreaks havoc with the environment. First, to expose the oil sands, large amounts of overburden, including forests and vegetation, need to be removed. Second, the tailings that are dumped into ponds contain potentially toxic material that may leach into streams and ground-water. Research and development is underway to find methods that can extract bitumen *in situ*. Such methods may include steam injection, solvent injection, or direct heating by radiofrequency electromagnetic waves. The idea is to make the bitumen less viscous, so it will flow toward the production wells.

### 2.7.4   Natural Gas Reserves

The combustible part of natural gas (NG) consists mainly of methane ($CH_4$) with some admixture of heavier gaseous hydrocarbons (ethane, propane, and butane). However, frequently noncombustible gases are found mixed with NG, namely, $N_2$ and $CO_2$. For example, the recently discovered gas fields off the coast of the Indonesian archipelago contain up to 70% by volume $CO_2$. On the average,

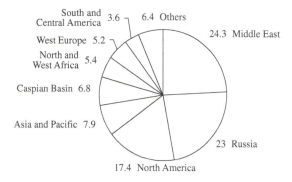

**FIGURE 2.15**  World's natural gas reserves, in percentages.[10]

NG contains 74.4% by weight of carbon, 24.8% hydrogen, 0.6% nitrogen, and 0.2% oxygen. On average, the heating value of NG is 55 MJ/kg, or 38.5 MJ/m$^3$.

Natural gas is a desirable fuel for several reasons. First, it is easy to combust because, being a gas, it readily mixes with air. Thus, the combustion is rapidly completed, and the boiler or furnace volume is smaller than that required for oil or coal combustion. Second, the combusted gas can directly drive a gas turbine with applications in power generation. Third, NG combustion does not produce particulate and sulfurous pollutants. Fourth, NG produces one half the amount of $CO_2$ per unit heating value as does coal and three quarters as much as oil.

The U.S. Geological Survey estimated the world's proven reserves of natural gas to be 190 E(12) m$^3$. This is equivalent to 7350 EJ. The world's rate of consumption of NG in 2008 amounted to 120 EJ. If that rate were to continue in the future, the world's proven gas reserves would be depleted in about 60–65 years. However, gas consumption worldwide is increasing at a rate of 2.45%/y. Thus, the proven reserves may not last longer than 35–40 years.

The gas reserves are distributed among the major reservoirs as follows (in percentages): Middle East (24.3), Russia, including Siberia (23), North America, including the United States, Canada, and Mexico (17.4), Asia and Pacific (7.9), Caspian Basin (6.8), North and West Africa (5.4), West Europe, including the North Sea (5.2), South and Central America (3.6), and others (6.4) (see Figure 2.15).

## 2.7.5  Unconventional Gas Resources

Natural gas exists also in unconventional reservoirs, such as: (a) gas trapped in sandstone, (b) gas trapped in shale rock, (c) gas trapped in coal seams, (d) gas trapped in pressurized underground water reservoirs (geopressurized methane), and (e) methane clathrates, also called hydrates, found at some locations at the bottom of oceans and ice caps. Methane hydrates are basically ice crystals in which a methane "guest" molecule is trapped. Hydrates are formed under the high pressures and low temperatures prevailing at the ocean bottom or ice caps. The first four reservoirs in the United States alone may yield 600–700 EJ, which is about 10% of the world's proven gas reserves. The amount of methane hydrates on the ocean bottoms and under the ice caps may range from 1 to 2 E(16) kg.[13] Taking the heating value of methane as 50.2 MJ/kg, the heating value stored in methane

---

[13]Kvenvolden, K.A. *Chem. Geol.* **71**, 41, 1988.

**TABLE 2.4**  World's Proven Fossil Fuel Reserves

| Fuel | Reserves (EJ) | 2008 consumption (EJ/y) | Rate of growth (%/y) | Lifetime, no growth | Lifetime, growth |
|------|---------------|-------------------------|----------------------|---------------------|------------------|
| Coal | 25,000 | 140 | 0.8 | 170–190 | 100–130 |
| Oil  | 10,000 | 180 | 1.1 | 50–60 | 40–45 |
| Gas  | 7,350 | 120 | 2.5 | 60–65 | 35–40 |

hydrates could amount to 5–10 E(5) EJ, two orders of magnitude larger than proven gas reserves. However, the amount of gas hydrates is speculative, and no technology yet exists to recover the methane from gas hydrates lying on the ocean bottom or under ice caps.

### 2.7.6  Summary of Fossil Reserves

The world's proven fossil fuel reserves are summarized in Table 2.4. Also listed are the estimated lifetimes of the reserves for the case that 2008 consumption continues without growth into the future and for the case that the present growth rates continue into the future. We emphasize the lifetime estimates are only approximate. The estimates include only the world's proven fossil fuel reserves, but not the potential resources and unconventional resources, such as oil shale, oil sands, methane hydrates, and others. Nevertheless, global fossil fuel reserves and resources are finite. Therefore, it is imperative for mankind to curtail the consumption of fossil fuels, not only to prolong their lifetimes, but also to minimize the environmental degradation that the consumption of fossil fuels entails, notably global warming and climate change. Throughout this book, we endeavor to describe the options and technologies for reducing fossil fuel consumption and carbon emissions, including conservation and efficiency improvements and shifting to noncarbon energy sources.

### 2.8  CONCLUSION

We reviewed the present and historic trends of energy consumption and supply patterns in the world as a whole, as well as in individual countries, by industrial sector, by end use, and by per capita use. The so-called "developed" countries consume a much larger amount of energy and emit a much higher rate of $CO_2$ per capita than the "less-developed" countries. However, the converse is true for energy use per GDP. The less developed countries have a higher ratio of energy consumption and a higher emission rate of $CO_2$ per \$GDP than the developed countries.

Measured by the available proven fossil energy reserves and present rate of consumption, coal may last 170–190 years, oil 50–60 years, and natural gas 60–65 years. Unconventional fossil energy resources, such as oil shale, oil sands, geopressurized methane, and methane hydrates, may extend the lifetime of fossil fuels several fold, but their exploitation will require greatly increased capital investment and improved technology. The price of the delivered product will be much higher than is currently paid for these commodities.

The major conclusion is that for the sake of husbanding the fossil fuel reserves and mitigating air pollution and the $CO_2$-caused global climate change, mankind ought to conserve these fuels, increase the efficiency of their uses, and shift to nonfossil energy sources.

## PROBLEMS

### Problem 2.1

From Figure 2.1 determine the rate of growth $r$ (%/y) of global energy consumption for the years 1980 to 2005. Because the rate of growth is exponential, use the equation $Q_t = Q_0 \exp(rt)$ to determine $r$.

### Problem 2.2

Obtain the latest trend of $CO_2$ concentrations in the atmosphere (ppmv) from 1958 to date from Carbon Dioxide Information Analysis Center (CDIAC), Oak Ridge National Laboratory (http://cdiac.ornl.gov). Because the rate of growth is exponential, use the equation as in Problem 2.1 to determine the rate of growth $r$ (%/y).

### Problem 2.3

Figure 2.4 depicts the trend of global energy consumption by fuel type (liquids, coal, natural gas, renewables, nuclear) from 1980 to 2008, with projections to 2030. Draw in a linear best-fit trend for the years 1980 to date. Determine the rate of growth by fuel type (%/y) using the year 1980 as a base.

### Problem 2.4

Figure 2.5 depicts the trend of global electric power generation from 1980 to 2008, with projections to 2030. Because the rate of growth is exponential, use the equation as in Problem 2.1 to determine the worldwide rate of growth $r$ (%/y).

### Problem 2.5

The world's coal reserves were estimated in Section 2.7.1 to have a total heating value of 2.5E(4) EJ. The world's coal consumption in 2008 amounted to 140 EJ/y. Estimate (a) the lifetime of the coal reserves assuming the yearly consumption is constant and (b) the lifetime of the coal reserves if the consumption rate increases exponentially with $r = 0.5, 0.8, 1\%/y$. For exponential growth, use the lifetime estimate equation $T = \ln[r(Q_T/Q_0) + 1]/r$, where $Q_T$ is the total reserve and $Q_0$ is the 2008 consumption rate. Tabulate results for (a) and (b).

### Problem 2.6

The world's petroleum reserves were estimated in Section 2.7.2 to have a total heating value close to 1E(4) EJ. The world's oil consumption in 2008 amounted to about 180 EJ/y. Estimate (a) the lifetime of the petroleum reserves assuming the yearly consumption is constant and (b) the lifetime of the petroleum reserves if the consumption rate increases exponentially with $r = 1, 1.5$, and $2\%/y$. Tabulate results for (a) and (b).

### Problem 2.7

The world's natural gas reserves were estimated in Section 2.7.4 to have a total heating value close to 7,350 EJ. The world's NG consumption in 2008 amounted to about 120 EJ/y. Estimate (a) the lifetime of the NG reserves assuming the yearly consumption is constant and (b) the lifetime of the NG reserves if the consumption rate increases exponentially with $r = 1, 2$, and $3\%/y$. Tabulate results for (a) and (b).

### Problem 2.8

It is estimated that the Alberta, Canada, oil sands contain 60 to 120 L petroleum per ton of oil sand (0.38 to 0.75 bbl/t). Suppose the petroleum (bitumen) is to be extracted *in situ* by heating the sandstone with injected steam to liquefy the bitumen. Assuming the heating value of the extracted liquid petroleum is 6.12 GJ/bbl and the heat capacity of sandstone is 1.1 J/g/°C, what percentage of the heating value of the extracted petroleum needs to be spent to heat the oil sand from 30 to 250°C to liquefy the bitumen?

## BIBLIOGRAPHY

Deffeyes, K. S. *Beyond Oil*. New York: Hill and Wang, 2005.

Ghosh, T. K., and M. A. Prelas. *Energy Resources and Systems*. New York: Springer, 2009.

Ngo, Ch., and J. Natowitz. *Our Energy Future: Resources, Alternatives and the Environment*. Hoboken, NJ: John Wiley and Sons, 2009.

Tester, J. F., E. M. Drake, M. W. Golay, M. J. Driscoll, and W. A. Peters. *Sustainable Energy: Choosing among Options*. Cambridge, MA: MIT Press, 2005.

U.S. National Academies. *America's Energy Future*. Washington, DC: National Academy Press, 2008.

# Thermodynamic Principles of Energy Conversion

## 3.1  INTRODUCTION

The development of the steam engine, an invention that powered the first two centuries of the industrial revolution, preceded the discovery of the scientific principle involved, namely, the production of mechanical work in a device that utilizes the combustion of fuel in air. The scientific understanding that explains the production of work in different kinds of combustion engines is derived from the laws of thermodynamics, developed in the 19th century. In the 20th century these principles aided the development of engines other than the steam engine, such as the reciprocating gasoline and diesel engines, the gas turbine, and the fuel cell. With the aid of this scientific hindsight, in this chapter we will review how the laws of thermodynamics determine the functioning of these sources of mechanical energy and especially how they limit the amount of mechanical work that can be generated from the burning of a given amount of fuel.

The source of mechanical power developed from the combustion of fossil fuels or the fission of nuclear fuel in an engine is the energy released by the change in molecular or nuclear composition. This released energy is never lost but is transformed into other forms, appearing as some combination of mechanical or electric energy, internal energy of the molecular or nuclear products of the reactions, or energy changes external to the engine caused by heat transfer from it. This conservation of energy is explicitly expressed by the first law of thermodynamics.

The principle of energy conservation, or first law, places an upper limit on the conversion of chemical or nuclear energy to mechanical work; namely, the work of an engine cannot exceed the energy available. Experience shows that the work is significantly less than the energy released by fuel reactions in engines, a matter having important practical consequences. The scientific principle that explains why, and by how much, there is a work shortfall is called the second law of thermodynamics. In combination with the first law, it enables us to understand the limits to producing work from fuel and to improve the engines that have been invented to accomplish this task.

The laws of thermodynamics cannot substitute for invention, but they do inform us of the performance limits of a perfected invention. In this chapter we review the principles of operation of the major inventions that transform chemical to mechanical energy: the steam engine, the gasoline and diesel engines, the gas turbine, and the fuel cell. In each case we show that the laws of thermodynamics provide limits on how much of the fuel's energy can be converted to work and describe how various debilitating factors reduce the work output of practical devices to values below this limit. Thermodynamic analyses of this type provide guidance for improving the energy efficiency of mechanical power production.

In the next five sections of this chapter we summarize the relevant principles of thermodynamics as embodied in its two laws, including the concepts of energy, work and heat, the definition of useful thermodynamic functions, and their application to the steady flow of working fluids, such as air and combustion gases. We then proceed to the special application of the combustion of fuels and the various thermodynamic cycles that explain how common heat and combustion engines function. A subsequent section treats the fuel cell separately, a more recent development that operates on a different principle than heat engines, producing work directly in electrical form. The fuel efficiencies of various power-producing cycles are then summarized. The chapter concludes with a short discussion of the energy efficiency of synthetic fuel production.

## 3.2   THE FORMS OF ENERGY

The concept of energy, which originated with Aristotle, has a long history both in science and as a colloquial term. It is a central concept in classical and quantum mechanics, where it appears as a constant of the motion of mechanical systems. In the science of thermodynamics, energy has a distinct definition that distinguishes it from heat, work, or power. In this section we define thermodynamic energy as a quantity that is derived from an understanding of the physical and chemical properties of matter.

### 3.2.1   The Mechanical Energy of Macroscopic Bodies

Newtonian mechanics identifies two forms of energy, the *kinetic energy* of a moving body and the *potential energy* of the field of force to which the body is subject. The kinetic energy $KE$ is equal to the product of the mass $M$ of the body times one half of the square of its velocity $V$,

$$KE \equiv \frac{1}{2}MV^2.$$

The potential energy $PE$ of a body subject to a force $\mathbf{F}\{\mathbf{r}\}$ at a location $\mathbf{r}$ in space is equal to the work done in moving the body to the location $\mathbf{r}$ from a reference position $\mathbf{r}_{\text{ref}}$,

$$PE \equiv -\int_{\mathbf{r}_{\text{ref}}}^{\mathbf{r}} \mathbf{F}\{\mathbf{r}\} \cdot d\mathbf{r}.$$

While the kinetic energy has always a positive value with a zero minimum, the potential energy's value is measured with respect to the reference value and may be positive or negative.

It is one consequence of Newton's laws of motion of a body in a force field that the sum of the kinetic and potential energies is a constant of the motion (i.e., it is not a function of time). Calling this sum the total energy $E$, we have

$$E \equiv KE + PE = \text{constant}.$$

Absent any other force applied to the body, its energy $E$ is unchanged despite its movement within the region of space available to it. We may call this the *principle of conservation of energy*.

A simple example of the motion of a body that possesses kinetic and potential energy is that of a satellite moving in orbit around the earth. The potential energy is that of the satellite mass in the earth's gravitational field, which increases inversely in proportion to the distance of the satellite from the center of the earth. If the satellite is in an eccentric orbit about the earth, the conservation of energy requires that the satellite speed (and kinetic energy) is a maximum where it dips closest to the earth's surface.

It is possible to change the energy $E$ of a body by acting upon it with an external force. In the case of the satellite, a quick impulse from its rocket engine can change its velocity and kinetic energy, thereby changing its total energy $E$. The amount by which the total energy $E$ is changed is directly related to the amount of the rocket impulse. It is thereby possible to extend the statement of the conservation of energy by taking into account the changes in $E$ brought about by external impulses applied to the moving body. This is the most general form of the principle of conservation of energy.

### 3.2.2   The Energy of Atoms and Molecules

The matter of a macroscopic body is composed of microscopic atoms and/or molecules (themselves aggregates of atoms). Sometimes, as for gases, these molecules are so widely separated in space that they may be considered to be moving independently of each other, each possessing a distinct total energy. Otherwise, in the case of liquids or solids, each molecule is under the influence of forces exerted by nearby molecules, and we can only distinguish the aggregate energy of all the molecules of the body. We call this energy the *internal energy* and give it the symbol $U$. Even though the motion of microscopic molecules is not describable by Newtonian mechanics, it is still possible to consider their total energy to be the sum of the kinetic energies of their motion and the potential energies of their intermolecular forces.

It is not possible to observe directly the energies of individual atoms of a thermodynamic substance, but changes in its internal energy are indirectly measurable by changes in temperature, pressure, and density. These observables, called thermodynamic state variables, are the surrogates for specifying internal energy.

### 3.2.3   Chemical and Nuclear Energy

Molecules are distinct stable arrangements of atomic species. Their atoms are held together by strong forces that resist rearrangement of the atoms. To disassemble a molecule into its component atoms usually requires the expenditure of energy, so that the molecules of a body may be considered to possess an energy of formation related to how much energy was involved in assembling them from their constituent atoms. If the internal energy $U$ of a material body is changed, but the individual molecules remain intact, then their chemical energy of formation remains unchanged and contributes nothing to the change in $U$. On the other hand, if a chemical change occurs, so that new molecular species are formed from the atoms present in the original species, there will be a redistribution of energy among the components of the internal energy, of which the chemical energy of formation of the molecules must be taken into account.

A similar energy change accompanies the formation of new atomic nuclei in the fission of the nuclei of heavy elements or the fusion of light ones. Because the binding forces that hold nuclei together are so much larger than those that hold molecules together, nuclear reactions are much more energetic than molecular ones. Nevertheless, we can consider both molecules and atomic nuclei to possess energies of formation that must be taken into account in expressing the conservation of energy for material bodies that experience chemical or nuclear changes in composition.

### 3.2.4 Electric and Magnetic Energy

Molecules that possess a magnetic or electric dipole moment can store energy when they are in the presence of a magnetic or electric field, in the form of magnetic or electric polarization of the material. This energy is associated with the interaction of the molecular dipoles of the material body with the external electric charges and currents that give rise to the applied electric or magnetic field. Since capacitors and inductors are common components of electronic and electrical circuits, this form of energy is important to their functioning.

### 3.2.5 Total Energy

The various forms of energy that can be possessed by a material body, as described above, can be added together to define a total energy, to which we give the symbol $E$,

$$E \equiv KE + PE + U + E_{chem} + E_{nuc} + E_{el} + E_{mag}. \tag{3.1}$$

It is very seldom that more than just a few of these forms are significant in any practical process for which there are changes in the total energy. There are many examples. In a gasoline engine, the combustion of the fuel–air mixture involves $U$ and $E_{chem}$; in a steam and gas turbine, only $KE$ and $U$ change; in a nuclear power plant fuel rod, $U$ and $E_{nuc}$ are involved; and in a magnetic cryogenic refrigerator $U$ and $E_{mag}$ are important. Nevertheless, the manner in which the various forms of energy enter into the laws of thermodynamics is expressed through the total energy $E$, a result of great generality and consequence.

## 3.3 WORK AND HEAT INTERACTIONS

Thermodynamics deals with the interaction of a thermodynamic material system and its environment.[1] It is through such interactions that we are able to generate mechanical power or other useful effects in the environment. There are two quite different but important modes of interaction of a system with its environment, called the work interaction and the heat interaction. Each of these is a process in which, over time, the system and its environment undergo physical and/or chemical changes related to the kind of interaction taking place, either work or heat (or both simultaneously).

As we shall see below, work and heat interactions are distinguishable from each other by the character of the changes in the system and the environment. Both are quantifiable in terms of the interaction, being expressed in energy units. Neither is a form of energy, but only a transaction quantity that accounts for the character of the exchange of energy between a thermodynamic system and its environment.

---

[1] In thermodynamics, the environment of a thermodynamic system is that part of its material surroundings with which the system interacts.

### 3.3.1  Work Interaction

Newtonian mechanics employs the concept of work as the exertion of a force acting through a displacement or a couple acting through an angular displacement.[2] We say that the amount of work required to lift a mass $m$ through a vertical distance $r$ in the earth's gravitational field is the product of the magnitude of the gravity force, $mg$, times the distance $r$, where g is the magnitude of the local acceleration of gravity. In thermodynamics, by convention, positive work is defined to be the product of the force exerted by a system on the environment times any displacement of the environment that occurs while the force is acting. (By Newton's principle, the force that the system exerts on the environment is equal in magnitude but opposite in direction to the force that the environment exerts on the system.) If the force $F_{en}$ exerted on the environment by the system is accompanied by an incremental displacement $dr_{en}$ of the environment in the direction of $F_{en}$, the increment of work may be expressed as

$$dW \equiv F_{en}\, dr_{en}. \tag{3.2}$$

When the system does work on the environment, $dW$ is positive; when the environment does work on the system then $dW$ is negative.

There are many simple examples of a work interaction. If a gas is contained in a circular cylinder capped at one end and fitted with a movable piston at the other, then the force exerted by the gas on that portion of the environment that is the movable piston is $pA$, where $p$ is the gas pressure and $A$ is the piston face area. If the piston is displaced an incremental distance $dr_{en}$ in the direction of the pressure force $pA$, the positive work increment $dW$ in this displacement is

$$dW = pA\, dr_{en} = p(A\, dr_{en}) = p\, dV, \tag{3.3}$$

where $dV = A\, dr_{en}$ is the increment in the volume $V$ of the gas in the cylinder. Or the work interaction with the environment may involve the movement of an increment of electric charge $dQ_{en}$ through an increase in electric potential $\phi_{en}$, such as when a current flows from the system to and from an electric motor in the environment, for which the work increment is

$$dW = \phi_{en}\, dQ_{en}. \tag{3.4}$$

Another common example is the rotation of the shaft of a turbine (a material system) that applies a torque $\mathcal{T}_{en}$ (or couple) to an electric generator (the environment) attached to the turbine, rotating it through an increment of angle $d\theta_{en}$ in the same direction as the torque, giving rise to a positive work increment,

$$dW = \mathcal{T}_{en}\, d\theta_{en}. \tag{3.5}$$

These are but a few specific examples of the many possible kinds of work interaction between a thermodynamic system and its environment.

---

[2] A couple is the product of the distance separating two equal but opposite forces times the magnitude of the force.

### 3.3.2   Heat Interaction

We are familiar with the processes whereby substances are warmed or cooled. Cooking or refrigerating food requires increasing or decreasing its temperature by bringing it into contact with a warmer or cooler environment. A temperature difference between a system and its environment is required for a heat interaction to transpire. If the environment undergoes a temperature increase after a system warmer than the environment is brought into contact with it, then a heat interaction has taken place. The incremental amount of the heat interaction $dQ$, which in this case equals the energy transfer from the system to the environment, is equal to the product of the heat capacity $C_{en}$ of the environment times its temperature increase $dT_{en}$. But by convention the energy transferred *to* a system in a heat interaction is regarded as a positive quantity so that in this case the energy transfer is negative. Consequently,

$$dQ = -C_{en}\, dT_{en}. \tag{3.6}$$

We usually describe this interaction as *heat transfer*, although it is energy that is being exchanged in a process solely involving a heat interaction.

Both heat and work quantities involved in an interaction of a system and its environment are recognized by their effects in the environment, as described in Equations (3.2)–(3.6) above. Furthermore, both heat and work interactions may occur simultaneously, being distinguishable by their different physical effects in the environment (e.g., $F_{en}\, dr_{en}$ vs. $C_{en}\, dT_{en}$).

### 3.4   THE FIRST LAW OF THERMODYNAMICS

The first law of thermodynamics is an energy conservation principle. It relates the incremental change in energy $dE$ of a system with the increments of work $dW$ and heat $dQ$ recognizable in the environment during an interaction of the system with its environment. In words, the law states that the increment in system energy $dE$ equals the increment in heat $dQ$ transferred to the system minus the work $dW$ done by the system on the environment,

$$dE = dQ - dW. \tag{3.7}$$

It is an energy conservation principle in the sense that the sum of the system energy change $dE$, the work $dW$, and the heat $(-dQ)$ added to the environment is zero (i.e., this sum is a conserved quantity in any interaction with the environment).

Equation (3.7) expresses the first law in differential form. If many successive incremental changes are added to accomplish a finite change in the system energy $E$ from an initial state $i$ to a final state $f$, the first law may be expressed in integral form as

$$E_f - E_i = \int_i^f dQ - \int_i^f dW. \tag{3.8}$$

In this form, the first law expresses the finite change in energy of the system as equal to the sum of the heat transferred to the system minus the work done by the system on the environment during the process that brought about the change from the initial to the final state.

The integrals of the heat and work quantities on the right side of Equation (3.8) cannot be evaluated unless the details of the process that caused the change from the initial to the final state of the system are known. In fact, there may be many different processes that can bring about the same change in energy of the system, each distinguished by different amounts of heat and work, but all having in common that the sum of the heat and work quantities added to the environment is the same for all such processes that change the system from the same initial to final states.

In some power-producing and refrigeration systems, a working fluid undergoes a series of heating, cooling, and work processes that returns the fluid to its initial state. Since $E_f = E_i$ for such a cyclic process, the integral expression of the first law of thermodynamics, Equation (3.8), has the form

$$\oint dQ = \oint dW, \qquad (3.9)$$

where the symbol $\oint$ identifies the cyclic process for which the heat and work integrals are evaluated. In other words, in a cyclic process the net heat and work quantities are equal.

## 3.5    THE SECOND LAW OF THERMODYNAMICS

The goal of engineers who design power plants is to devise a system to convert the energy of a fuel into useful work. If we consider the combustion of a fossil fuel to provide a source of heating, then the desirable objective is to convert all of the fuel energy to work, as the first law, Equation (3.9), allows. However, the second law of thermodynamics states that it is not possible to devise a cyclic process in which heating supplied from a single source is converted entirely to work. Instead, only some of the heat may be converted to work; the remainder must be rejected to a heat sink at a lower temperature than the heat source. In that way the net of the heat added and subtracted in the cycle equals the work done, as the first law requires.

It is not possible to express directly this second law statement in the form of an equation. However, it is possible to deduce three important consequences of the second law. The first is that there exists an absolute temperature scale, denoted by $T$, which is independent of the physical properties of any substance and which has only positive values. The second is that there is a thermodynamic property called entropy, denoted by $S$, whose incremental change is equal to the heat interaction quantity $dQ$ divided by the system temperature $T$ for any incremental process in which the system temperature remains spatially uniform, called a reversible heat addition, or

$$dS \equiv \left(\frac{dQ}{T}\right)_{rev}. \qquad (3.10)$$

The third deduction is called the inequality of Clausius. It states that, in any process, $dS$ is equal to or greater than the ratio $dQ/T$,

$$dS \geq \frac{dQ}{T}. \qquad (3.11)$$

As a consequence, in a process for which $dQ = 0$, which is called an *adiabatic* process, the entropy may remain the same or increase, but may never decrease. An adiabatic process for which the entropy increases is an irreversible process because the reverse of this process, for which the entropy would decrease ($dS < 0$), violates the second law as expressed in Clausius' inequality.

There are many important consequences of Clausius' inequality that we will not examine in detail in this chapter, but will be identified as such at the appropriate occasion of use. Among them are the conditions for thermodynamic equilibrium, including thermochemical equilibrium, and the limits on the production of useful work in cycles or processes.[3]

## 3.6   THERMODYNAMIC PROPERTIES

The most common methods for utilizing the energy of fossil or nuclear fuels require the use of fluids as the means to generate mechanical power or to transport energy to a desired location. The thermodynamic properties of fluids thereby assume a great importance in the systems that transform energy.

We know that in a steam power plant the working fluid, water, undergoes large changes in temperature and pressure as it moves through the boiler, turbine, and condenser. The thermodynamic properties pressure $p$ and temperature $T$ are called *intensive* properties because their values are not proportionate to the mass of a fluid sample but are the same at all points within the sample. On the other hand, the energy $E$, volume $V$, and entropy $S$ are *extensive* properties in that their values are directly proportionate to the mass of a fluid sample.[4] But if we divide an extensive property, such as $E$, by the mass $M$ of fluid whose energy is $E$, then the ratio $E/M$, called the *specific energy*, is independent of the amount $M$ of fluid. Denoting specific extensive properties by a lowercase letter, we have for the specific energy $e$, volume $v$, and entropy $s$

$$e \equiv \frac{E}{M}; \qquad v \equiv \frac{V}{M}; \qquad s \equiv \frac{S}{M}. \tag{3.12}$$

The use of specific extensive properties simplifies the analysis of thermodynamic systems utilizing fluids and other materials to produce work or transform energy. By following the changes experienced by a unit mass of material as it undergoes a change within the system, the work and heat quantities per unit mass may be determined. The total work and heat amounts for the system may then be calculated by multiplying the unit quantities by the total mass utilized in the process.

The first and second law properties, energy and entropy, are sometimes not convenient to use in analyzing the behavior of thermodynamic systems. Rather, particular combinations of the properties $p$, $T$, $v$, $e$, and $s$ turn out to be more helpful. One of these useful properties is called the *enthalpy* and is defined as

$$h \equiv e + pv. \tag{3.13}$$

The enthalpy has a simple physical interpretation. Suppose a unit mass of material is surrounded by an environment in which the pressure is fixed and equal to the pressure $p$ of the system. If a small

---

[3]The third law of thermodynamics is an additional principle that is closely related to the second law. It states that the entropy of all thermodynamic systems is zero at the absolute zero of temperature. Among other things, it is important to the determination of the free energy change in combustion reactions.

[4]In the following discussion, we disregard the kinetic and potential energy components of the total energy as expressed in equation (3.1) as we are considering the properties of a system that is stationary in the earth's gravity field. When it is necessary to take this motion into account, as in a flow through a turbine, we will explicitly add these additional energy components at the appropriate point.

amount of heat, $dq$, is added to the system, its temperature will rise and it will expand, undergoing an increment of volume $dv$ and performing an amount of work $dw = p\,dv$ on the environment. According to the first law, Equation (3.7), the heat and work amounts change the energy $e$,

$$de = dq - p\,dv$$

$$dq = de + p\,dv = de + d(pv) = d(e + pv) = dh,$$

where the equality $p\,dv = d(pv)$ follows from the constancy of $p$ in this process. Thus, the amount of heat added in a constant pressure process is equal to the increase in enthalpy of the material. The ratio of the increase in enthalpy, at fixed pressure, to the increment of temperature experienced in this process is called the *constant-pressure specific heat* and is given the symbol $c_p$,[5]

$$c_p \equiv \left( \frac{\partial h\{p, T\}}{\partial T} \right)_p. \tag{3.14}$$

When we consider a similar heating at fixed volume, no work is done and the increase in energy $de$ is equal to the heat increment $dq$. The ratio of the energy increase to the concomitant temperature increase is called the *constant-volume specific heat*, $c_v$,

$$c_v \equiv \left( \frac{\partial e\{v, T\}}{\partial T} \right)_v. \tag{3.15}$$

A second property that will be found useful is *Gibbs' free energy*, given the symbol $f$, and defined by

$$f \equiv h - Ts = e + pv - Ts. \tag{3.16}$$

For a process that proceeds at constant temperature and pressure, the second law of thermodynamics requires that the amount of work done by a system cannot exceed the reduction of free energy $f$.[6] The free energy is a useful thermodynamic function in cases of chemical or phase change. For example, a sample of liquid water and water vapor can be held in equilibrium at the boiling temperature corresponding to the sample pressure. If heat is added while the pressure remains fixed, some liquid is converted to vapor but the temperature remains unchanged. For this heat transfer process at fixed temperature and pressure, the free energy $f$ is unchanged.

It is possible to express the relationship between these properties in differential form by noting that, for a reversible process, $dq = Tds$ by Equation (3.10), so that

$$Tds = de + p\,dv \tag{3.17}$$

$$= dh - v\,dp \tag{3.18}$$

$$= dh - s\,dT - df. \tag{3.19}$$

---

[5]In this expression, $h\{p, T\}$ is considered a property depending upon the pressure $p$ and temperature $T$. The partial derivative $\partial h / \partial T$ is taken with respect to $T$ holding $p$ fixed. A similar constraint is implied in Equation (3.15), where $e\{v, T\}$ is a function of volume $v$ and temperature $T$.

[6]See Section 3.11.

## 3.7   STEADY FLOW

Many thermodynamic systems incorporate components through which a fluid flows at a mass flow rate $\dot{m}$ that is invariant in time, which we call *steady flow*. That is true for the compressor, combustor, and turbine of a gas turbine power plant; and for the boiler, steam turbine, condenser, and feed pump of a steam power plant; but not for the cylinder of an automobile engine, where the flow is intermittent.[7] If the flow is steady, the first law may be expressed in a form that relates the thermodynamic properties of the inflowing and outflowing fluid streams with the rates $\dot{Q}$ and $\dot{W}$ at which heat is added to and work is done by the fluid within the component in question,

$$\dot{m}\,h_{\text{out}} = \dot{m}\,h_{\text{in}} + \dot{Q} - \dot{W}, \tag{3.20}$$

where the subscripts out and in identify the thermodynamic state of the fluid at the outlet and inlet of the component.[8] For a boiler, $\dot{Q}$ would be the rate at which heat is added to change the water flow to steam; for a steam turbine, $\dot{W}$ is the mechanical power delivered by the turbine as its shaft rotates to drive an electric generator or other mechanical load. If we divide Equation (3.20) by $\dot{m}$, then $q \equiv \dot{Q}/\dot{m}$ and $w \equiv \dot{W}/\dot{m}$ are the heat and work quantities per unit mass of fluid flowing through the device; the change in fluid enthalpy $h_{\text{out}} - h_{\text{in}}$ is then equal to the sum of these terms,

$$h_{\text{out}} - h_{\text{in}} = q - w. \tag{3.21}$$

Many power plant components belong to one of two categories: adiabatic ($\dot{Q} = 0$) devices that deliver or absorb mechanical power (e.g., turbines, pumps, compressors) or workless ($\dot{W} = 0$) heat exchangers in which a fluid is heated or cooled. The combustion chamber of a gas turbine power plant is an exception to this rule because it is both adiabatic and workless.

## 3.8   HEAT TRANSFER AND HEAT EXCHANGE

While the laws of thermodynamics tell us how much work can be generated by adding and subtracting heat from a working fluid, they don't tell us how quickly we can accomplish this task, a matter of great practical consequence because the time rates of exchange of work and heat quantities determine the mechanical or thermal power that can be produced. The more power that can be generated from a given mass of material, at a given cost, the more desirable the power system becomes.

Depending upon the circumstances, we may want to augment or diminish the rate at which heat flows from hot to cold environments. For example, a steam power plant boiler is designed to facilitate the rapid heating of the circulating water by the hot combustion gases, with the heat being transferred through the wall of the metal tubes within which the water flows and outside of which the hot gases circulate. On the other hand, when heating a building's interior space in

---

[7]When operated at a steady speed, the inflows of fuel and air and outflow of exhaust gas from an automobile engine may be regarded as steady, so that steady flow Equation (3.20) below may be applied.

[8]We have omitted in this expression the contributions of the kinetic and potential energies of the fluid, which in many cases are negligible compared with the other quantities.

winter months, the loss of heat to the cold exterior environment is minimized by installing thermal insulation in the walls.

In most cases of steady heat transfer from a hot to a cold environment, the time rate of heat transfer $\dot{Q}$ may be represented by[9]

$$\dot{Q} = \mathcal{U}A(T_h - T_c), \tag{3.22}$$

where $T_h - T_c$ is the temperature difference between the hot and cold environments, $A$ is the surface area of the material that separates the two environments and across which the heat flows, and $\mathcal{U}$ is the *heat transfer coefficient*, a property of the material separating the two environments.[10] To attain high values of $\mathcal{U}$, one should use a thin layer of a material such as copper, which is a good heat conductor, and provide vigorous motion of the hot and cold fluids with which it is in contact. To obtain low thermal conductances, one needs thick layers of thermally insulating material, like foamed plastics.

Heat exchangers are passive devices that accomplish a transfer of heat, usually between two streams of fluids, one hot and the other cold. Typically, one fluid flows inside parallel cylindrical tubes and the other outside of them. In a steam boiler, for example, the cold water (or steam) flows inside the tubes while the hot combustion gases flow around them. Similarly, in a steam power plant condenser the cold cooling water flows through the tubes while the hot exhaust steam from the turbine passes outside the tubes, condensing on the cold surface. The use of heat exchangers is often necessary to the functioning of a power plant, as in these examples, but they necessarily exact penalties in the form of loss of mechanical power, increased economic cost, and reduced thermodynamic efficiency. As an example of the latter, consider the design of a condenser that must transfer a fixed amount of heat per unit time, $\dot{Q}$. According to Equation (3.22), we could reduce its size ($A$), and thereby its cost, by increasing the temperature difference ($T_h - T_c$) and thereby reduce the steam cycle efficiency. Alternatively, we could increase the heat transfer coefficient $\mathcal{U}$ by pumping the cooling water through the tubes at a higher speed, but that would incur an extra pumping power loss. As a consequence, the transfer of heat at finite rates in thermodynamic systems inevitably incurs performance penalties that cannot be reduced to zero except by the expenditure of infinite amounts of capital. Fortunately, these performance penalties can be limited to acceptable levels at a cost commensurate with that of other components of the system.

## 3.9  IDEAL HEAT ENGINE CYCLES

Generating mechanical power from the combustion of fossil fuel is not a straightforward matter. One must utilize the combustion process to change the temperature and/or pressure of a fluid and then find a way to use the fluid to make mechanical work by moving a piston or turning a turbine. The first and second laws of thermodynamics limit the amount of work that can be generated for

---

[9]This expression is not a thermodynamic law, although it is in agreement with the requirement of the second law that heat can be transferred only from a hot to a cold body, not the reverse.

[10]The product $\mathcal{U}A$ is called the thermal conductance, in analogy with the electrical conductance of an electric circuit, which is the ratio of the electric current (analogous to the heat flux $\dot{Q}$) to the voltage difference (analogous to the temperature difference $T_h - T_c$).

each unit mass of fuel used, and those limits depend upon the details of how the fuel is used to create power.

To understand the implications of the thermodynamic laws for the conversion of fuel energy to mechanical power, it is convenient to analyze ideal devices in which a fluid is heated and cooled, and produces or absorbs work, as the fluid moves through a cycle. Such a device can be called a *heat engine* in that it exchanges heat with its environment while producing work in a cyclic process. The combustion of fuel is represented in this idealized cycle by the addition of heat from a high-temperature source. Some practical engines, like the gas turbine and the automobile engine, are not heated from an external source. These are termed *internal combustion engines* (ICE). Nevertheless, most of their features can be modeled as ideal heat engine cycles to help us understand their chief attributes.

In this section we consider simple models of heat engines that illustrate the principal features of several practical devices. Of particular importance is the amount of work produced ($w$) in proportion to the amount of heat that is added ($q$) to represent the combustion of fuel, the ratio of which is called the *thermodynamic efficiency* $\eta_{th}(\equiv w/q)$. But other features are of practical consequence as well, and these are displayed in the analysis.

In all these analyses, we assume that the fluid that produces the mechanical work undergoes reversible processes, so that the entropy change is related to the heat addition by Equation (3.10).

### 3.9.1  The Carnot Cycle

The Carnot cycle is a prototype cycle that has little practical importance but is beautifully illustrative of the second law limits on the simplest of heat engine cycles. It is sustained by two heat reservoirs, a hot one of temperature $T_h$ and a cold one of temperature $T_c$. (We may think of the hot reservoir as one that is kept warm by heat transfer from a burning fuel source and the cold one as the atmosphere.) Consider the heat engine to be a cylinder equipped with a movable piston and enclosing a fluid of unit mass. The cycle consists of four parts, as illustrated in Figure 3.1: an isothermal expansion during which an amount of heat $q_h$ is added to the engine (1 → 2 in Figure 3.1); an adiabatic isentropic additional expansion during which the fluid decreases in temperature from $T_h$ to $T_c$ (2 → 3); an isothermal compression while the system adds a quantity of heat $q_c$ to the cold reservoir (3 → 4);

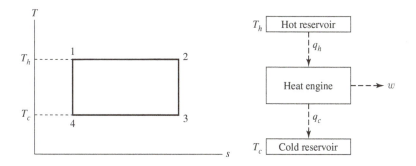

**FIGURE 3.1**  The Carnot cycle consists of isothermal and isentropic expansions (1 → 2, 2 → 3) and compressions (3 → 4, 4 → 1) of a fluid in a cylinder while absorbing heat $q_h$ from a hot reservoir (1 → 2), rejecting heat $q_c$ to a cold reservoir (3 → 4), and producing work $w$.

and finally an isentropic compression to the initial state $(4 \rightarrow 1)$. For this cycle the net work $w$ of the piston per cycle is $\oint p\,dv$ and by the first law Equation (3.9),

$$q_h - q_c = w = \oint p\,dv. \tag{3.23}$$

In Figure 3.1 a temperature–entropy plot of the Carnot cycle shows that the entropy increase $(s_2 - s_1)$ during heating by the hot reservoir is equal in magnitude to the decrease $(s_3 - s_4)$ during cooling, and by the second law (3.10) it follows that

$$\frac{q_h}{T_h} = \frac{q_c}{T_c}. \tag{3.24}$$

By combining these two relations, we may find the thermodynamic efficiency $\eta_{th}$ of the Carnot cycle to be

$$\eta_{th} \equiv \frac{w}{q_h} = 1 - \frac{T_c}{T_h} \tag{3.25}$$

The most remarkable aspect of this result is that the thermodynamic efficiency depends only upon the temperatures of the two reservoirs and not at all upon the properties of the fluid used in the heat engine. On the other hand, the amount of net work $w$ that the heat engine delivers does depend upon the fluid properties and the amount of expansion,

$$w = \oint p\,dv = \oint T\,ds, \tag{3.26}$$

where we have used Equation (3.17) to show that the net work is equal to the area enclosed by the cycle path in the $T, s$ plane of Figure 3.1.

The conclusion to be drawn from the example of the Carnot cycle is that the thermodynamic efficiency is improved by supplying the heat $q_h$ to the engine at the highest possible temperature $T_h$. But for a fuel burning in ambient air and supplying this heat to the hot reservoir so as to maintain its temperature, $T_h$ could not exceed the adiabatic combustion temperature $T_{ad}$. Furthermore, for any $T_h < T_{ad}$ only a fraction of the fuel heating value could be added to the hot reservoir, that fraction being approximately $(T_{ad} - T_h)/(T_{ad} - T_c)$. The resulting thermodynamic efficiency based upon the fuel heating value would then be

$$\eta_{th} = \left(1 - \frac{T_c}{T_h}\right)\left(\frac{T_{ad} - T_h}{T_{ad} - T_c}\right), \tag{3.27}$$

which has a maximum value, when $T_h = \sqrt{T_{ad}T_c}$, of

$$\eta_{th} = \frac{\sqrt{T_{ad}/T_c} - 1}{\sqrt{T_{ad}/T_c} + 1}. \tag{3.28}$$

For example, if $T_{ad} = 1{,}900°C = 2{,}173$ K and $T_c = 25°C = 298$ K, then the maximum thermal efficiency would be 46% when $T_h = 805$K $= 532°C$. This is considerably less than the Carnot efficiency of 86.3% when $T_h = T_{ad}$ for this case.

A possible plan for increasing the efficiency above the value of Equation (3.28) would be to employ a large number of Carnot engines, each operating at a different hot reservoir temperature $T_h$ but the same cold reservoir temperature $T_c$. The combustion products of a constant-pressure burning of the fuel would then be brought into contact with successively cooler reservoirs, transferring heat amounts $dh$ from the combustion gases to produce work amounts $dw$, where

$$dw = \left(1 - \frac{T_c}{T_h}\right) dh = d(h - T_c s), \tag{3.29}$$

and where we have made use of the fact that $dh = T_h\,ds$ for a constant-pressure process. The thermodynamic function $h - T_c s$ that appears in Equation (3.29) is called the *availability*. In this case the change in availability of the combustion products is the amount of the fuel heat that is available to convert to work by use of an array of Carnot engines. It is the maximum possible work that can be generated if the fuel is burned at constant ambient pressure when the ambient environment temperature is $T_c$.

If we assume that the constant-pressure specific heat of product gas is constant, then Equation (3.29) may be integrated to find the thermodynamic efficiency $\eta_{th}$,

$$\eta_{th} = 1 - \frac{\ln(T_{ad}/T_c)}{(T_{ad}/T_c) - 1} \tag{3.30}$$

For the values of $T_{ad} = 2{,}173$ K and $T_c = 298$ K used above, $\eta_{th} = 68\%$. While this is an improvement on the previous case (Equation (3.28), $\eta_{th} = 46\%$), it is still lower than the Carnot efficiency of 86.3% for $T_h = 2{,}173$ K. Evidently, considerable complication is required to boost ideal thermodynamic efficiencies for fossil-fueled cycles above 50%, and there is little hope of approaching the Carnot efficiency at the adiabatic flame temperature. The Carnot cycle is an important guide to understanding how a simple heat engine might work, but it is not a very practical cycle.

### 3.9.2 The Rankine Cycle

From the beginning of the industrial revolution until the eve of the 20th century, most mechanical power generated by the burning of fossil fuel utilized the steam cycle, called the Rankine cycle. In a steam power plant, fuel mixed with air is burned to heat water in a boiler to convert it to steam, which then powers a turbine. This is an external combustion system where the working fluid, water/steam, is heated in pipes that are contacted by hot flue gas formed in the combustion chamber of the furnace. In an efficient steam plant, nearly all the fuel's heating value is transferred to the boiler fluid, but of course only part of that amount is converted to turbine work. The steam cycle is mechanically robust, providing mechanical power even when the boiler and turbine are not perfectly efficient, accounting for its nearly universal use through the end of the 19th century.[11]

---

[11]Among the earlier American automobiles was the Stanley steamer, powered by a Rankine cycle engine. Although it established a speed record in its day, it was soon surpassed by the more practical gasoline engine.

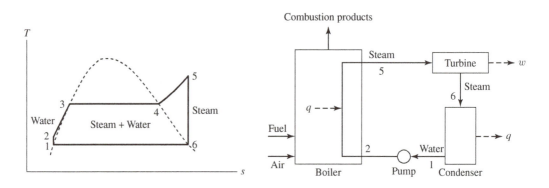

**FIGURE 3.2** The Rankine cycle is a steady flow cycle where water is turned to steam at high pressure in a boiler (2 → 5) and then passes through a turbine to generate mechanical power (5 → 6), after which the steam is condensed to water (6 → 1) and pumped back into the boiler (1 → 2) . (In the $T$, $s$ diagram on the left, the area underneath the dashed line delineates the conditions where both steam and liquid water coexist, in contrast to steam only to the right and liquid water to the left.)

The thermodynamic processes of the Rankine cycle are illustrated in Figure 3.2, showing (on the left) the changes in temperature and entropy that the working fluid undergoes. In a steam power plant, ambient temperature water is pumped to a high pressure and injected into a boiler (1 → 2 in Figure 3.2), whereupon it is heated to its boiling point (3), completely turned into steam (4), and then usually heated further to a higher temperature (5). This heating within the boiler occurs at a constant high pressure $p_b$. The stream of steam flows through a turbine (5 → 6) undergoing a pressure reduction to a much lower value, $p_c$, while the turbine produces mechanical power. The low-pressure steam leaving the turbine is cooled to an ambient temperature liquid in the condenser (6 → 1) and then pumped into the boiler to complete the cycle.

In the idealized Rankine cycle of Figure 3.2, the adiabatic steady flow work per unit mass of steam $w_t$ produced by the turbine is equal to the enthalpy change $h_5 - h_6$ across the turbine, by virtue of the first law Equation (3.20). Because this is ideally an isentropic process, the enthalpy change may be expressed, through Equation (3.18), as

$$w_t = h_5 - h_6 = \int_6^5 dh = \int_6^5 v\, dp.$$

There is a similar expression for the work required to operate the pump. The net work $w$ produced in the cycle may then be expressed as

$$w = \int_{p_c}^{p_b} (v_s - v_w)\, dp, \tag{3.31}$$

where $v_s$ and $v_w$ are the specific volumes of the steam in the turbine and water in the pump and $p_b$ and $p_c$ are the boiler and condenser pressures. Because the specific volume of liquid water is so much smaller than that of steam, the power to run the pump is only a tiny fraction of the power produced by the turbine, a mechanically robust attribute of the Rankine cycle.

Because the heating and cooling processes of the ideal Rankine cycle (2 → 5, 6 → 1) occur at constant pressure while the work processes are isentropic, the thermodynamic efficiency may be expressed as

$$\eta_{th} = \frac{(h_5 - h_6) - (h_2 - h_1)}{h_5 - h_2} = 1 - \frac{h_6 - h_1}{h_5 - h_2} = \frac{\oint T \, ds}{\int_2^5 T \, ds}. \tag{3.32}$$

There are several aspects of the Rankine cycle that deserve notice. First, unlike the Carnot cycle, its thermodynamic efficiency depends explicitly upon the properties of the working fluid, water, as may be seen in Figure 3.2 and Equation (3.32). Second, the cycle efficiency is increased if the boiler pressure (and steam temperature) is increased. At the same time, high boiler pressures increase the amount of work produced per unit mass of water flowing through the system, reducing the cost of the turbine per unit of power output. The basic cycle is capable of improvements in efficiency by use of internal heat exchange at intermediate pressure levels.

For Rankine cycles using water as the working fluid, the temperature of the steam from the boiler seldom exceeds 550°C and then only for boilers that operate at a high pressure. A high-pressure and -temperature steam cycle is one for which the steam pressure and temperature exceed the values at the *critical point* of water.[12] The thermodynamic efficiency of a steam cycle is improved if a high pressure and temperature are used. But there are additional steps that may be taken to improve the efficiency of the basic Rankine cycle. One, called *superheating*, is illustrated in Figure 3.2. Here the steam evolved by boiling the water is further heated in the superheater section of the boiler to a temperature $T_5$ that is higher than the boiling point $T_4$ corresponding to the boiler pressure $p_4 = p_3$. Another step is to reroute steam leaving the high-pressure turbine (see Figure 3.3) back to the boiler for *reheating* to $T_5$, whereupon it returns to the lower-pressure turbine stages, producing more turbine power than if it had not been reheated. A third improvement involves extracting a fraction of the steam from the high-pressure turbine exhaust and using it to heat the water leaving the feedwater pump, at temperature $T_2$. This latter is called *regenerative feed water heating*. The net effect of any or all of these alternatives is to increase the average temperature of the working fluid during which heat is added in the boiler. Using the Carnot cycle as the paradigm, this increases the Rankine cycle thermodynamic efficiency. Depending upon the circumstances, employing all these measures can add up to 10 points of efficiency to the basic cycle.

The thermodynamic efficiency of the ideal Rankine cycle is in the range of 30–45%, depending upon the details of the cycle complexity. But actual steam plants have lower than ideal efficiencies for several reasons. The steam turbine and feed water pumps are not 100% efficient, resulting in less net work than in the ideal cycle. Other mechanical power is required to operate the boiler fans and condenser cooling water pumps, reducing the net power output. The boiler does not transfer to the working fluid (water, steam) all of the fuel higher heating value (HHV), because the flue gases exit from the boiler at higher than fuel input temperature and excess air, above that required for stoichiometric combustion, is used. Even the best steam electric power plants seldom exceed 40% thermal efficiency based upon the ratio of the net mechanical power output divided by the heating value of the fuel supply.

The steam turbine for an electric power plant experiences a large change in pressure between entrance and exit, during which the steam density decreases greatly requiring ever longer turbine

---

[12]The critical point is the condition for which the liquid and vapor phases are indistinguishable. For water the critical pressure and temperature are 221.3 bar (= 22.13 MPa = 3,210 psi) and 374.2°C (= 705.6°F).

**FIGURE 3.3**    A multimegawatt steam turbine rotor, showing the rotor stages. From left to right, the high-pressure, intermediate-pressure, and low-pressure stages. (Copyright Norbert Millauer/Getty Images.)

blades to extract power from the flow. Figure 3.3 shows the rotor of a multimegawatt steam turbine. It is divided into three stages: high-, intermediate-, and low-pressure, from left to right in the photograph. In the high-, intermediate-, and low-pressure stages the steam pressure is reduced from 71 to 10, 10 to 3, and 3 to 0.1 bar, respectively.[13]

### 3.9.3    The Otto Cycle

The most ubiquitous fossil-fueled engine is that in the automobile. Unlike the steam plant, the automobile engine does not depend upon heat transfer to the working fluid from an external combustion source. Instead, the fuel is burned adiabatically inside the engine and the products of combustion produce more work during the expansion stroke than must be invested in the compression stroke, giving a net power output. The combustion products, which are exhausted to the atmosphere, are replaced by a fresh air–fuel charge to begin the next combustion cycle. The working fluid flows through the engine and is not recycled. This is termed an *open cycle*, in contrast to the steam cycle, which is closed.

The thermodynamic heat engine closed cycle that replicates the pressure–volume characteristics of the reciprocating ICE is called the Otto cycle. As sketched in the $T$, $s$ plane of Figure 3.4, it consists of an isentropic compression from a volume $v_e$ at the beginning of the compression stroke

---

[13]For a more detailed description of Rankine cycle power plants using fossil or nuclear fuels, see Chapters 6 and 7.

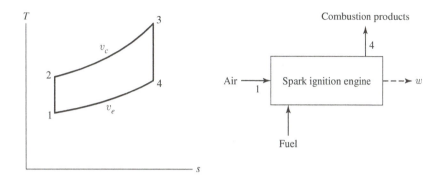

**FIGURE 3.4** The Otto cycle comprises two isentropic compression and expansion strokes of a piston in a cylinder ($1 \rightarrow 2, 3 \rightarrow 4$) interspersed with two constant-volume heating and cooling processes ($2 \rightarrow 3, 4 \rightarrow 1$). It is a model for the spark ignition engine.

to $v_c$ at the end ($1 \rightarrow 2$), followed by a constant-volume heating ($2 \rightarrow 3$) to replicate the adiabatic combustion of the fuel–air mixture, then an isentropic expansion to the maximum volume $v_e$, after which a constant-volume cooling ($4 \rightarrow 1$) completes the cycle (a process that is absent in the open cycle ICE).

The net cyclic work $w$, equivalent heat added $q$, and thermodynamic efficiency $\eta_{th}$ for the Otto cycle are

$$w = \oint p \, dv = \oint T \, ds$$

$$q = \int_2^3 T \, ds$$

$$\eta_{th} = \frac{\oint T \, ds}{\int_2^3 T \, ds}. \tag{3.33}$$

The Otto cycle efficiency is a monotonically increasing function of the volumetric compression ratio, $v_e/v_c$, and, of course, the thermodynamic properties of the working fluid. For simplifying assumptions about the working fluid, it may be expressed as

$$\eta_{th} = 1 - \frac{1}{(v_e/v_c)^{(c_p/c_v)-1}} \tag{3.34}$$

For a typical gasoline engine compression ratio of 9 and $c_p/c_v = 1.26$, $\eta_{th} = 43.5\%$.

In gasoline engines, the compression ratio is limited by the tendency of the fuel–air mixture to combust spontaneously (called *knock*). A higher compression ratio is used in the diesel engine, where the fuel is injected after the air is compressed, providing a greater thermodynamic efficiency for the diesel engine compared with the gasoline engine.

Because the Otto cycle is a model for the reciprocating ICE, the amount of equivalent heat $q$ is limited by the amount of fuel that can be burned in the fuel–air charge in the cylinder at the end of the compression stroke. The combustible fuel is a maximum when the air/fuel ratio is stoichiometric, which is the normal condition for gasoline engines.[14] The maximum temperature $T_3$ at the end of combustion is thus the adiabatic combustion temperature of the compressed mixture in the cylinder.[15] Because the reciprocating ICE engine experiences this peak temperature only momentarily and is otherwise exposed to a much lower average temperature, it is able to operate successfully with peak temperatures that exceed the melting point of most materials and secure the favorable thermodynamic efficiencies that accompany such high temperatures.

The thermodynamic efficiencies of automotive engines are noticeably less than the ideal efficiency of Equation (3.34). Friction of pistons and bearings, power required to operate valves, cooling pump and the fuel supply system, pressure losses in intake and exhaust systems, and heat loss to the cylinder during the power strokes all combine to reduce the power output compared with the ideal cycle. For four-stroke cycle gasoline engines, the low intake pressure experienced at part load is an additional loss that has no counterpart in diesel engines. The best thermal efficiencies for automotive engines are about 28 and 39% for the gasoline and diesel engine, respectively.[16]

The thermodynamic analysis of the Otto cycle does not explain the most salient feature of the reciprocating ICE, that it can be constructed in useful sizes between about 1 kilowatt to 10 megawatts. This is in marked contrast to the steam power plant, which, for the generation of electric power, is usually built in units of 100 to 1,000 MW. These differences are a consequence of mechanical factors related to the limiting speed of pistons versus turbine blades and other factors unrelated to the thermodynamics of the cycles.[17]

## 3.9.4   The Brayton Cycle

Since the middle of the 20th century the gas turbine has become the dominant propulsive engine for large aircraft because of its suitability to high subsonic speed propulsion, light weight, fuel economy, and reliability. But it has made inroads into other uses such as naval vessel propulsion, high-speed locomotives, and, more recently, electric power production. In the latter case, the gas turbine is often used with a Rankine cycle steam plant that is heated by the gas turbine exhaust, with the coupled plants being termed a *combined cycle*.

In its simplest form the gas turbine plant consists of a compressor and turbine in tandem, both attached to the same shaft that delivers mechanical power. Situated between the compressor and turbine is a combustion chamber within which injected fuel burns at constant pressure, raising the temperature of the compressed air leaving the compressor to a higher level prior to its entering the turbine. In passing through the turbine, the hot combustion gas is reduced in pressure and temperature, generating more turbine power than is consumed in compressing the air entering the compressor and making available a net mechanical power output from the shaft. The compression,

---

[14]In gasoline engines the load is varied by changing the pressure $p_1$ and thus the amount of stoichiometric fuel–air mixture in the cylinder. In diesel engines, the air pressure is fixed but the amount of fuel injected is varied.

[15]In the Otto cycle model of the ICE, combustion occurs at constant volume. The constant-volume fuel heating value and adiabatic combustion temperature are not exactly the same as those for constant-pressure combustion, but the differences are small.

[16]Under average operating conditions, the thermal efficiency is less than these maximum values because the engine operating conditions are selected to optimize vehicle performance rather than efficiency.

[17]For a detailed description of the Otto cycle, see Chapter 9.

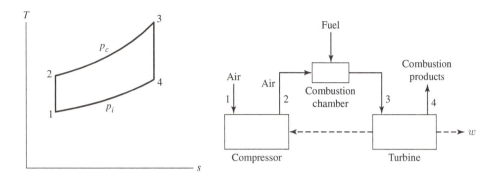

**FIGURE 3.5** The Brayton cycle models the gas turbine cycle, where air is compressed from the inlet pressure $p_i$ in a compressor to the outlet pressure $p_c$ ($1 \rightarrow 2$), then burned with fuel to a higher temperature at the constant pressure $p_c$ ($2 \rightarrow 3$), and subsequently expanded in the turbine ($3 \rightarrow 4$), producing net work $w$.

combustion, and expansion processes are adiabatic, and like the reciprocating engine cycles the gas turbine is an open cycle.

The ideal thermodynamic cycle that models the thermal history of the air and combustion gas flow through a gas turbine power plant is called the *Brayton cycle*. Illustrated in Figure 3.5, it consists of an isentropic compression of air in the compressor from the intake pressure $p_i$ to the compressor outlet pressure $p_c$ ($1 \rightarrow 2$ in Figure 3.5), followed by a constant-pressure heating ($2 \rightarrow 3$) that raises the gas temperature to the value $T_3$ at the turbine inlet. The combustion gas expands isentropically while flowing through the turbine, its pressure being reduced from $p_c$ to $p_i$ ($3 \rightarrow 4$).

In the Brayton cycle, the steady flow work produced by the turbine and that absorbed by the compressor are each equal to the change in enthalpy of the fluid flowing through them. Per unit mass of fluid, the net work $w$ of the gas turbine plant is the difference between the turbine work and the compressor work,

$$w = (h_3 - h_4) - (h_2 - h_1). \tag{3.35}$$

The heat $q$ added to the fluid leaving the compressor, which replicates the temperature rise caused by adiabatic combustion, is just the increase in enthalpy in the constant-pressure process,

$$q = h_3 - h_2. \tag{3.36}$$

As a consequence, the thermodynamic efficiency $\eta_{th}$ of the Brayton cycle is

$$\eta_{th} = \frac{w}{q} = \frac{(h_3 - h_4) - (h_2 - h_1)}{h_3 - h_2} = 1 - \frac{h_4 - h_1}{h_3 - h_2} = \frac{\oint T \, ds}{\int_2^3 T \, ds}, \tag{3.37}$$

where the final expression on the right of Equation (3.37) follows from the equality of constant-pressure heat addition and $\int T \, ds$.

The thermodynamic efficiency of the ideal Brayton cycle depends upon the pressure ratio $p_2/p_1 = p_3/p_4$ and the thermodynamic properties of air and combustion products. For simplifying assumptions about these properties, this efficiency may be expressed as

$$\eta_{th} = 1 - \frac{1}{(p_2/p_1)^{(1-[c_v/c_p])}},$$
(3.38)

showing that the efficiency increases with increasing pressure ratio. For example, if the pressure ratio $p_2/p_1 = 10$ and $c_p/c_v = 1.3$, then $\eta_{th} = 41.2\%$.

There are several practical problems in building a successful gas turbine power plant. The power produced by the turbine, and usually that absorbed by the compressor, is each greater than the net power output, so that the total power of this machinery is considerably greater than the net output power. The aerodynamic efficiencies of the compressor and turbine need both be high so that as much net power is produced as possible. The thermodynamic efficiency of the cycle can be improved by increasing the turbine inlet temperature, but the latter is limited by the high temperature strength of the turbine blades. It was not until the development of efficient aerodynamic blade designs and high temperature turbine materials that the gas turbine plant became practical and economically feasible in the 20th century.

For the simple Brayton cycle, the best thermodynamic efficiencies are about 33%. By use of heat exchange between the hot exhaust gas and and the compressed gas entering the combustion chamber, this efficiency may be increased by about 4 percentage points. Economic factors may require that the plant operate at maximum power output, for which the efficiency would be somewhat lower than these values.

The compressor and turbine of a gas turbine power plant are usually built into a single rotor, with the combustion chamber sandwiched between the compressor and turbine. This is the arrangement for aircraft gas turbines, which must be as light as possible. The rotor shaft delivers the net power difference between the turbine and compressor powers to the electric generator.

### 3.9.5  Combined Brayton and Rankine Cycles

The combustion products gas stream leaving the gas turbine carries with it that portion of the fuel heating value that was not converted to work. This hot stream of gas may be used to generate steam in a boiler and produce additional work without requiring the burning of more fuel. The use of a gas turbine and steam plant to produce more work from a given amount of fuel than either alone could produce is called a combined cycle.

The thermodynamic efficiency $\eta_{cc}$ of a combined cycle power plant may be determined as a function of the component efficiencies, $\eta_g$ and $\eta_s$, of the gas turbine and steam cycles. For the gas turbine, the work $w_g$ is equal to $\eta_g q_f$, where $q_f$ is the fuel heat added per unit mass of combustion products. The amount of heat that can be utilized in the steam cycle is just $q_f - w_g = q_f(1 - \eta_g)$. The steam cycle work output $w_s$ is therefore $\eta_s$ times this heat, or $\eta_s q_f(1 - \eta_g)$. Thus we find the combined cycle efficiency,

$$\eta_{cc} \equiv \frac{w_g + w_s}{q_f} = \eta_g + \eta_s(1 - \eta_g) = \eta_g + \eta_s - \eta_g \eta_s.$$
(3.39)

The efficiency of the combined cycle is always less than the sum of the efficiencies of the component cycles. Nevertheless, the combination is always more efficient than either of its components. For example, if $\eta_g = 30\%$ and $\eta_s = 25\%$, then $\eta_{cc} = 47.8\%$.

In the combined cycle gas plus steam power plant, the thermal efficiency of the steam cycle component is considerably lower than that for the most efficient steam-only power plant, because the gas turbine exhaust gas is not as hot as the combustion gas in a normal boiler and because the gas turbine requires much more excess air than does a steam boiler. Both restrictions limit the steam cycle efficiency, but nevertheless the combined cycle plant provides an overall fuel efficiency that is higher than that for any single cycle plant.

The combined cycle power plant burning natural gas or jet fuel is often the preferred choice for new electric generating plants, rather than coal-fueled steam plants, for a variety of reasons that overcome the fuel price differential in favor of coal. These reasons are mostly financial and environmental, with the latter including the reduced air pollutant emissions, especially carbon dioxide.[18]

## 3.10   THE VAPOR COMPRESSION CYCLE: REFRIGERATION AND HEAT PUMPS

The use of mechanical power to move heat from a lower temperature source to a higher temperature sink is the thermodynamic process that underlies the functioning of refrigerators, air conditioners, and heat pumps. The process is the reverse of a heat engine in that power is absorbed, rather than being produced, but it still observes the restrictions of the first and second laws of thermodynamics that the heat and work quantities balance (Equation 3.9) and that the ratio of the heat quantities is related to the temperatures of the heat source and sink.

The most common form of refrigeration system employs the *vapor compression cycle.* The refrigeration equipment consists of an evaporator, a vapor compressor, a condenser, and a capillary tube connected in series in a piping loop filled with the refrigerant fluid. The refrigerant is chosen so as to undergo a change of phase between liquid and vapor at the temperatures and pressures within the system. The fluid flows through these components in the order listed, being propelled by the pump. The purpose of the capillary tube is to reduce the pressure of the refrigerant fluid flowing from the condenser to the evaporator, resulting in a lowering of its temperature. The fluid states in the ideal vapor compression cycle are illustrated in the temperature–entropy diagram on the left of Figure 3.6. Vapor leaving the evaporator (1) is compressed isentropically ($1 \rightarrow 2$) by the compressor to a higher pressure, where it enters the condenser at a temperature $T_2$ that is higher than the environment to which heat will be transferred from the condenser. The condenser, a heat exchanger, condenses the vapor to liquid form ($2 \rightarrow 4$) by transferring heat to the atmosphere or other environmental sink. The liquid refrigerant leaving the condenser (4) passes through a small-diameter capillary tube, undergoing a viscous pressure drop to enter the evaporator at a lower pressure (5). In this adiabatic, constant-enthalpy process, the fluid temperature decreases and some of the liquid changes to the vapor form. The liquid–vapor mixture then passes through the evaporator, a heat exchanger that absorbs heat from the refrigerated space while changing the liquid portion of the refrigerant to a vapor, completing the cycle.

In heat engine cycles that produce work from the combustion of fuel, the thermodynamic efficiency measures the ratio of the output (work) to the fuel input (heat). The second law of thermodynamics assures that the output is always less than the input, so that the thermodynamic efficiency is less than 100%. For refrigerators and air conditioners, however, the desired output

---

[18]For further discussion, see Section 6.3.

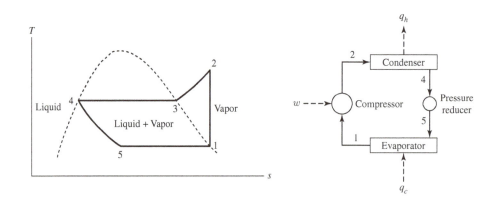

**FIGURE 3.6**   The vapor compression cycle for refrigeration begins with an isentropic compression $(1 \rightarrow 2)$ of the vaporized fluid leaving the evaporator, followed by constant-pressure cooling in the condenser $(2 \rightarrow 4)$ to form a saturated liquid (4). The liquid leaving the condenser undergoes an adiabatic pressure decline $(4 \rightarrow 5)$, entering the evaporator as a cold liquid–vapor mixture (5), whereupon it absorbs heat from the refrigerated space $(5 \rightarrow 1)$. (In the $T$, $s$ diagram on the left, the area beneath the dashed line delineates the conditions where both vapor and liquid refrigerant coexist, in contrast to vapor only to the right and liquid to the left.)

(heat removed from the refrigerated space) is not necessarily less than the input (compressor work). Nevertheless, we may form the ratio of output to input, which is called the *coefficient of performance (COP)*, using this as a figure of merit for the performance of these devices.

The COP of the vapor compression cycle may be determined in terms of the changes in the thermodynamic states of the refrigerant fluid, such as those illustrated in Figure 3.6. The work $w$ required to compress a unit mass of refrigerant equals its change in enthalpy $h_2 - h_1$. The heat $q_c$ absorbed by the refrigerant from the refrigerated space is equal to its change in enthalpy $h_1 - h_5$, which also equals $h_1 - h_4$ since the process $4 \rightarrow 5$ is one of unchanging enthalpy. As a consequence, the coefficient of performance is

$$\text{COP} \equiv \frac{q_c}{w} = \frac{h_1 - h_4}{h_2 - h_1} = \frac{h_1 - h_4}{(h_2 - h_4) - (h_1 - h_4)}. \tag{3.40}$$

The COP is greatest when the temperature difference between the refrigerated space and the environment is least. It decreases monotonically as this temperature difference becomes larger and becomes zero at a sufficiently large value of the temperature difference, depending upon the characteristics of the refrigerant fluid.

A heat pump is a refrigerator operating in reverse (i.e., it delivers heat to an enclosed space by transferring it from an environment having a lower temperature). Heat pumps are commonly used to provide wintertime space heating in climates where there is need for summertime air conditioning. The same refrigeration unit is used for both purposes by redirecting the flow of air (or other heat transfer fluid) between the condenser and evaporator.

For a heat pump, the $(\text{COP})_{\text{hp}}$ is defined as the ratio of the heat $q_h$ transferred to the higher temperature sink, divided by the compressor work $w$,

$$(\text{COP})_{\text{hp}} \equiv \frac{q_h}{w} = \frac{q_c + w}{w} = 1 + \frac{q_c}{w} = \frac{h_2 - h_4}{h_2 - h_1} = \frac{h_2 - h_4}{(h_2 - h_4) - (h_1 - h_4)}, \tag{3.41}$$

where we have used the first law relation that $q_h = q_c + w$ and Equation (3.40) to simplify the right side of (3.41).

The heat pump's coefficient of performance is always greater than unity, because the heat output $q_h$ always includes the energy equivalent of the pump work $w$. But in very cold winter climates, $(COP)_{hp}$ may not be very much greater than unity and the heat delivered would not be much greater than that from dissipating the compressor's electrical power in electrical resistance heating of the space, a much less capital-intensive system. It is the year-round use of refrigeration equipment for summertime air conditioning and wintertime space heating that justifies the use of this expensive system.

## 3.11   ENERGY PROCESSING: FIRST AND SECOND LAW CONSTRAINTS

In previous sections we considered the major technologies for transforming the energy of fossil fuels to mechanical or electrical power: stream power plants, internal combustion engines, and gas turbines. Despite these quite different schemes for generating mechanical power, all of them have the same limitation: only a fraction of the fuel energy is transformed to mechanical energy output. This is a consequence of the second law of thermodynamics, which expresses the limitations of the physical and chemical properties of the substances that bring about the intended energy transformation. There are other power systems that generate electrical power, such as fuel cells and electric storage batteries, that are also limited by these thermodynamic principles. In addition, physicochemical processes used to transform fuels to new forms or separate chemical species, used in advanced power systems, also have limitations circumscribed by thermodynamics. This section explains how the laws of thermodynamics have practical consequences for such systems that may be expressed in general terms.

Most of these energy devices can be modeled as a steady flow of matter into and out of a control volume, as illustrated in Figure 3.7. The inflow stream has an enthalpy per unit mass of $h_{in}$, while the outflow stream enthalpy is $h_{out}$. The electrical/mechanical work produced by the device, per unit mass of flow, is $w$ and the heat transfer from the surrounding environment, per unit mass of flow, is $q$. As explained in Section 3.7, the first law of thermodynamics requires that

$$h_{out} - h_{in} = q - w. \tag{3.42}$$

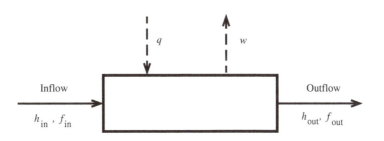

**FIGURE 3.7**   Steady flow of reacting material through an energy processing device, with heat input $q$ from, and work output $w$ to, the environment.

The constraint of the second law is more subtle. It may be explained best in terms of the free energy $f \equiv h - Ts$ defined in (3.16), for which the first law takes the form

$$f_{\text{out}} - f_{\text{in}} = q - w - T(s_{\text{out}} - s_{\text{in}}).  \tag{3.43}$$

But the change in entropy is related to $q$ and $T$ by the inequality of Clausius (3.11),

$$T\,ds \geq ds$$
$$T(s_{\text{in}} - s_{\text{out}}) \geq q$$
$$w \leq f_{\text{in}} - f_{\text{out}},  \tag{3.44}$$

where (3.44) incorporates both the first and the second law restrictions. The constraint of (3.44) is a useful way of expressing the limits of converting free energy to work. We shall apply it to several examples of energy processes in Chapter 5.

### 3.11.1   Fuel Heating Value

When a mixture of fuel and air is burned, the temperature of the combustion products formed is much higher than that of the fuel–air mixture. In some instances heat may be transferred from the hot combustion products to a colder fluid; for example, in a steam boiler, this heat causes the water to warm and then boil to steam. The amount of heat available for this purpose is called the *fuel heating value* and is usually expressed in energy units per unit mass of fuel.

Consider a combustion chamber that is supplied with a steady flow of a fuel–air mixture (the reactants) at a pressure $p_r$ and temperature $T_r$. If the fuel is burned at constant pressure $p_r$, and if no heat is lost from the combustion chamber ($\dot{Q} = 0$), then the product gas temperature $T_p$ will be higher than $T_r$, but the product gas enthalpy $h_p\{T_p, p_r\}$ will exactly equal the reactant stream enthalpy $h_r\{T_r, p_r\}$, by Equation (3.20), there being no work done ($\dot{W} = 0$). This process may be illustrated by identifying the reactant and product states as points in the enthalpy–temperature diagram of Figure 3.8, in which the enthalpies of the reactants ($h_r$) and products ($h_p$) are shown as functions of temperature, at the pressure $p_r$, as the upper and lower curve, respectively. The reactant enthalpy can be identified as the point $R$ at the intersection of the upper (reactant) curve and the vertical line at the reactant temperature $T_r$. The horizontal line through this point then intersects the lower (product) curve at the point $P$, where the product temperature is $T_p$, assuring that $h_p\{T_p, p_r\} = h_r\{T_r, p_r\}$. $T_p$ is called the *adiabatic combustion temperature*.

We are now in a position to determine the fuel heating value. If the hot product gases are subsequently cooled at constant pressure to the reactant temperature $T_r$ at the point $P'$, then the heat removed per unit mass of product gas will be equal in magnitude to the reduction in enthalpy of the product gas between $T_p$ and $T_r$, or $h_p\{T_p, p_r\} - h_p\{T_r, p_r\} = h_r\{T_r, p_r\} - h_p\{T_r, p_r\}$. Multiplying this by the mass flow rate of products, $\dot{m}_p$, divided by the mass flow rate of fuel, $\dot{m}_f$, we obtain the fuel heating value FHV$\{T_r, p_r\}$,

$$\text{FHV}\{T_r, p_r\} = \left(\frac{\dot{m}_p}{\dot{m}_f}\right)(h_r\{T_r, p_r\} - h_p\{T_r, p_r\}).  \tag{3.45}$$

Common hydrocarbon fuels, such as gasoline or diesel fuel, are mixtures of many hydrocarbons of varying molecular structure. The fuel heating value for such fuels is measured using a calorimeter,

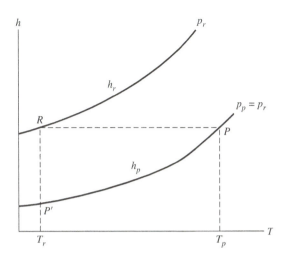

**FIGURE 3.8**    The enthalpy $h$ of the reactants (upper curve) and the products (lower curve) of a combustion process, as functions of the temperature $T$, are related by the fuel heating value. For adiabatic, constant-pressure combustion, the products temperature $T_p$ is greater than the reactant temperature $T_r$.

to which Equation (3.45) is directly applicable. But for pure compounds, such as methane ($CH_4$), the fuel heating value may be calculated from values of the enthalpy of formation of the fuel and the products of combustion.[19]

## 3.11.2    Free Energy Change

In Section 3.11.1, the fuel heating value per unit mass of fuel FHV was defined in (3.45) in terms of the enthalpy change per unit mass of products, $(h_r - h_p)$. A similar relationship exists for the free energy change per unit mass of fuel, $\Delta f$:

$$\Delta f \equiv (\dot{m}_p/\dot{m}_f)(f_r - f_p). \tag{3.46}$$

Expressed in terms of the work per unit of fuel mass, $w_f$, Equation (3.44) becomes

$$w_f \leq \Delta f. \tag{3.47}$$

Traditionally, the maximum amount of work $w_f$ that could be obtained from burning fuel in an engine is the fuel heating value FHV, that is, the maximum amount of heat that can be transferred to an engine cycle, such as a steam engine. For internal combustion engines, such as a gas turbine or automobile engine, where the combustion of fuel occurs within the open cycle flow of reactants through the engine, (3.47) sets $\Delta f$ as the upper limit of $w_f$. There is some, but not great, difference

---

[19]The enthalpy of formation of a compound is the difference between the enthalpy of the compound and that of its elemental constituents (in their stable form), all evaluated at the same reference temperature and pressure.

between these values for most fuels. For a hydrogen fuel cell, it will be seen in Chapter 5 that $\Delta f$ does indeed constitute the upper limit of $w_f$.

### 3.11.3  Separating Gases

An important process in some energy processing plants is the separation of undesirable chemical components from the discharge streams to air or water, so as to prevent pollution of the environment. This is an energy-intensive process, especially the removal of $SO_2$ and $CO_2$, covered in Chapters 10 and 12. Here we will consider a simple example, separation of two gases in a stream in which the temperature and pressure are fixed at ambient values.

We consider a stream of a mixture of two perfect gases, in which one has a partial pressure $xp$, with $x$ being the mole fraction and $p$ being the total pressure. The other component has a partial pressure $(1 - x)p$. In the separation process, for a perfect gas, the free energy increase per mole of separated gas is $RT \ln(p/xp) = RT \ln(1/x)$ for one gas and $RT \ln(p/(1-x)p) = RT \ln(1/(1-x))$ for the other, where R is the universal gas constant. Adding these increases, the free energy increase per mole of mixture is $RT[x \ln(1/x) + (1-x) \ln(1/(1-x))]$. This has a minimum value of $RT \ln 2$ when $x = 1/2$. For very small values of $x$ the free energy increase per mole of separated gas is $RT \ln(1/x)$.

A separation process of this type requires the input of mechanical or electrical work of amount that is not less than the increase of free energy of the separated gases and in general is much greater than this amount. As the initial mole fraction $x$ of the component to be removed is reduced, the required work per unit mass increases in proportion to $\ln(1/x)$.

In removing $CO_2$ from power plant stack gases, it is less costly to remove it in a fuel gasification process, where its concentration is high, than after combustion in air, where its concentration is much lower (see Section 12.5.1.5). By the same token, removing sulfur from a fuel is less expensive than removing $SO_2$ from combustion products formed by burning the fuel in air.[20]

### 3.12   FUEL (THERMAL) EFFICIENCY

The energy embodied in fuels can serve many purposes: generating mechanical or electrical energy, propelling vehicles, heating working or living spaces, creating new materials, refining ores, cooking food, etc. In this chapter we have focused on the production of mechanical work as an example of the constraints imposed on fuel energy use by the laws of thermodynamics. Nevertheless, these laws apply universally to all energy transactions.

The use of fuel to produce mechanical or electrical power, for whatever purpose, accounts for more than half of all fossil fuel consumption. It is for this reason that the considerations of this chapter provide important information needed to evaluate the environmental consequences of current use of this type and especially of options for reducing fossil fuel use in the future. Whatever the beneficial use of the mechanical energy produced, be it vehicle propulsion, material processing, fluid pumping, and so forth, the initiating step of converting fuel energy to mechanical form is an essential ingredient that has important economic and environmental consequences. A measure of

---

[20]If we seek to reduce the concentration of toxic or global warming gases in the atmosphere, it is less costly in energy consumption to capture them before they are emitted into the atmosphere than to capture them after they have been greatly reduced in concentration by mixing throughout the atmosphere.

**TABLE 3.1**   Fuel (Thermal) Efficiencies of Current Power Technologies

| Type | Efficiency (%) |
| --- | --- |
| Steam electric power plant | |
|     Steam at 62 bar, 480°C | 30 |
|     Steam at 310 bar, 560°C | 42 |
| Nuclear electric power plant | |
|     Steam at 70 bar, 286°C | 33 |
| Automotive gasoline engine | 25 |
| Automotive diesel engine | 35 |
| Gas turbine electric power plant | 30 |
| Combined cycle electric power plant | 45 |
| Fuel cell electric power | 45 |

the influence upon these consequences is the efficiency with which the fuel energy is converted to mechanical form.

A practical measure of the efficiency of converting fuel energy to work is the ratio of the work produced to the heating value of the fuel consumed, which we may call the *fuel efficiency* $\eta_f$ (or alternatively, *thermal efficiency*). Usually we use the lower heating value (LHV)[21] of the fuel for this purpose, as this measures the practical amount of fuel energy available. Fuel efficiency is useful because we can readily calculate the fuel mass consumption rate $\dot{m}_f$ of an engine of given power output $\mathcal{P}$ if we know its fuel efficiency,

$$\dot{m}_f = \frac{\mathcal{P}}{\eta_f\,(\mathrm{FHV})}. \tag{3.48}$$

Sometimes the fuel efficiency is expressed differently. The ratio of fuel consumption rate $\dot{m}_f$ to engine power $\mathcal{P}$ is called the *specific fuel consumption*. From Equation (3.48), we can conclude that the specific fuel consumption is the inverse of the product $\eta_f$ (FHV).

Table 3.1 summarizes the range of values of fuel (thermal) efficiencies of current technologies for producing mechanical or electrical power. It can be seen that none of these exceeds 50%. These efficiencies reflect the constraints of the laws of thermodynamics, the limitations of materials, and the compromises inherent in achieving economical as well as efficient systems. While there is room for improvement, only modest increases above the values in Table 3.1 can be expected from extensive development efforts.

## 3.13   CONCLUSION

Nearly 90% of the world's energy is supplied by the combustion of fossil fuel. While the processes by which the energy of this fuel is made available for human use in the form of heat or mechanical power are circumscribed by the principles of thermodynamics, the technologies employed are the consequences of human invention.

---

[21]LHV is the heating value assuming that the $H_2O$ formed is in the vapor phase.

In this chapter we have described the implications of the first and second law of thermodynamics for the functioning of selected technologies for producing mechanical power, with particular attention to the efficiency of conversion of fuel energy to useful work. We found that the fuel energy that can be made available by burning fuel in air, called the fuel heating value, appears as the sum of work produced by a heat or combustion engine and heat rejected to the surrounding environment, as required by the first law of thermodynamics. But only a fraction of the fuel energy can be converted to work, according to the second law of thermodynamics, with the magnitude of that fraction depending upon the detailed operation of the technology being used. It is extremely difficult to convert more than half of the fuel heating value to work, but very easy to convert all of it to heat alone.

Most mechanical power is produced in steam power plants, where it is converted to electrical form for distribution to end consumers. Water/steam is circulated within a closed loop, being heated in a boiler by the combustion of fuel and then powering a steam turbine. The most efficient steam plants convert about 40% of the fuel energy to mechanical power.

The gas turbine engine, developed initially for aircraft propulsion and utilizing combustion of fuel in the air steam that flows through the engine, is one prominent form of internal combustion engine. Unlike the steam engine, it does not require the exchange of heat with an external combustion system. Its performance is limited by the strength of the turbine blades that must endure high combustion temperatures. When used to generate electricity, its efficiency is only about 30%. By combining with a steam plant, called a combined cycle, the efficiency of the combination is about 45%.

The other prominent form of internal combustion engine is the common automobile engine, either gasoline (spark ignition) or diesel (compression ignition). Here the cyclic nature of the engine allows much higher combustion temperatures than in the gas turbine and higher efficiencies of 25–35%.

The science of thermodynamics provides a guide for improvements in the performance of combustion systems, but cannot substitute for the invention and ingenuity needed to bring them about.

# PROBLEMS

### Problem 3.1

It takes 4 million tons of coal per year to fuel a 1,000-MW steam-electric plant that operates at a capacity factor (ratio of annual average electric power to rated electric power) of 70%. If the heating value of coal is 12,000 Btu/lb, calculate the plant's thermal efficiency and heat rate. (The latter is defined as the ratio of fuel heat in Btu to electric output in kWh.)

### Problem 3.2

Given a pressure ratio $p_2/p_1 = 12$ across a gas turbine and a specific heat ratio $c_p/c_v = 1.35$ of the working gas fluid, calculate the ideal thermal efficiency of a Brayton cycle gas turbine plant. Explain why real gas turbine plants achieve only 25–35% thermal efficiency.

## Problem 3.3

A rankine cycle ocean thermal power plant is supplied with heat from a flow of warm ocean surface water, which is cooled by 15°C as it passes through the boiler. If the plant electrical output is 10 MW and the thermal efficiency is 5%, calculate the volume flow rate of warm water, in m$^3$/s, supplied to the boiler.

## Problem 3.4

A combined cycle power plant has a gas turbine cycle thermodynamic efficiency of 30% and a steam cycle efficiency of 30%. Calculate the combined cycle thermodynamic efficiency, the ratio of gas turbine power to steam turbine power, and the fraction of the fuel heat that is removed in the condenser of the steam plant.

# BIBLIOGRAPHY

Aschner, F. S. *Planning Fundamentals of Thermal Power Plants.* New York: John Wiley & Sons, 1978.

Black, William Z., and James G. Hartly. *Thermodynamics.* New York: Harper and Row, 1985.

Flagan, Richard C., and John H. Seinfeld. *Fundamentals of Air Pollution.* Englewood Cliffs, NJ: Prentice Hall, 1988.

Haywood, R. W. *Analysis of Engineering Cycles*, 4th ed. Oxford: Pergamon Press, 1991.

Holman, J. P. *Thermodynamics*, 3rd ed. New York: McGraw–Hill Book Co., 1980.

Horlock, J. H. *Combined Power Plants Including Combined Cycle Gas Turbine (CCGT) Plants.* Oxford: Pergamon Press, 1992.

Saad, Michael A. *Thermodynamics; Principles and Practice.* Upper Saddle River, NJ: Prentice Hall, 1997.

Spalding, D. B., and E. H. Cole. *Engineering Thermodynamics.* London: Edward Arnold (Publishers) Ltd., 1973.

Wark, Kenneth, Jr. *Thermodynamics*, 5th ed. New York: McGraw–Hill Book Co., 1988.

# Thermodynamics of Fossil, Biomass, and Synthetic Fuels

## 4.1 INTRODUCTION

Chapter 2 summarizes the sources and uses of global energy. It reveals that the three principal fossil fuels, coal, oil (petroleum), and natural gas, supply about 85% of the current annual global energy consumption. These fuels lie below the surface of the earth (including the shallow ocean bottom of the continental shelves) in amounts that could be exhausted within centuries if used at current rates. They are the residues of dead organic matter (principally plants) trapped for millions of years in sedimentary deposits. As the name implies, they are mostly burned in air to produce heat, electrical or mechanical power, or other useful products. In this chapter we examine the properties of these fuels and how they are related to the fuels' atomic and molecular structure. We also consider the thermochemical reforming of these fuels into synthetic forms that are more useful for certain purposes than the parent fuel.

A second class of fuels is biomass, the dried remains of vegetation harvested from agricultural and silvicultural crops, including residues of plant and animal foods. While their contribution to energy fuels is much less than fossil fuels, they are renewable and free of fossil carbon emissions. Synthetic forms of biomass fuel are especially important as vehicle fuels.

Finally, we show how electrochemical reactions can generate electric power directly. The technology of fuel cells illustrates the advantages of direct transformation of fuel energy to electricity without the use of electromechanical machinery.

## 4.2 FOSSIL FUELS

The atomic constituents of fossil fuels are the residues of their plant parents: mostly the light atoms of carbon, hydrogen, and oxygen that constitute carbohydrates, with lesser amounts of nitrogen, sulfur, and phosphorus and trace amounts of minor elements found in soils. Over the long periods of transformation to fossil fuels, the oxygen content is greatly reduced and the dominant molecular form is that of a mixture of hydrocarbons, especially for fluid fuels. In addition, there is considerable variability in the atomic and molecular composition of these fuels as they are recovered from geologic deposits in different regions of the earth's surface. Even so, the energy released in their combustion depends primarily upon their carbon and hydrogen content.

The molecular composition of fossil fuels, predominantly hydrocarbons, can be quite various. Natural gas has the fewest molecular components: mostly low-molecular-weight hydrocarbons (1–4 carbon atoms per molecule) having boiling points below standard temperature. Petroleum has a myriad of hydrocarbons, a mixture of liquid components having elevated boiling points and carbon numbers[1] of 1 to 3 digits, although a smaller range of carbon numbers comprise most of the liquid mass. In contrast, coal is a heterogeneous solid, mostly amorphous carbon, with inclusions of hydrocarbons of lower carbon numbers.[2] As recovered from the well or mine, gaseous, liquid, and solid nonhydrocarbon impurities of various amounts are common as well.

Hydrocarbon molecules are composed of subunits linked to each other by carbon–hydrogen and carbon–carbon bonds. In increasing order of size, these are the acetyl ($CH$), formyl ($CH_2$), and methyl ($CH_3$) radicals. Carbon has four bonds; the carbon atoms are linked to each other by single ($-$), double ($=$), or triple ($\equiv$) carbon–carbon bonds and to hydrogen atoms by single bonds. The only single carbon molecule is methane, composed of a methyl radical and hydrogen atom: $H-CH_3$. There are three two-carbon molecules: ethane ($H_3C-CH_3$), ethylene (($H_2C=CH_2$)), and acetylene ($HC\equiv CH$).[3] Linear molecules of the form ($H_3C-[HCH]_n-CH_3$) are called alkanes; some of them are listed in Table 4.1. As the carbon number increases, many more possible forms of hydrocarbon molecules are possible, both in atomic composition and in geometric configuration.[4]

The energy and economic value of fossil fuels reside in their fuel heating values (FHV), defined in Section 4.3. Some values are listed in Table 4.1, expressed in units of J/kg. The FHV of raw fuel at the wellhead or minemouth is variable, depending upon its molecular composition, and must be determined empirically in a fuel calorimeter. Furthermore, all raw fuel streams are processed into forms more suitable for end-use consumption, in which case the processed product has a different FHV than the raw product. For natural gas, which is predominantly methane, this processing may involve separating ethane, propane, and butane (called natural gas liquids) and impurities such as carbon dioxide, nitrogen, hydrogen sulfide, and water. For petroleum, fractional distillation will provide a continuous spectrum of increasingly higher boiling point fractions having a narrower range of carbon numbers and correspondingly better defined FHV. These fractions may be directly usable as commercial fuels, such as vehicle fuels (gasoline, diesel, and jet fuel) or residential and commercial space heating fuels. By contrast, coal has the widest range of constituents and thereby FHV, especially in its ash content. Mostly used for electric power production, coal is commonly processed to reduce its ash and sulfur content so as to limit air pollutant emissions from these plants.

Because of the chemical complexity of fossil fuels and their commercial variants, it is helpful to represent them as simple pure fuel substances. For example, natural gas can be modeled as pure methane ($CH_4$), coal as pure carbon (C), gasoline as octane ($C_8H_{18}$), diesel fuel as decane

---

[1]The carbon number of a molecule is the number of carbon atoms per molecule.

[2]See Table 2.3.

[3]Methane is a common gaseous fuel, ethylene is not used as a fuel but as a feed stock for polyethylene plastic, and acetylene is the fuel for the oxy-acetylene torch used for welding and cutting metals.

[4]By adding an oxygen atom to an alkane, we can form alcohols, such as methanol or ethanol, which are useful liquid fuels.

**TABLE 4.1**  Thermodynamic Properties of Fuel Combustion in Air at 25°C and One Atmosphere Pressure[a]

| Fuel | Symbol | Mol wt (g/mol) | FHV[b] (MJ/kg fuel)[c] | $(A/F)_{st}$ | $(h_r - h_p)$[b] (MJ/kg product) | $\Delta f$ (MJ/kg fuel) | FHV[b] (MJ/kg C) |
|------|--------|----------------|------------------------|--------------|----------------------------------|-------------------------|------------------|
| Pure compounds[d] | | | | | | | |
| Hydrogen | $H_2$ | 2.016 | 119.96 | 34.28 | 3.400 | 117.63 | na |
| Carbon (graphite) | $C_{(solid)}$ | 12.01 | 32.764 | 11.51 | 2.619 | 32.834 | 32.764 |
| Methane | $CH_4$ | 16.04 | 50.040 | 17.23 | 2.745 | 51.016 | 66.844 |
| Carbon monoxide | CO | 28.01 | 10.104 | 2.467 | 2.914 | 9.1835 | 23.564 |
| Ethane | $C_2H_6$ | 30.07 | 47.513 | 16.09 | 2.780 | 48.822 | 59.480 |
| Methanol | $CH_4O$ | 32.04 | 20.142 | 6.470 | 2.696 | 22.034 | 53.739 |
| Propane | $C_3H_8$ | 44.10 | 46.334 | 15.67 | 2.779 | 47.795 | 56.708 |
| Ethanol | $C_2H_6O$ | 46.07 | 27.728 | 9.000 | 2.773 | 28.903 | 53.181 |
| Isobutane | $C_4H_{10}$ | 58.12 | 45.576 | 15.46 | 2.769 | | 53.142 |
| Hexane | $C_6H_{14}$ | 86.18 | 46.093 | 15.24 | 2.838 | | 54.013 |
| Octane | $C_8H_{18}$ | 114.2 | 44.785 | 15.12 | 2.778 | | 53.246 |
| Decane | $C_{10}H_{22}$ | 142.3 | 44.599 | 15.06 | 2.778 | | 52.838 |
| Dodecane | $C_{12}H_{26}$ | 170.3 | 44.479 | 15.01 | 2.778 | | 52.567 |
| Hexadecane | $C_{16}H_{34}$ | 226.4 | 44.303 | 14.95 | 2.778 | | 52.208 |
| Octadecane | $C_{18}H_{38}$ | 254.5 | 44.257 | 14.93 | 2.778 | | 52.102 |
| Commercial fuels | | | | | | | |
| Natural gas | | | 36–42 | | | | |
| Gasoline | | | 47.4 | | | | |
| Kerosene | | | 46.4 | | | | |
| No. 2 oil | | | 45.5 | | | | |
| No. 6 oil | | | 42.5 | | | | |
| Anthracite coal | | | 32–34 | | | | |
| Bituminous coal | | | 28–36 | | | | |
| Subbituminous coal | | | 20–25 | | | | |
| Lignite | | | 14–18 | | | | |
| Biomass fuels | | | | | | | |
| Wood (fir) | | | 21 | | | | |
| Grain | | | 14 | | | | |
| Manure | | | 13 | | | | |

[a] Data from Lide, David R., and H. P. R. Frederikse, eds. *CRC Handbook of Chemistry and Physics.* 75th ed. Boca Raton, FL: CRC Press, 1994. Probstein, Ronald F., and R. Edwin Hicks. *Synthetic Fuels.* New York: McGraw–Hill Book Co., 1982. Flagan, Richard C., and John H. Seinfeld. *Fundamentals of Air Pollution.* Englewood Cliffs, NJ: Prentice Hall, 1988.

[b] $H_2O$ product in vapor phase; heating value is the lower heating value (LHV).

[c] 1 MJ/kg = 429.9 Btu/lb mass.

[d] Gas phase, except carbon.

($C_{10}H_{22}$), etc. The properties of these pure fuels, listed in Table 4.1, are proxies for the properties of natural gas, coal, gasoline, diesel fuel, etc., especially those properties listed in the last five columns of the table.

## 4.3  COMBUSTION OF FOSSIL FUEL

The source of energy that is utilized in fossil-fueled power systems is the chemical energy that is released when a fuel is oxidized by burning in air. The most common fossil fuels are hydrocarbons (i.e., mixtures of molecules composed of carbon and hydrogen).[5] Upon their complete combustion, the carbon in the fuel is oxidized to carbon dioxide and the hydrogen to water vapor. The energy made available in this oxidation is the net amount released when the carbon and hydrogen atoms are separated from each other and subsequently combined with oxygen to form carbon dioxide and water.

Denoting a hydrocarbon fuel molecule as $C_nH_m$, where the integers $n$ and $m$ denote the number of carbon and hydrogen atoms in a fuel molecule, the molecular rearrangement accompanying complete oxidation of the carbon and hydrogen may be represented by the reaction

$$C_nH_m + \left(n + \frac{m}{4}\right)O_2 \rightarrow nCO_2 + \left(\frac{m}{2}\right)H_2O. \tag{4.1}$$

For each hydrocarbon molecule, $n + m/4$ diatomic oxygen molecules are required to convert the carbon and hydrogen to $n$ molecules of $CO_2$ and $m/2$ molecules of $H_2O$. The ratio of the number of oxygen molecules to the number of fuel molecules, $n + m/4$, is called the *stoichiometric ratio*. It may be expressed alternatively as a mass ratio by multiplying the number of molecules by their molecular masses, yielding

$$\frac{\text{oxygen mass}}{\text{fuel mass}} = \frac{32n + 8m}{12n + 1.008m}. \tag{4.2}$$

This mass ratio lies in the range between $8/3 = 2.667$ (for pure carbon, $m = 0$) and 7.937 (for pure hydrogen, $n = 0$), being a function of the molar ratio $m/n$ only.

Since fossil fuels invariably are burned in air, the stoichiometric proportions are more usefully expressed in terms of the ratio of air mass to fuel mass by multiplying Equation (4.2) by the ratio of the mass of air to the mass of oxygen in air, which is 4.319,

$$\frac{\text{air mass}}{\text{fuel mass}} \equiv (A/F)_{st} = 4.319 \left(\frac{32n + 8m}{12n + 1.008m}\right). \tag{4.3}$$

If less air is available than is required for a stoichiometric proportion, then not all of the carbon or hydrogen will be fully oxidized and some amount of CO, solid C, or $H_2$ may be present in the products of combustion. In such "rich" mixtures not all of the available chemical energy is released in the (incomplete) combustion process. On the other hand, if extra or excess air is available, then not all of the oxygen available is needed and some will remain unconsumed in the combustion

---

[5] Synthetic fuels made from hydrocarbons may include oxygen containing components such as alcohols and carbon monoxide.

products, but all of the fuel's chemical energy will have been released in the combustion of this "lean" mixture.

## 4.3.1 Fuel Heating Value

When a mixture of fuel and air is burned, the temperature of the combustion products formed is much higher than that of the fuel–air mixture. In some instances heat may be transferred from the hot combustion products to a colder fluid; for example, in a steam boiler, this heat causes the water to warm and then boil to steam. The amount of heat available for this purpose is called the fuel heating value and is usually expressed in energy units per unit mass of fuel.

Consider a combustion chamber that is supplied with a steady flow of a fuel–air mixture (the reactants) at a pressure $p_r$ and temperature $T_r$. If the fuel is burned at constant pressure $p_r$, and if no heat is lost from the combustion chamber ($\dot{Q} = 0$), then the product gas temperature $T_p$ will be higher than $T_r$, but the product gas enthalpy $h_p\{T_p, p_r\}$ will exactly equal the reactant stream enthalpy $h_r\{T_r, p_r\}$, by Equation (3.20). This process may be illustrated by identifying the reactant and product states as points in the enthalpy–temperature diagram of Figure 3.9, in which the enthalpies of the reactants ($h_r$) and products ($h_p$) are shown as functions of temperature, at the pressure $p_r$, as the upper and lower curve, respectively. The reactant enthalpy can be identified as the point $R$ at the intersection of the upper (reactant) curve and the vertical line at the reactant temperature $T_r$. The horizontal line through this point then intersects the lower (product) curve at the point $P$, where the product temperature is $T_p$, assuring that $h_p\{T_p, p_r\} = h_r\{T_r, p_r\}$. $T_p$ is called the *adiabatic combustion temperature*.

Table 4.1 lists the FHV of some common fuels at 25°C and one atmosphere of pressure, assuming that the $H_2O$ formed in the product is in the vapor phase. (This is called the *lower heating value*.) Also listed is the stoichiometric air–fuel ratio $(A/F)_{st}$, the enthalpy difference per unit mass of combustion product, $h_r - h_p$, at the reference temperature and pressure and the free energy difference $\Delta f$ per unit mass of fuel,[6]

$$\Delta f \equiv (\dot{m}_p/\dot{m}_f)(f_r - f_p). \tag{4.4}$$

The fuel heating values of Table 4.1 cover a wide range, from about 10 to 120 MJ/kg fuel. For the saturated hydrocarbons listed, $CH_4$ to $C_{18}H_{38}$, the range is much smaller, about 44 to 50 MJ/kg fuel. The partially oxygenated fuels, CO, $CH_4O$, and $C_2H_6O$, have lower heating values than their parents, C, $CH_4$, and $C_2H_6$, because they have less oxidation potential and greater molecular weight. The low value for solid carbon reflects the considerable energy required to convert the carbon atoms from solid to gaseous form.

A different aspect of the fuels is made evident in the fifth column of Table 4.1, which compares the enthalpy difference $h_r - h_p$ (measured in MJ/kg product) at the reference conditions. This enthalpy difference is the chemical energy that is made available in a constant-pressure, adiabatic, stoichiometric combustion process to increase the product temperature to the adiabatic value. If the product gases of the different fuels possessed the same specific heat, then the temperature rise would be proportional to this value. It can be seen that most of these fuels would have approximately the same adiabatic combustion temperature, which turns out to be about 1900°C, with the exception of hydrogen, which has a somewhat higher temperature.

---

[6] See Section 3.11.2.

The last column of Table 4.1 lists the FHV per unit mass of fuel carbon. Its reciprocal is the carbon emissions to the atmosphere (in the form of $CO_2$) per unit of FHV realized when the fuel is burned and the products of combustion are released into the atmosphere. Of the pure hydrocarbon fuels listed, methane provides the most FHV per unit mass of carbon, although all other hydrocarbon fuels possess only about 20% less than methane.[7] Pure carbon, a component of coal, has substantially less FHV per unit mass of carbon than methane, the principal component of natural gas.

The principal commercially available fuels are coal (anthracite, bituminous, and subbituminous), liquid petroleum fuels (gasoline, diesel fuel, kerosene, home heating fuel, commercial heating fuel), petroleum gases (natural gas, ethane, propane, butane), and wood (hardwood, softwood). The heating values of these fuels vary according to the fuel composition, with none of them having a pure molecular composition and some of them including inert components. The unit selling price of these fuels may be based upon the volume (liquids, gases, and wood) or the mass (coal), but the heating value may be a factor in the price. Their heating values are listed in Table 4.1.

In virtually all combustion systems, the water molecules in the products of combustion leaving the device are in the form of vapor, not liquid, because the effluent temperature is high enough and the concentration of water molecules is low enough to prevent the formation of liquid droplets. As a practical matter, the heat of condensation of the water vapor is not available for partial conversion to work and the effective FHV should be based upon the water product as a vapor, as assumed in Table 4.1. Nevertheless, sometimes a *higher heating value* (HHV) is used in the sale of fuel, based upon the assumption that the water product is in the liquid form. To determine this HHV for the fuel, we should add to the LHV (FHV of Table 4.1) the heat of vaporization of water at the reference temperature, expressed as enthalpy per unit mass of hydrogen in water,[8] times the mass fraction of hydrogen in the fuel.[9]

The distinction between HHV and LHV is primarily a matter of convention. Sellers of fuel like to quote their price in terms of dollars per million Btu of HHV, a lower price than that per million Btu of LHV. On the other hand, users of fuel who generate electricity prefer to rate their plant efficiency in terms of electrical energy produced per unit of fuel LHV consumed, leading to higher efficiencies than when using the HHV. As long as the basis of the price or plant efficiency is stated, no confusion should result.

## 4.4  BIOMASS FUELS

Biomass fuels are the dried residues of plant matter such as food plants, grasses, trees, algae, and animal waste. They are the remains of living cellular material, heterogeneous in composition and structure, and composed of the great variety of organic molecules that form biological systems.

As such, they are directly combustible, as occurs in forest and grassland fires or when used for cooking or heating. Their approximate atomic composition is $CH_2O$ and the FHV is that of their carbon content, or FHV $\sim$ 33 MJ/kg C = 13.2 MJ/kg fuel. But the molecular composition is

---

[7]For a discussion of the significance of carbon emissions, see Chapter 11.

[8]At 25°C, this value is 21.823 MJ/kg H.

[9]The difference in heating values is a maximum for hydrogen (21.823 MJ/kg fuel), but approaches 3.136 MJ/kg fuel for the heaviest hydrocarbons of Table 4.1.

a combination of of saccharides (glucose, sucrose), polysaccharides (starch, cellulose, lignocellulose), and lignin, with the latter two having molecular masses of 5,000 to 100,000 atoms. As long as the raw fuel is burned, its molecular composition is not of great consequence, but if it is used as the raw material for synthesizing biofuels, especially for fermenting to ethanol, the molecular form is significant.

The combustion of biomass fuel, modeled as the carbohydrate of average composition $CH_2O$, can be described as

$$CH_2O + O_2 \rightarrow CO_2 + H_2O; \quad FHV \sim 33MJ/kgC; \quad \Delta f \sim 33MJ/kgC, \qquad (4.5)$$

where the values of FHV and $\Delta f$ are those of carbon.


## 4.5   SYNTHETIC FUELS

A synthetic fuel is one that is manufactured from raw fuel so as to enhance its usefulness while retaining as much of the original heating value as possible. Typical examples are oil produced from coal, oil shale, or tar sands; gas from coal, oil, or biomass; alcohols from biomass or natural gas; and hydrogen from coal, oil, natural gas, or biomass. Some liquid fuels, such as gasoline, are partially synthetic in that the refining process produces components that are synthesized from petroleum constituents and added to the natural fractions of petroleum that ordinarily comprise the liquid fuel. The major advantages of a synthetic fuel, other than its form as liquid, gas, or solid that might enhance its transportability and convenience of storage, are the removal of base fuel constituents such as sulfur, nitrogen, and ash that lead to harmful air pollutants and the ability to burn the fuel in special devices such as gas turbines and fuel cells. A major disadvantage is the economic cost of synthesizing the fuel and the lessening of its heating value, both of which raise the economic cost of synthetic FHV. This cost factor has been the major obstacle to widespread production and use of synthetic fuels.

Liquid synthetic fuels for use in highway vehicles have significant advantages over traditional fuels made from petroleum in nations that seek to reduce their dependence on imported oil, especially if they can be made from domestic coal or natural gas. If made from biomass, such vehicle fuels also reduce fossil carbon emissions from these vehicles.

Synthetic fuels are formed in steady flow reactors supplied with fuel and other reactants, usually at elevated pressures and temperatures. The chemical reactions that generate the synthetic fuels may be aided by the use of catalysts that enhance the reaction rates, lowering the size and cost of the reactor vessel to economically practical levels. Synthesizing reactions are usually slightly exothermic; heat is rejected to the ambient environment, maintaining a constant reactor temperature. Thus the synthesis proceeds at nearly fixed temperature and pressure.

The second law of thermodynamics requires that the free energy of the product gas leaving the reactor not exceed that of the reactant mixture ($\Delta f \geq 0$) since no mechanical work, or its equivalent, is invested in the process.[10] As a consequence, practical synthesis results in some reduction of heating value in the synthesized fuel.[11]

———————

[10] See Section 3.11.

[11] Synthetic fuel may be produced within advanced power plant cycles. See Section 6.3.

**TABLE 4.2**    Thermal Efficiencies of Synthetic Fuel Production

| Fuel | Product | Efficiency[a] (%) |
|---|---|---|
| Coal | Synthesis gas | 72–87 |
| Coal | Methane | 61–78 |
| Coal | Methanol | 51–59 |
| Coal | Hydrogen | 62 |
| Oil | Hydrogen | 77 |
| Methane | Hydrogen | 70–79 |
| Coal, oil, or gas | Hydrogen (electrolytic) | 20-30 |
| Oil shale | Oil and gas | 56–72 |
| Methanol | Oil and gas | 86 |
| Wood | Gas | 90 |
| Corn | Ethanol | 46 |
| Manure | Gas | 90 |

[a] Thermal efficiency is the ratio of the heating value of the synthetic product divided by the heating value of the parent fuel.

Table 4.2 summarizes the thermal efficiencies (the ratio of synthetic FHV to that of the parent fuel) for several synthetic fuel production processes. With but few exceptions, these efficiencies lie within the range of 60–90%. Most conversion processes require high process temperatures and pressures, need catalytic support to improve the production rate, and consume mechanical power to provide for the requisite pressurization and heat transfer processing. The economic and energy costs of synthetic fuel production can only be justified when there are compensating gains attending the use of synthetic fuels, such as the suitability for use in fuel cells or convenience of storage and transport.

Synthetic nuclear fuels can be produced in nuclear reactors. Uranium-238, which is not a fissionable nuclear fuel, can be converted to plutonium-239, which can be used to fuel a nuclear fission reactor. See Section 7.2 for a description of this process.

### 4.5.1    Examples of Fossil Fuel Synthesis

In this section we consider the conversion of coal, oil, and natural gas to other synthetic forms, modeling these raw fuels as simple pure fuels: solid carbon (C), alkanes ($C_nH_m$), and ($CH_4$). We start by considering the combustion reactions of four pure fuels: C, CO, $H_2$, and $CH_4$, whose relevant properties are given in Table 4.1 and listed below.

$$C + O_2 \rightarrow CO_2; \quad \Delta h = 8.941 \, \text{MJ/kg r/p}; \quad \Delta f = 8.960 \, \text{MJ/kg r/p} \qquad \textbf{(4.6)}$$

$$CO + \frac{1}{2}O_2 \rightarrow CO_2; \quad \Delta h = 6.431 \, \text{MJ/kg r/p}; \quad \Delta f = 5.845 \, \text{MJ/kg r/p} \qquad \textbf{(4.7)}$$

$$H_2 + \frac{1}{2}O_2 \rightarrow H_2O; \quad \Delta h = 13.424 \, \text{MJ/kg r/p}; \quad \Delta f = 13.163 \, \text{MJ/kg r/p} \qquad \textbf{(4.8)}$$

$$CH_4 + 2O_2 \rightarrow CO_2 + 2H_2O; \quad \Delta h = 10.01 \, \text{MJ/kg r/p}; \quad \Delta f = 10.203 \, \text{MJ/kg r/p} \qquad \textbf{(4.9)}$$

Here $\Delta h$ and $\Delta f$ are the differences in the enthalpy and free energy, respectively, between the reactants (the molecules to the left of $\rightarrow$) and the products (the molecules to the right of $\rightarrow$), per unit of mass of reactants or products[12] and denoted in units of MJ/kg r/p.

The combustion reactions (4.6)–(4.9) are *oxidation reactions* in which the fuel molecules are oxidized to $CO_2$ and $H_2O$ and are characterized by large reductions in enthalpy $h$ and free energy $f$ ($\Delta h \sim \Delta f \gg 0$). When occurring in a reactor at fixed temperature and pressure, these reactions are exothermic, transferring heat to the surrounding environment in the amount $\Delta h$.

On the other hand, consider the reverse of these reactions, in which $\rightarrow$ is replaced by $\leftarrow$ and in which the $CO_2$ and $H_2O$ are converted to the fuels C, CO, $H_2$, and $CH_4$. They are accompanied by corresponding negative values of $\Delta h$ and $\Delta f$ ($\Delta h \sim \Delta f \ll 0$). These are reactions in which oxygen is removed from the $CO_2$ and $H_2O$, converting them to fuel and oxygen, and are called *reduction reactions*. In these reduction reactions the fuel/oxidant mixtures created from combustion products have a higher free energy that necessitates an equivalent input of work as a consequence of the second law of thermodynamics. For example, making $H_2$ by reduction of water in an electrolytic cell requires 13.42 MJ/kg water of electrical energy, a much more expensive process than production of hydrogen from coal or natural gas.

In what follows, we investigate combinations of oxidation and reduction reactions that will convert fossil fuels to synthetic fuels without loss of heating value and not require the consumption of work. These reactions are isoenergetic ($\Delta f = 0$) and therefore thermodynamically reversible. They constitute the most energy-efficient synthetic fuel process permitted by the second law of thermodynamics, an upper limit to what is possible in practical synthetic fuel reactors.[13]

### 4.5.1.1  Coal to Gas

The conversion of coal to *synthesis gas* is one of the oldest of fuel syntheses. Coal reforming is a major resource for synthesizing liquid and gaseous hydrocarbon fuels. We first consider the formation of synthesis gas, a combination of CO and $H_2$, by reforming coal with steam. We choose a combination of the oxidation reaction (4.6) and the reduction (reverse) reaction (4.7) and (4.8) such that $\Delta f = 0$:

$$C + 0.6807\,H_2O + 0.1579\,O_2 \rightarrow 0.6807\,H_2 + CO;$$

$$\Delta h = -0.197\,\text{MJ/kg r/p}; \qquad \Delta f = 0. \tag{4.10}$$

This reaction is slightly endothermic; the FHV of the synthesis gas is greater than that of the coal by 2.2%. Synthesis gas can be used as a gaseous fuel or further reformed to various synthetic products.

Coal may also be used to generate methane:

$$C + 0.964\,H_2O + 0.036\,O_2 \rightarrow 0.518\,CO_2 + 0.482\,CH_4;$$

$$\Delta h = 0.10\,\text{MJ/kg r/p}; \qquad \Delta f = 0. \tag{4.11}$$

For pure methane production, the $CO_2$ must be separated from the product stream.[14]

---

[12]Equations (4.6)–(4.9) imply the conservation of mass of the labeled molecules.

[13]See Table 4.2.

[14]See Section 12.5.1.3.

Hydrogen is often the fuel of choice for fuel cells. It may be formed from coal:

$$C + 1.664\,H_2O + 0.168\,O_2 \rightarrow CO_2 + 1.664\,H_2; \qquad \Delta h = 0; \qquad \Delta f = 0, \qquad \textbf{(4.12)}$$

which is isothermic. The $CO_2$ must be removed to obtain pure hydrogen.[15]

Reactions similar to (4.11) may be constructed in which coal is reformed to liquid alkanes suitable for vehicle fuels by appropriate proportions of C, $H_2O$, and $O_2$ that yield $\Delta f = 0$, as in (4.10)–(4.12). Furthermore, similar reforming reactions utilizing liquid hydrocarbons and natural gas as raw fuels to create synthetic gaseous or liquid hydrocarbon products are common in the petrochemical and oil refining industries.

These simplified examples show how the atoms of solid, liquid, or gaseous hydrocarbon raw fuels may be rearranged to form other gaseous or liquid fuels that have more desirable properties without encountering significant loss of FHV. Nevertheless, the details of how this can be accomplished in an efficient and rapid manner and the synthesis processing facilities needed may present significant technological and economic challenges.[16]

## 4.5.2  Examples of Biochemical Synthesis

Biomass is produced by photochemical processes in plant cells, resulting in a biofuel of approximate atomic composition $CH_2O$. It is possible to rearrange the biomass molecules into other useful fuel forms utilizing unicellular microbes such as bacteria or yeast. These microbes act as catalysts and do not invest their metabolic energy in the fuel molecule processing. The synthesizing proceeds at concentration and temperature ranges within which the microbes can sustain their viability. The microbes may be naturally present in the raw biomass or may be added to the biomass. Biochemical processing is inherently a slow process, much like metabolic processing.

A common example of a biosynthetic process is the generation of methane by anaerobic bacteria in bogs or municipal solid or liquid organic waste facilities and by enteric fermentation in the digestive tracts of animals. In this process carbohydrate fragments $(CH_2O)$[17] are reconfigured into methane and carbon dioxide:

$$CH_2O \rightarrow \frac{1}{2}CH_4 + \frac{1}{2}CO_2; \qquad \Delta h = 1.34\,\text{MJ/kg fuel}; \qquad \Delta f = 1.258\,\text{MJ/kg fuel}, \qquad \textbf{(4.13)}$$

which is an exothermic process.

A more important biochemical process is the fermentation of carbohydrate sugars to form ethanol $(C_2H_6O)$:

$$CH_2O \rightarrow \frac{1}{3}C_2H_6O + \frac{1}{3}CO_2; \qquad \Delta h = 3.351\,\text{MJ/kg fuel};$$

$$\Delta f = 3.258\,\text{MJ/kg fuel}. \qquad \textbf{(4.14)}$$

---

[15]This process is used in coal-fired electric power plants with carbon capture (see Section 12.5.1.5 ).

[16]Reactions (4.10)–(4.12) are global reactions linking the reactants to the products formed. The path from reactants to products may require more than one step, increasing the values of $\Delta h$ and $\Delta f$.

[17]We assume that $CH_2O$ has the thermochemical properties of [C + $H_2O$].

This process forms the basis of the production of alcoholic beverages, a food product, but can be used to produce pure ethanol, a gasoline additive or substitute. The latter industrial scale process is discussed in Section 4.6.

## 4.6   BIOCHEMICAL PRODUCTION OF ETHANOL FROM BIOMASS

The production of vehicle-fuel ethanol from biomass has been spurred within the past decade for two reasons: the desire to replace imported petroleum with bioethanol so as to enhance national fuel supply security and to reduce fossil carbon emissions from road vehicles. While there are direct gains to both objectives, there are also indirect losses stemming from the production and processing of biomass for this purpose.

The major biomass feeds for biochemical conversion to ethanol are sugar cane, corn grain, and cellulosic plant matter (wood and crop waists), all of which are agricultural food or fiber crops in limited supply, nationally and globally. In addition, the global growth of biomass-derived ethanol production will have adverse indirect effects on food supply and land use that will tend to offset the desired reduction in fossil carbon emissions.

The global biochemical process for converting biomass to ethanol is that of Equation (4.14), but the steps required to convert biomass carbohydrate to methanol increase in complexity with the three fuels: cane sugar, corn grain, and cellulosic plant matter. For cane sugar, the carbohydrate content of the cane juice pressed from the cane stalk is almost entirely soluble sugars that can be fermented to ethanol, after which the ethanol is separated by distillation. Corn grain contains mostly starch, which must be hydrolyzed to lower carbon soluble sugars and then fermented to ethanol. Cellulosic plant matter must be pretreated to remove the lignin that prevents hydrolysis, after which it is hydrolyzed and fermented, as for corn. Each of these steps requires the addition of microbes (bacteria or yeast) and enzymes, and the process must be carried out at elevated temperatures that promote a rapid reaction rate but not so high as to kill the microbes. Ethanol is toxic to living cells in high enough concentrations, around 15% by volume, so large volumes of water are needed and processed in ethanol production. The ethanol separation process is energy intensive, as discussed in Section 3.11.3.

There are several aspects of this process that deserve mention. First, not all of the biomass FHV is converted to ethanol fuel, only that which is fermentable. The remaining residue constitutes a by-product with some heating value that can be used to generate process power and heat or, in the case of corn grain, has food energy value for food animal consumption. Second, the energy and economic costs of growing, fertilizing, reaping, and transporting the biomass feedstock to the refinery are integral parts of converting sunlight to feed biomass and thence to ethanol, and these require some use of fossil fuels and their attendant fossil carbon emissions. Third, productive arable land is a scarce resource, a storehouse for atmospheric carbon, and an irreplaceable reservoir of primary productivity, which can be damaged by destructive agricultural practices that induce soil erosion. Finally, utilizing food crops to produce biofuel has the potential to threaten the integrity and sufficiency of the human food supply. For all of these reasons, the conversion of food grains to ethanol and food oils to biodiesel fuel is not considered a long-term contributor to the reduction of fossil carbon emissions from vehicles.

## 4.7   ELECTROCHEMICAL REACTIONS

In Sections 4.2–4.4 we focused attention on combustion reactions, in which the burning of fuels in air provided the fuel energy needed to make mechanical power. This process is rapid, allowing quick combustion in but a small fraction of the power plant equipment such as the combustion chamber of a gasoline engine, a gas turbine combustion chamber, or the burners in a steam boiler. This energy is then utilized by moving engine pistons, whirling turbine blades, or boiling water changed to steam in a power plant. In many cases, the mechanical power is converted to electrical power sent out over transmission lines to users. This transformation of fuel energy to electrical form entails much more electrical/mechanical structure, and thereby expense, than just the combustion equipment that frees the fuel energy in the form of hot combustion gases.

Is it possible to convert the fuel energy into electric power directly without the use of the complex electromechanical equipment? In principle, and in practice, it is possible to intervene at the molecular level in the oxidation of the fuel molecules to force electrons released from the fuel molecules to flow through an external electric circuit, rejoining the oxidant at a higher electric potential as it forms product molecules. In that process the current flow in the external circuit generates electric power "without any moving parts." The device that accomplishes this magical feat is called a *fuel cell*. It consumes fuel and air (or oxygen) and emits combustion products while generating a direct current (DC) electric power stream.

A fuel cell is a particular example of a general device, the electrochemical cell. In Chapter 5 we consider another example, the electric storage battery. There are other important industrial uses for electrochemical reactions, such as the electrolytic refining of metal ores that utilize electric energy to extract pure metals from their ores.

## 4.7.1 Fuel Cells

In Section 3.10 we considered several different systems for converting the energy of fuel to mechanical energy by utilizing direct combustion of the fuel with air, each based upon an equivalent thermodynamic cycle. In these systems, a steady flow of fuel and air is supplied to the "heat engine," within which the fuel is burned, giving rise to a stream of combustion products that are vented to the atmosphere. The thermal efficiency of these cycles, which is the ratio of the mechanical work produced to the heating value of the fuel, is usually in the range of 25–50%. This efficiency is limited by the combustion properties of the fuel and mechanical limitations of the various engines. Thermodynamically speaking, the combustion process itself is an irreversible one and accounts for a large part of the failure to convert more of the fuel energy to work.

Is there a more efficient way to convert fuel energy to work? The second law of thermodynamics places an upper limit on the amount of work that can be generated in an exothermic chemical reaction, such as that involved in oxidizing a fuel in air. In a chemical change that proceeds at a *fixed temperature and pressure*, the maximum work that can be extracted is equal to the decrease in free energy of the reactants as they form products in the reaction. For most fuels the change of free energy $f$, defined in Equation (3.46), is only slightly different from the FHV (see Table 4.1), but this limit is certainly much greater than the work produced by practical heat or combustion engines. But thermodynamics alone does not tell us how it might be possible to capture this much greater amount of available energy in fuels.

As explained above, a fuel cell is a device that can convert chemical to electrical energy. It consists of an electrolyte[18] filling the space between two electrodes. In a *battery*, the electrodes are

---

[18]An electrolyte is a fluid or solid in which some atoms or molecules are dissociated into positive and negative ion pairs that are free to move through the electrolyte, carrying electric current. Pure water is an electrolyte composed of $H^+$ and $OH^-$ ions in low concentrations. By dissolving salt or strong acids in water, the ion concentration and electrical conductivity is increased above that of pure water.

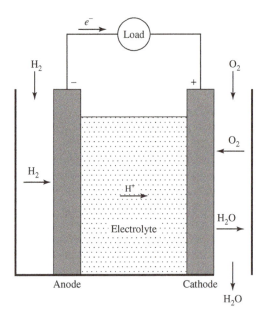

**FIGURE 4.1**    A sketch of a hydrogen–oxygen fuel cell (not to scale). Hydrogen and oxygen are supplied to porous electrodes separated by an electrolyte in which the electric current is carried by hydrogen ions. In the external electric circuit, current is carried by a matching electron flow. The product of oxidation, water, evolves from the cathode.

chemically dissimilar, which causes an electric potential difference between them. In a *fuel cell*, the electrodes are chemically similar but one is supplied with a fuel and the other with an oxidant, generating an electric potential difference between them. By closing the electric circuit external to an electrochemical cell, a current may be drawn from the cell, generating electrical power. The electrical energy consumed in the external circuit is generated by chemical changes within the cell. The second law of thermodynamics requires that the electrical energy delivered to the external electric circuit is never greater than the reduction of free energy of the accompanying chemical changes within the cell.[19]

The structure of a fuel cell, sketched in Figure 4.1, is amazingly simple, albeit complicated in detail. Two porous metal electrodes are separated by a space filled with an electrolyte, a fluid or solid in which positive or negative ions can move readily under the influence of electric fields, much as electrons do in an electrically conducting metal. Fuel and oxidant are supplied to separate electrodes, diffusing through the porous material to the electrolyte. At the anode (the negative electrode), electrons are transferred to the electrode from the electrolyte as positive ions are formed from fuel molecules; at the cathode (the positive electrode), electrons are emitted to the electrolyte to form negative ions or to neutralize positive ions. If the electrodes are connected by electrically

---

[19] See Section 3.11.2.

conducting wires to an external load, as shown in Figure 4.1, an electric current will flow (in the counterclockwise direction in Figure 4.1) and electrical work will be expended on the load because the cathode electric potential is greater than that of the anode. Inside the fuel cell, the electric current completing the circuit is carried by ions moving through the electrolyte.

The chemical reaction that generates the electrical energy expended in the external load occurs partially at each electrode. Taking as an example the hydrogen-oxygen fuel cell shown in Figure 4.1, the surface reaction at the anode when a hydrogen fuel molecule is ionized in a *reducing* reaction and enters the electrolyte can be expressed as

$$H_2 \rightarrow 2H^+\{\Phi_{el}\} + 2e^-\{\Phi_a\} \tag{4.15}$$

where $\Phi_{el}$ is the electric potential of the electrolyte and $\Phi_a$ is that of the anode. In this reaction the electron moves into the anode at its potential $\Phi_a$ while the hydrogen ion moves into the electrolyte at the potential $\Phi_{el}$. At the cathode, the *oxidizing* reaction is

$$2H^+\{\Phi_{el}\} + 2e^-\{\Phi_c\} + \frac{1}{2}O_2 \rightarrow H_2O \tag{4.16}$$

where $\Phi_c$ is the cathode electric potential. The net effect of both of the electrode reactions is the production of water and the movement of a charge through the external circuit, found by adding (4.15) and (4.16):

$$H_2 + \frac{1}{2}O_2 \rightarrow H_2O + 2e^-\{\Phi_a\} - 2e^-\{\Phi_c\} \tag{4.17}$$

In this overall reducing–oxidizing reaction, the hydrogen and the oxygen molecules are changed to water molecules. In the process, for each hydrogen molecule two electrons flow from the low anode potential to the high cathode potential in the external circuit, producing electrical work. If $q_e$ is the magnitude of the charge of an electron and $m_{H_2}$ is the mass of a hydrogen molecule, then the electrical work per unit mass of fuel in the reaction is $w_e = (2q_e/m_{H_2})(\Phi_c - \Phi_a)$. Multiplying the numerator and denominator of the first factor of this expression by Avogadro's number (see Table A.3), we find

$$w_e = \left(\frac{2\mathcal{F}}{\mathcal{M}_{H_2}}\right)(\Phi_c - \Phi_a), \tag{4.18}$$

where $\mathcal{F}$=9.6487E(4) coulomb/mol is the Faraday constant and $\mathcal{M}_{H_2}$ is the molecular mass (g) of a mole of diatomic hydrogen ($H_2$).

The second law limits the electrode potential difference $\Phi_c - \Phi_a$ that can be achieved, because the work $w$ cannot exceed the free energy change $\Delta f$ available in the oxidation reaction

$$w_e \leq (\Delta f)_{H_2}$$
$$(\Phi_c - \Phi_a) \leq \frac{(\Delta f)_{H_2} \mathcal{M}_{H_2}}{2\mathcal{F}}, \tag{4.19}$$

where (4.18) has been used to eliminate $w_e$ in the second line of (4.19). The right hand side of (4.19) is thus the maximum possible electrode potential difference. For a hydrogen–oxygen fuel cell at 20°C and one atmosphere of pressure, using the values of Table 4.1, this is calculated to be 1.225 V.

The maximum potential difference of a fuel cell, determined by the free energy change of the fuel oxidation reaction as in Equation (4.19), is only reached when the cell is operated in a reversible manner by limiting the current to extremely low values, in effect zero current or open external circuit. For finite current draw, there will be a voltage drop at the electrode surfaces and within the electrolyte that is needed to move the fuel ions to the cathode at a finite rate and a corresponding decline in the electrode potential difference. The thermodynamic efficiency $\eta_{fc}$ of a fuel cell may then be defined as the ratio of the actual electric work $w_e$ delivered by the cell to the maximum work $\Delta f$,

$$\eta_{fc} \equiv \frac{w_e}{\Delta f}. \tag{4.20}$$

The high efficiencies of fuel cells, compared with heat engines utilizing the direct combustion of fuel with air, stems from the electrode processes where the electrostatic energy binding molecules can be converted directly to electrostatic energy of the ions and electrons that move in the cell circuit. In contrast, in an adiabatic combustion process the fuel energy is converted to random kinetic and potential energy of product molecules, which cannot be fully recovered in subsequent flow processes.

Figure 4.2a illustrates how the fuel cell voltage and power vary with the current when the electrolyte provides the only resistance to current flow within the cell. As the current increases, the cell voltage drops linearly from its maximum value given in Equation (4.19). The power output, which is the product of the current times the voltage, reaches a maximum when the voltage has fallen to 50% of its maximum value. The fuel cell efficiency, given in Equation (4.20), declines

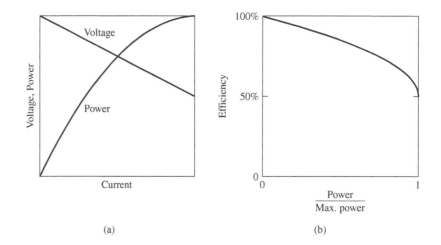

(a)                                          (b)

**FIGURE 4.2**   Idealized sketches of (a) fuel cell voltage and power as a function of current and (b) efficiency as a function of power.

with increasing power, as shown in Figure 4.2b, to 50% at maximum cell power. When operating at part load, fuel cells can have significantly higher efficiencies than do combustion engines.

To maintain the fuel cell temperature at a fixed value, heat must be removed. The magnitude of the heat removed per unit mass of fuel, $|q_{fc}|$, is determined from the first law of thermodynamics applied to this steady flow process to be

$$|q_{fc}| = \text{FHV} - w_e \geq \text{FHV} - \Delta f, \tag{4.21}$$

where FHV is the fuel heating value and the inequality follows from the second law constraint. In fuel cells used to generate electric utility power, the heat removed from the fuel cell may be used to generate additional electricity in a Rankine cycle plant provided the fuel cell operating temperature is sufficiently high. This combined cycle plant can achieve high thermal efficiencies.

### 4.7.2  Practical Fuel Cell Systems

Fuel cells can be constructed from a variety of materials and utilize different fuels. The objective is to maximize the output power per unit area of electrode surface and the cell efficiency of conversion of fuel energy to electric energy, while minimizing the capital cost per unit output of power. The type of fuel cell can be categorized by the choice of the electrolyte, the electrolyte ion current carrier, the type of electrode catalyst, the fuel utilized, and the operating temperature.

The most common types of fuel cells, designated by their electrolyte composition, are the polymer (P), alkaline (A), phosphoric acid (PA), solid oxide (SO), and molten carbonate (MC). Their other characteristics are described below.[20]

*Polymer* electrolyte fuel cells utilize a solid perfluorosulfonic acid polymer and electrodes coated with noble metal catalysts. The fuel is hydrogen and the oxidant is oxygen or air. The ion carrier is $H^+$. Operating temperature is 30–100°C.

*Alkaline* fuel cells utilize an electrolyte of potassium hydroxide solution in water and electrodes coated with non-noble metal catalysts. The fuel is hydrogen and the oxidant is oxygen. The ion carrier is $OH^-$. Operating temperature is 60–250°C.

*Phosphoric acid* fuel cells utilize an electrolyte of phosphoric acid ($H_3PO_4$) solution in a porous silicon carbide matrix and electrodes coated with noble metal catalysts. The fuel is hydrogen and the oxidant is oxygen. The ion carrier is $H^+$. Operating temperature is 160–210°C.

*Solid oxide* fuel cells utilize an electrolyte of phosphoric acid ($H_3PO_4$) solution in a porous silicon carbide matrix and electrodes coated with noble metal catalysts. The fuel is hydrogen and the oxidant is oxygen. The ion carrier is $O^+$. Operating temperature is 800–1,000°C.

*Molten carbonate* fuel cells utilize an electrolyte of molten alkali metal (Li/K or Li/Na) in a porous matrix of $LiAlO_2$. The electrode catalysts are nickel based. The fuel is hydrogen (or CO) and the oxidant is air plus $CO_2$. The ion carrier is $CO_3^{2-}$. Operating temperature is 600–800°C.

Each of these types of fuel cells has advantages and disadvantages and generally is appropriate for different services, such as road vehicles, auxiliary power sources, and space vehicles.

Practical fuel cells have cell potentials and power outputs that are less than those of the idealized cell shown in Figure 4.2, primarily because of internal electrical losses at the two electrodes. An example of a typical fuel cell performance is shown in Figure 4.3. The cell potential difference $\Delta\Phi$,

---

[20]For a more complete description, see Mench, Matthew M. *Fuel Cell Engines.* Hoboken: John Wiley & Sons, 2008.

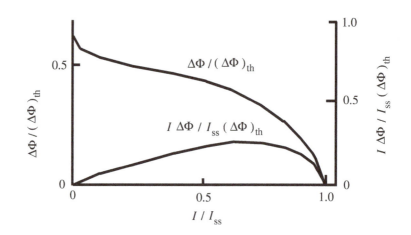

**FIGURE 4.3** A plot of typical fuel cell performance showing the dimensionless electrode potential $\Delta\Phi/(\Delta\Phi)_{th}$ and electrical power output $I\Delta\Phi/I_{ss}(\Delta\Phi)_{th}$ versus the dimensionless current $I/I_{ss}$.

divided by the maximum theoretical cell potential $(\Delta\Phi)_{th}$ of Equation (4.19), is plotted as a function of the cell current $I$ divided by the maximum short circuit current $I_{ss}$. Also shown is the electrical power $I\Delta\Phi$, divided by the theoretical power $I_{ss}(\Delta\Phi)_{th}$, as a function of the cell current ratio $I/I_{ss}$. It can be seen that the cell potential and power are less than expected, which is a consequence of the losses at the electrodes. In Figure 4.3, the cell efficiency $\eta_{fc}(=\Delta\Phi/(\Delta\Phi)_{th})$ is about 35% at maximum power. Typical cell maximum power per unit of cell electrode area is about 1 W/m$^2$.

The electromechanical structures of fuel cell power systems are complex. Each cell must be supplied with separate fuel, oxidant, and product streams and a method of cooling to remove heat at a rate comparable to the electrical power output. Cells are stacked in series, with electrical connections that pass the same current through each but at increasing electric potential. Fuel and oxidant composition may have to exclude contaminants that can poison electrode catalysts. Needless to say, there are many possibilities for ingenuity in material selection and design geometry that will need to be developed to secure robust and efficient fuel cell systems.

## 4.8   THE HYDROGEN ECONOMY

Hydrogen is a carbon-free fuel that can substitute directly for fossil fuels in automobile engines and gas turbines or in fuel cells that directly generate electric power. But it is not a naturally occurring fuel like fossil fuel or biofuel; it is instead a synthetic fuel whose fuel energy is supplied by other sources, such as fossil fuels (coal, petroleum, natural gas) and biomass or by electric power. From an energy viewpoint, hydrogen fuel energy is less, and more costly, than that of any of its sources, but can possess other desirable characteristics that may offset these deficiencies.

We consider here several hydrogen fuel systems that possess some of these advantages and their offsetting disadvantages:

- Hydrogen in the form of a compressed gas or refrigerated liquid, stored aboard an automobile or other highway vehicle, and burned with air in a fuel cell or reciprocating engine.

- A carbon-free fuel synthesized from carbon-containing fossil fuel in which the carbon is captured as $CO_2$ and stored underground and can be used as non-carbon-emitting fuel for any purpose.
- An energy storage fuel produced by an electric power source, stored for a period of time, and then later used to generate electric power when needed.
- Compressed gas distributed by pipeline to stationary users as a natural gas substitute.

In all of these examples hydrogen serves as an intermediate carrier of fuel energy whose ultimate consumption results in no $CO_2$ emissions to the atmosphere.

### 4.8.1  Hydrogen Fuel for Vehicle Propulsion

The use of hydrogen-fueled cells for generating electric power aboard highway vehicles has been under development for several decades. The original purpose was to eliminate toxic tailpipe emissions from such vehicles and to secure the higher energy efficiency of fuel cells feeding electric traction motors. If the hydrogen is generated by electrolysis powered by renewable or nuclear power plants or synthesis of fossil fuels incorporating carbon capture with subsurface storage, then the climate warming emissions reduction becomes an additional and potentially more valuable benefit.

Offsetting these benefits are problems associated with vehicle cost and performance and problems associated with the fuel supply system. For the vehicle, the weight and volume of the fuel storage system and fuel cell are greater than that of the traditional system it replaces, which adversely affects the vehicle energy efficiency (distance traveled per unit of fuel energy consumed) and vehicle range. In addition, the capital and operating costs are also greater than that of comparably sized conventional vehicles.

The fuel supply system also presents problems. Filling stations would need pressure vessels to store hydrogen, whatever the source, sufficient to meet vehicle demand between storage refills and to maintain vehicle fuel storage pressure. If supplied by pipeline from a central hydrogen-generating station or on site by an electrolysis plant, storage requirements would be minimal. For isolated filling stations, supply by highway tank trucks may be more practical, but the pressurized storage would be more costly than for an equivalent amount of gasoline.

Finally, compressed hydrogen gas storage, whether aboard a vehicle or at a filling station, presents fire and explosion hazards in the event of an accidental leakage of gas to the atmosphere, of greater hazard than comparable amounts of compressed natural gas, for example.[21]

### 4.8.2  Synthetic Hydrogen from Fossil Fuels with Carbon Capture and Storage

In Section 4.5.1 we discussed the generation of hydrogen from coal, as in Equation (4.12), in which coal plus steam is changed to hydrogen and $CO_2$ in a synfuel reactor. This reaction permits the separation of the fossil carbon from the reactor products, leaving a pure hydrogen fuel; the $CO_2$ may then be stored underground as waste material. This coal-based synthetic hydrogen fuel is the fuel preparation step in coal gasification combined cycle power plants discussed in Section 12.5.1.4. In this type of power plant, synthetic hydrogen is the gaseous fuel that powers a combined cycle (gas

---

[21] See Chapter 5 for a discussion of compressed gas storage.

turbine plus steam cycle turbine) to generate electric power. Of course, this hydrogen fuel could otherwise be used to generate electricity in a hydrogen–air fuel cell.

There are other uses for hydrogen than electric power production. Hydrogen is used in the catalytic cracking of heavy hydrocarbons to synthesize lighter-weight fuels.[22] It is also used in the synthesis of ammonia ($NH_3$), a component of agricultural fertilizer. By combining a synthetic hydrogen plant that incorporates carbon capture and storage with these other uses for hydrogen, the energy content of fossil fuels may be utilized without generating $CO_2$ emissions to the atmosphere.

A drawback of the synthetic hydrogen process is the difficulty of varying its output rate; this is especially bothersome for variable load electric power plants, some of which need to be started and stopped to coincide with the diurnal cycle of electric power demand. It is difficult to operate the synthesis plant to provide a variable hydrogen supply that meets the requirements of the electrical demand. For this reason synthetic hydrogen can only be used for base load electricity or steady load petrochemical synthesis plant operations.

Synthetic hydrogen with carbon capture and storage can be produced from coal or natural gas. It could also be used with biomass fuels, providing a net sink for biomass carbon.

### 4.8.3   Hydrogen as Energy Storage for Intermittent Electric Power Plants

Some renewable energy electric power plants produce electric power only when the energy source is available, but not always when it is needed.[23] It would be desirable to provide storage of this energy when it is available in periods of low demand for use during high demand when the renewable resource may not be available.

Conventional electric power systems tie together base load plants, which operate continuously at their rated power with variable load power plants that operate part time to provide the time-variable demand. The installed power of the system is greater than the average power output so as to provide for the highest demand encountered.

To provide this storage and reuse function, it would be necessary to convert the renewable energy to hydrogen fuel, to store this fuel for future use, and then generate electricity from the draw-down of the stored hydrogen. For the hydrogen generation, water could be electrolyzed to hydrogen and oxygen, a standard process for supplying highly purified forms of these gases. The hydrogen would be stored under high pressure in tanks or in underground storage caverns. This hydrogen could then be used to generate electric power in fuel cells or as fuel for combined cycle gas/steam turbine plants.

There are difficulties to be overcome in such a scheme. There are energy losses in both electricity-to-hydrogen conversion and its inverse, hydrogen to electricity; it would be difficult to reach more than about 50% conversion efficiency overall. A considerable capital investment would be required, perhaps as large as that for the renewable energy source of electric power. Until a cheap and efficient technology can be developed, it is unlikely that intermittent renewable energy will be deployed in this manner.

---

[22] See Section 4.5.1.

[23] For example, see Section 8.6.

### 4.8.4   Hydrogen as a Substitute for Pipeline Natural Gas

Residential and commercial users of natural gas and highway vehicles using compressed natural gas comprise a plethora of end users distributed over a wide region. To reduce fossil carbon emissions from such numerous small users, the current natural gas supply could be converted to hydrogen gas with carbon capture and storage at a central large-scale plant and the hydrogen distributed through existing distribution pipelines. This change might be economically justified if carbon emission reduction allocations can be sold to emitters that are unable to make regulated emission reductions.

## 4.9   CONCLUSION

Continued consumption of fossil fuel, the major source of global energy, presents an increasing threat to the global environment because of its effect on climate change. This chapter discussed the many ways that the energy locked up in these fuels can continue to be utilized by technologies that prevent the current atmospheric emissions of global warming gases, principally carbon dioxide, which can be stored safely underground. The energy-containing residue of the fossil fuel, mainly hydrogen, can be efficiently utilized by known technologies to replace the traditional fossil fuel uses.

The technology needed to accomplish this transformation will incur additional expense and some lessening of the useable energy originally in the fossil fuel. New and more energy efficient technologies will be needed to utilize the synthetic fuels created from the natural fossil fuels that provide the energy source.

The physicochemical principles that underlie the development of such energy technologies are well understood and provide a confident basis for the needed transformation of fossil fuel use. There will be adverse consequences to national and global economies, but over time these may be manageable.

There are current technologies for accomplishing these transformations in fuel use, but improvements in their energy efficiencies and economic costs are needed to provide the energy supply and services required by global development.

# PROBLEMS

### Problem 4.1

A synthetic fuel consists of 50% by weight CO and 50% by weight $H_2$. Calculate its FHV, MJ/kg fuel and MJ/kg product, and stoichiometric air–fuel ratio using the data of Table 4.1.

### Problem 4.2

In synthesizing hydrogen for use in a fuel cell, methane may be used in the overall reaction

$$CH_4 + 2H_2O \rightarrow CO_2 + 4H_2,$$

which is endothermic. Using the data of Table 4.1, calculate the amount of heat that must be supplied for this synthesis, per unit mass of methane consumed, if it proceeds at 25°C and one atmosphere of pressure. If this heat is supplied by burning additional methane in air, calculate the total kilograms of methane consumed per kilogram of hydrogen produced.

### Problem 4.3

The potential difference $\Delta\Phi$ of a hydrogen fuel cell, as a function of the cell current density $i$ amp/cm$^2$, is found to be

$$\Delta\Phi = (1.1 \text{ volt})(1 - 0.5\,i).$$

Calculate the maximum power output of the cell in W/m$^2$ and the values of $i$, $\Delta\Phi$, and the cell efficiency at maximum power.

### Problem 4.4

A PEM fuel cell using methane as fuel is supplied with a methane–water mixture at the anode. The anode reaction produces hydrogen ions in the electrolyte and anode electrons according to the reaction

$$CH_4O + H_2O \rightarrow 6H^+\{\Phi_{el}\} + 6e^-\{\Phi_a\} + CO_2.$$

At the cathode, the cathodic electrons and electrolytic hydrogen ions combine with oxygen to form water molecules,

$$6H^+\{\Phi_{el}\} + 6e^-\{\Phi_c\} + 3O_2 \rightarrow 3H_2O.$$

The combination of these reactions is the oxidation of methanol while producing a current flow in the external circuit across the electrode potential difference, $\Phi_a - \Phi_c$,

$$CH_4O + 3O_2 \rightarrow 2H_2O + CO_2 + 6e^-\{\Phi_a\} - 6e^-\{\Phi_c\}.$$

(a) Using the data of Table 4.1, calculate the maximum potential difference $\Phi_a - \Phi_c$ that the fuel cell can generate. (b) Calculate the ratio of the maximum electrical work per unit of fuel mass (J/kg) to the FHV.

# BIBLIOGRAPHY

Bartok, William, and Adel F. Sarofim, eds. *Fossil Fuel Combustion.* New York: John Wiley & Sons, 1991.
Khartchenko, Nikolai V. *Advanced Energy Systems.* London: Taylor & Francis, 1998.
Kordesch, Karl, and Gunter Simader. *Fuel Cells and Their Applications.* New York: VCH Publishers, 1996.
Mench, Matthew M. *Fuel Cell Engines.* Hoboken: John Wiley & Sons, 2008.
National Research Council. *Electricity from Renewable Resources: Status, Prospects, and Impediments.* Washington, D.C.: National Academies Press, 2010.
Probstein, Ronald F., and R. Edwin Hicks. *Synthetic Fuels.* New York: McGraw–Hill, 1982.
Twidell, John, and Tony Weir. *Renewable Energy Resources*, 2d. ed. London: Taylor & Francis, 2006.

# Electrical Energy Generation, Transmission, and Storage

## 5.1  INTRODUCTION

Prior to the industrial revolution, human and animal power had provided the bulk of the mechanical work needed in an agricultural society to provide food, clothing, and shelter for human settlements. Through invention and technological development, wind, tidal, and river flows provided mechanical power for milling of grain, sawing of timber, and ocean transportation of goods. But the invention of the steam engine in the early years of the industrial revolution greatly expanded the amount of mechanical power available in industrializing countries for the manufacture of goods and the transportation of people and freight, giving rise to economic growth. By the late 19th century the forms of mechanical power generation had evolved to include the steam turbine and gasoline and diesel engines and their uses in ocean and land vehicles. By that time the dominant fuels that produced mechanical power were coal and oil rather than wood. Although human and animal power, as well as the renewable power of wind and stream, were still significant at the dawn of the 20th century, fossil-fueled mechanical engines were clearly the major and rapidly growing sources of industrial energy.

A technological development that greatly augmented the usefulness of mechanical power in manufacturing and, eventually, commercial and residential settings was the 19th century invention of the electric generator and motor that converts mechanical and electrical power from one form to the other with little loss of energy. Unlike mechanical power, which is generated from fossil fuel at the site of power use, as in a manufacturing plant, railroad locomotive, or steamship, electric power generation made possible the transmission of electrical power from a central location to distant consumers via electric transmission lines, which greatly increased the usefulness of the electrical form of work. Together with the end-use inventions that make electrical power so useful, such as the electric light and electric communication devices, the production of electric power has grown so that it constitutes nearly one third of the energy use in current industrialized societies.

Today in the United States, most electric power is generated in large power plants where fossil or nuclear fuel provides the heat needed to generate mechanical work in a steam cycle, with mechanical power being converted to electrical power in the electric generator, or alternator, as

60-cycle synchronous alternating current (AC) electricity.[1] In transmitting the power to distant consumers, the voltage is increased by electrical transformers and the power is merged into a network of high-voltage transmission lines that joins the outputs of many power plants to supply the myriads of end users who are connected to the transmissions lines by a network of distribution lines at lower voltage. More recently, electric power is being generated in smaller plants, often employing gas turbine engines, either alone or in combination with steam turbines (called combined cycle power plant[2]), as the mechanical power source. In addition, some electricity is generated in industrial or commercial plants that utilize the waste heat from the power production process for process or space heating (called cogeneration). The transmission and distribution systems that connect all these sources to each other and to the users of electric power are usually owned and managed by public utility companies.[3]

Some electric power is produced from nonthermal sources of energy. The most prominent of these is hydropower, where mechanical power is produced by hydroturbines supplied with high-pressure water from a reservoir impounded by the damming of a river. The energy source is the difference in gravitational energy of the higher-level water behind the dam compared with the lower-level water downstream of the dam, this difference in level being called the head.[4] Hydropower may be generated as well from the rise and fall of ocean tides. Some electric power is now generated by the mechanical power of wind turbines extracting energy from the wind.

Solar insolation is being used to generate electric power either directly, utilizing photovoltaic cells that convert the energy of solar photons to electric power, or indirectly by supplying heat to thermal steam or vapor engines that drive electric generators. Solar insolation provides the energy needed to grow plants and may thereby be utilized indirectly to generate electric power when biomass crops are used as fuel in thermal power plants.

Nonthermal and solar sources of electric power are termed renewable, in contrast to mineral fuels, such as fossil and nuclear, that are extracted from the earth or ocean. Renewable energy systems are discussed in Chapter 8.

Electrochemical systems, such as batteries and fuel cells, convert the chemical energy of reactant molecules directly to electricity without an intermediate step where mechanical power is generated. While they currently contribute very little to the amount of electric power generation, they are obviously important in such applications as portable communication devices and in the development of electric-drive road vehicles.[5]

The physical principles of electricity and magnetism, which explain how stationary and moving electric charges generate electric and magnetic fields, how mechanical forces are exerted on electric currents flowing in electrical conductors in the presence of magnetic fields, and how electric current is induced to flow by electric fields in conducting materials, provide the basis for the mechanism

---

[1]Electric utility plants are interconnected with each other by transmission lines so that electric power may be reliably supplied to all customers. This requires synchronization of the generators and standardization of voltages among participating plants. For a description of these plants, see Chapters 6 and 7.

[2]See Sections 3.9.5 and 6.3.

[3]In many U.S. states, the ownership of transmission and distribution lines is being separated from the ownership of generation facilities as a part of the government's deregulation of the electric utility industry.

[4]The ultimate source of this power is solar insolation that evaporates ocean water, which is subsequently precipitated to the land drainage basin of the reservoir.

[5]See Section 9.6 for a discussion of electric-drive vehicles.

by which electrical power is generated, transmitted, and utilized.[6] These interactions are those of thermodynamic work, using the parlance of the science of thermodynamics explained in Chapter 3. Insofar as electromagnetic physics deals with forces on charges and currents, it may be regarded as explaining a form of work or power, reasonably called electrical work or power, to distinguish it from mechanical work or power, in which the forces are caused by mechanical contact or a gravitational field. Correspondingly, we may regard that the work required to move a charge or current-carrying conductor in an electric or magnetic field results in a change of electric or magnetic energy of the field, so that energy may be said to be stored in the field, in analogy with the change in potential energy of a mass in the earth's gravity field.

There are circumstances where it is desirable to store mechanical or electrical work generated from mineral fuel consumption for use at later times. A common example is the storage of electrical energy in an automobile's electric storage battery to supply the power needed to start the engine in a subsequent use. In the automobile engine itself a flywheel stores rotational energy produced by the power stroke of the pistons, returning it during the compression stroke.

One of the most prominent uses of electric energy storage is that employed by electric utility systems to even out diurnal variations in the demand for electric power by storing energy during night-time hours when demand is low and excess power is available and then restoring this energy to the system during daytime hours of peak demand. This permits the central electric generating plants to operate at constant power and best efficiency all day long, lowering the cost of electricity generation.

The principle involved is illustrated in Figure 5.1 depicting the diurnal variation of electric power demand in a typical electric utility system. Starting at midnight, the demand declines to a minimum in the early morning hours and then increases to a daily maximum in the late afternoon or early evening hours, after which it declines again to its midnight level. The utility power suppliers usually match this demand by turning on individual plants during the rising portion of

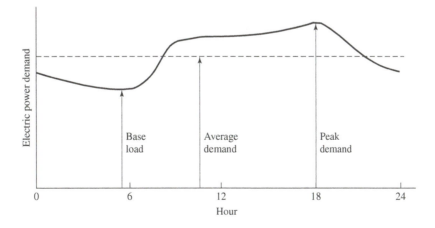

**FIGURE 5.1**    A typical diurnal demand for electric power has an early morning minimum and a late afternoon maximum, with the former defining the base load demand that is met by continuously operating plants.

---

[6]Electric communication systems consume only a small proportion of electrical power in modern economies. Despite their great importance, we will not explore here their interesting technologies.

the demand and then taking them off line as the demand declines. However, a sufficient number of plants must be run continuously to supply the minimum demand, called the base load. Supplying the base load can be accomplished more efficiently than supplying the variable load between the minimum and peak load where plants operate at less-than-optimal conditions, both economically and thermodynamically. If electric storage systems are used to store energy during times of less-than-average demand and return this energy during times of above-average demand, then fewer power plants will be needed to satisfy the diurnal demand and those fewer plants may be operated at optimal conditions.

The most common form of storing energy for use in central electric power systems is the pumped storage system where water is pumped from a lower- to a higher-level reservoir, with the energy being stored in the earth's gravitational field. This energy is later recovered when the stored water flows through a turbogenerator to the lower-level reservoir. Other schemes have been proposed as well, including the use of electric storage batteries, electric capacitors and inductors, and flywheels. The latter are also possible sources of emergency electric power for essential purposes, like hospital operating room power and computer power.

Electric power produced from some sources of renewable energy, such as photovoltaic cells, wind turbines, and tidal power systems, may not be available at times that synchronize with electric power demand, as illustrated in Figure 5.1. Energy storage may be necessary for the successful use of these systems, especially where they cannot be tied into an electric power grid that will provide power at times when the renewable source is absent.

In recent years a renewed interest in electrically propelled automobiles and other vehicles has arisen as a consequence of a desire to improve vehicular energy efficiency and reduce air pollutant emissions from ground transportation vehicles. Most of the proposed systems incorporate in part some energy storage in the form of electric storage batteries, flywheels, or electric capacitors. For such applications, the weight, volume, and cost of sufficient energy storage to provide acceptable vehicle performance is generally much greater than that for conventional hydrocarbon-fueled vehicles, posing obstacles to their widespread use. Compared with stationary energy storage systems, mobile ones must meet much more stringent requirements.

This chapter treats the physical principles that determine how electrical power is generated, transmitted, and stored. We first consider how mechanical work is converted to electrical work in an electrical generator, along with the inverse of that process in an electric motor. We next show how electric power flows in a transmission line that connects the electric generator with the end user of the electric power and explain how some of that power is lost in resistance to the flow of current in electric wires. Finally, we describe the kinds of energy storage systems that are used to provide electrical or mechanical power for use at times when it is not otherwise available. Emphasis is placed on the energy and power per unit mass and volume of these systems, the cost per unit energy stored, and the efficiency of recovery of the stored energy, as these are the parameters that determine their use to replace the conventional systems where mineral fuel is consumed as needed to supply electrical or mechanical power and heat.

## 5.2   ELECTROMECHANICAL POWER TRANSFORMATION

The electric generator and electric motor are the principal devices by which mechanical and electrical power are converted from one form to the other. In the United States almost all utility electric power is generated by steam, gas, hydro, or wind turbines driving an electric generator. About 60%

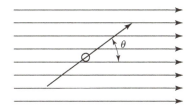

**FIGURE 5.2**   A magnetized needle in a magnetic field requires a counterclockwise torque to hold it in place at an angle $\theta$.

of this electric power is converted by electric motors to mechanical power for residential, commercial, and industrial use. Fifty percent of all mechanical power produced by fossil and nuclear fuel consumption is used to generate electric power.

The magnetic interaction that underlies the operation of an electric motor or generator is illustrated in Figure 5.2, showing a magnetized compass needle placed in a magnetic field. The needle seeks to align itself with the magnetic field lines, as does a compass needle in the earth's magnetic field. If we hold the needle stationary at an angle $\theta$ from the direction of the magnetic field, we must apply a torque $T$ that is equal to the product of the needle's magnetic dipole moment $M$, the strength of the magnetic field $B$, and $\sin\theta$. Alternatively, work can be done by the needle if it rotates to align itself with the magnetic field, in the amount $MB(1 - \cos\theta)$, but work can be generated only for a half revolution of the needle, at most. To make an electric motor of this device, we must reverse the direction of the needle's magnetization every half revolution. This can be accomplished by surrounding the needle with a coil of wire through which a current flows from an external circuit, with the current being reversed each half revolution. The basic elements of both motor and generator are a magnetizable rotor and a stationary magnetic field, either or both of which are connected to external electric circuits that adjust the amount and direction of the magnetic fields.

The physical principles underlying both the electric motor and the electric generator are illustrated in Figure 5.3a. A wire of length $L$ carrying a current $I$ in the presence of a magnetic field $B$ is subject to a restraining force $F$,

$$F = IBL. \tag{5.1}$$

In addition, if the wire moves at a velocity $V$ in the direction of the force, it experiences an electric field $E$ in the direction opposite to that of the current in the amount[7]

$$E = VB. \tag{5.2}$$

In a generator or motor, wires attached to a rotating armature move through a magnetic field established by a field coil or permanent magnet. The force $F$ applied to a generator armature wire delivers mechanical power $\mathcal{P}$ at a rate $FV$ while the electrical power produced when the current $I$ flows in the direction of the potential increase $EL$ is $IEL$. Utilizing Equations (5.1) and (5.2), these

---

[7]The general forms of Equations (5.1) and (5.2), using boldface characters to represent vector quantities, are, respectively, $\mathbf{F} = -(\mathbf{I} \times \mathbf{B})L$ and $\mathbf{E} = -\mathbf{V} \times \mathbf{B}$.

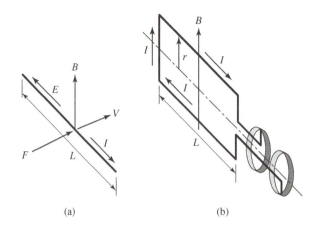

(a)                                        (b)

**FIGURE 5.3**   (a) A wire carrying a current $I$ and moving perpendicular to a magnetic field $B$ at a speed $V$ is subject to a restraining force $F$ and experiences an electric field $E$. (b) A sketch of a simple armature circuit showing how the current loop is connected to slip rings that deliver the current to an external circuit.

powers are found to be equal,[8]

$$\mathcal{P} = FV = IBLV = IEL. \tag{5.3}$$

Neglecting any electrical and mechanical losses, the mechanical power input *(FV)* to an ideal generator then equals the electrical power output *(IEL)*.

If we reverse the direction of the velocity $V$, the direction of the mechanical power is also reversed (i.e., there is a mechanical power output from the device). Simultaneously, the electric field is reversed and electric power is now an input. In this mode the device is an electric motor. Neglecting losses, Equation (5.3) defines the equality of input electrical power to output mechanical power for an electric motor.

Figure 5.3b shows how a rectangular loop of wire attached to a rotating armature and connected to slip rings can deliver the current to an external stationary circuit via brushes that contact the rotating slip rings. For this geometry, the peripheral velocity $V$ of the armature wires is $2\pi rf$, where $f$ is the armature's rotational frequency and $r$ is the distance of the wire from the armature axis, so that the ideal output and input power $\mathcal{P}$ is, using Equation (5.3),

$$\mathcal{P} = 2IBL(2\pi rf) = I(4\pi frLB), \tag{5.4}$$

where the factor of 2 arises from the return leg of the circuit. Since the electrical power equals the product of the current $I$ times the potential difference $\Delta\phi$ across the electrical output terminals, the factor $(4\pi frLB)$ in Equation (5.4) is equal to $\Delta\phi$.[9]

---

[8]Here we are ignoring any resistive loss in the wire.

[9]The potential increase $\Delta\phi$ of an electric generator is called an *electromotive force*. The generator's internal current $I$ flows in the direction of increasing electric potential $\phi$, whereas the current in an external circuit connected to the generator flows in the direction of decreasing potential. A source of electromotive force, such as a generator or battery, is a source of electric power for the attached external circuit.

In the case of the armature circuit of Figure 5.3b, the electric potential changes algebraic sign as the armature loop rotates through 180 degrees, thus producing alternating current (AC) power (in the case of a generator), provided the magnetic field $B$ maintains its direction. Various forms of armature and magnetic field circuits give rise to the several types of AC and direct current (DC) motors and generators.

The AC synchronous generator in an electric utility power plant rotates at a precise speed so as to produce 60-cycle AC power. All power plants that feed into a common transmission line must adhere to the same frequency standard. Because generators have an integral number of magnetic poles, their rotational speeds are an integral submultiple of 60 Hz. The generators maintain a fixed voltage in the transmission and distribution systems, while the current varies to suit the power needs of electricity customers.

Because of electrical and mechanical losses in electric motors and generators, the output power is less than the input power. The ratio of the output to the input powers is the efficiency $\eta$ of the device,

$$\eta \equiv \frac{\text{output power}}{\text{input power}}. \tag{5.5}$$

According to the first law of thermodynamics, the difference between the input and output powers appears as a heat flow from the device to the environment, in the amount of $(1 - \eta)$ times the input power. Both electric motors and generators must be cooled when operating to prevent the overheating of internal parts.

The electrical resistance $R$ of the armature wire is a source of inefficiency in both generators and motors. The electrical power lost in overcoming this resistance, $I^2R$, increases as $I^2$, whereas the power increases as $I$, as in Equation (5.3). This electrical loss thereby becomes a larger fraction of the power output as the latter is increased. As a consequence, the electrical efficiency of motors and generators is least at full power. For economic efficiency we would like to obtain the maximum power for a given investment, which means operating at maximum power, and thereby least electrical efficiency.

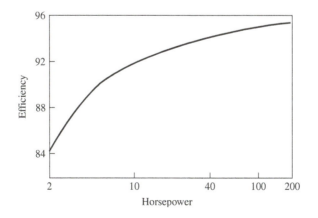

**FIGURE 5.4**   The efficiency of energy-efficient induction motors as a function of motor power. (Data from Andreas, John C. *Energy-Efficient Electric Motors*, 2d ed. New York: Marcel Dekker, 1992.)

The efficiency of electric motors and generators is greater for large than for small machines. Figure 5.4 exhibits this increase of efficiency with size for energy-efficient induction motors. The size effect is related to the lower rotational speed of larger machines and the necessarily more complex structure of the magnetic and electric circuits.

To achieve good efficiency and meet the practical limits on magnetic field strength, the power per unit volume or mass of a motor or generator is limited, being of the order of 1E(5) $W/m^3$ and 30 W/kg. Because of the similarity of the structure of these devices, the material amounts per unit volume are approximately independent of size and hence power output.

## 5.3   ELECTRIC POWER TRANSMISSION

The vast growth of electric power production and consumption in the 20th century was a consequence of the ability to transmit electric power from the source of production, the central power plant, to the many consumers in residential, commercial, and industrial locations far removed from the site of the power plant. Even though there is little energy storage in the transmission and distribution system and the power plant must produce sufficient power to meet the instantaneous demands of the many consumers, this system works remarkably well, especially when many power plants are connected in a network that permits sharing the power demand in an economical and reliable manner.

The components of an electric power system are sketched in Figure 5.5. A gas or steam turbine powers an electric generator supplying AC power to a transformer that steps up the line voltage to a high value in the transmission line, usually 60–500 kilovolts. After transmission from the power plant to the vicinity of load centers, a step-down transformer reduces the voltage to a lower value, 12–35 kilovolts, for the distribution system that delivers power to the end users. For residential use, a further reduction to 120–240 volts is required.

Electric power is transmitted via electric cable consisting of two wires that conduct electrical current at different electric potentials.[10] In a DC system the electric current in each wire flows in one direction only. On the other hand, in an AC system, where the potential difference and current are sinusoidal functions of time, the currents in each wire reverse direction every half cycle. In

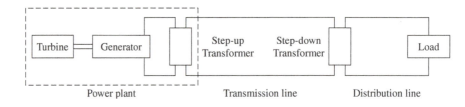

FIGURE 5.5   A sketch of the elements of an electric power system for generating, transmitting, and distributing power to end users.

---

[10]Electric utility transmission and distribution lines usually include more than two wires, one of which is a common ground return.

either case, the instantaneous electrical power $\mathcal{P}_{el}$ transmitted in the cable is the product of the potential difference $\Delta\Phi$ between the wires and the current $I$ flowing in them,

$$\mathcal{P}_{el} = (\Delta\Phi)I. \tag{5.6}$$

The time-averaged power $\overline{\mathcal{P}_{el}}$ is

$$\overline{\mathcal{P}_{el}} = \sqrt{\overline{(\Delta\Phi)^2}\ \overline{(I)^2}}\cos\phi, \tag{5.7}$$

where the overline indicates a time-averaged value and $\phi$ is the phase angle between the current $I$ and the potential difference $\Delta\Phi$ for AC power transmission.

In both DC and AC transmission systems there is a loss of electric power in the transmission line in the amount $I^2R$, where $R$ is the electrical resistance of the line. To minimize this loss, the line resistance can be reduced by using large-size copper wire and the current minimized, for a given power, by increasing the potential difference. Long-distance transmission lines operate at hundreds of thousands of volts to reduce the transmission loss. But high voltages are impractical and unsafe for distribution to residences and commercial users so the voltage in AC distribution systems is reduced to much lower levels by transformers. Since DC power cannot be transformed easily to a different voltage, its use is restricted to special applications, such as electric rail trains. The use of AC power predominates in modern electrical power systems. Nevertheless, high-voltage DC ($\sim$500 kilovolt) transmission lines may be used for very-long-distance transmission ($\sim$1500 kilometers) to reduce energy loss.

The energy efficiency of electric power transmission and distribution is almost entirely determined by economic choices. To increase the efficiency, more money must be invested in copper wire and transformer cores, which is only justified if the value of the saved electric energy exceeds the amortization costs of the increased investment. Transmission and distribution losses in electric power systems are usually held to 5–10%.

### 5.3.1  AC/DC Conversion

Although nearly all electric power is generated, transmitted, and utilized in AC form, there are important uses for DC power that usually require the conversion of an AC power input to DC form. The most prominent of these are communication systems, such as the telephone and computer, where digital circuits use DC power. Another common DC system is that of the automobile, where AC power generated in the engine driven alternator is converted to 12-volt DC power that charges the battery and supplies power for lights, fans, radio, and so on.

A *rectifier* is an electrical circuit device that converts AC to DC power. It consists of diodes that permit current to flow in one direction only, thereby transforming the AC current to DC form. As in most electrical devices, some power is lost in this transformation. Every computer has an internal or external power supply that converts household AC to the DC power needed for digital circuits.

There are some sources of electric power that are DC in nature, most notably photovoltaic cells and fuel cells. Where this power is fed to electric utility AC transmission or distribution lines, the DC power first must be transformed to AC form. The electric device that accomplishes this conversion is called an *inverter*. While it is possible to achieve this conversion mechanically by employing a DC motor to power an AC generator, inverters are generally electrical circuit components that accomplish the same purpose. There is some power loss that accompanies this conversion.

## 5.4  ENERGY STORAGE

There is very little energy stored in the electric utility system that supplies electric power to consumers. The electric power must be supplied at the same rate that it is being utilized by the utility's customers. Although there is some energy temporarily stored (and removed) each half cycle in the transmission and distribution lines and transformers, it is not available for supplying power when demand exceeds supply during a sustained period. However, there are systems that will fulfill this need and that of other applications, such as storing energy that could be used for electric-drive vehicles. In this section we will describe the principles that form the basis for the construction of such devices.

### 5.4.1  Electrostatic Energy Storage

A capacitor is a device for storing electric charge at an elevated potential. As sketched in Figure 5.6a, it consists of two electrically conducting plates of area $A$, separated by an electrically insulating dielectric medium of equal area and thickness $h$. Positive and negative charges, in equal amounts $Q$, stored on the plates induce an electric potential difference $\Delta\phi$ between them. The charge and potential difference are related by Coulomb's law for this configuration,

$$Q = \left(\frac{\epsilon A}{h}\right) \Delta\phi \equiv C \, \Delta\phi, \tag{5.8}$$

where $\epsilon$ is the *electric permittivity* of the dielectric medium and $C \equiv \epsilon A/h$ is the *capacitance* of the device.[11] If an increment $dQ$ of charge is moved from the negative plate to the positive one through the potential increase $\Delta\phi$, via the circuit external to the capacitor, an amount of electrical

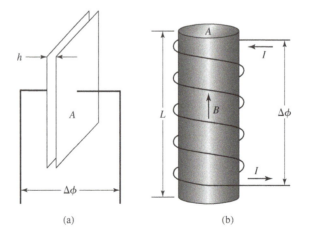

(a)                                  (b)

**FIGURE 5.6**  The electric capacitor (a) and inductor (b) are devices for storing electrical energy.

---

[11]The dimension of the electrical charge is the coulomb and that of the capacitance is the farad = coulomb/volt. The electric permittivity has the dimensions of farad/meter = coulomb/volt meter. (See Table A.1.)

work $\Delta\phi\, dQ$ is done in this process, increasing the capacitor free energy by the amount $dF$,

$$dF = \Delta\phi\, dQ = \frac{Q\, dQ}{C} = \frac{1}{2C}\, d(Q^2)$$

$$F = \frac{Q^2}{2C} = \frac{C\,(\Delta\phi)^2}{2}. \tag{5.9}$$

We may determine the free energies per unit volume and mass in terms of the electric field $E = \Delta\phi/h$ in the capacitor as

$$\frac{F}{\mathcal{V}} = \frac{(\epsilon A/h)\Delta\phi^2}{2Ah} = \frac{\epsilon E^2}{2}$$

$$f = \frac{\epsilon E^2}{2\rho}, \tag{5.10}$$

where $f$ is the free energy per unit mass and $\rho$ is the mass density (kg/m$^3$) of the dielectric medium, and we neglect the mass of the electrodes.

To obtain high energy-storage densities we should choose a material with a high electric permittivity $\epsilon$ and an ability to withstand a high electric field $E$ without breakdown (i.e., without conducting a current that would short circuit the capacitor internally). Such materials are composed of molecules that have a permanent electric dipole moment and that are not easily ionized in the presence of strong electric fields. For example, polypropylene has a permittivity $\epsilon = 2E(-11)$ farad/meter, a dielectric strength of $E = 6.5E(9)$ V/m, and density $\rho = 1E(3)$ kg/m$^3$, so that the free energy per unit volume is $F/\mathcal{V} = 4.3E(8)$ J/m$^3$ and that per unit mass is $f = 4.3E(5)$ J/kg.

High-energy-density capacitors are being developed for potential electric vehicle use, with energy densities of about 10 Wh/kg $= 0.036$ MJ/kg. The electric power input and output requires high and variable electric potentials requiring power conditioning equipment to deliver the lower voltage and higher current needed for traction motors. Electric failure of the capacitor dielectric can present safety problems.

### 5.4.2  Magnetic Energy Storage

It is possible to store energy in the magnetic field produced by a current flowing in a conducting wire. In the sketch of Figure 5.6b, a magnetic inductor consists of a long cylinder of material, of cross-sectional area $A$ and length $L$, around which a coil of electric wire having $N$ turns carries a current $I$. Ampere's law relates the *magnetic induction $B$*[12] in the material to the current flowing in the wire,

$$B = \frac{\mu NI}{L}, \tag{5.11}$$

where $\mu$ is the magnetic permeability of the material.[13] To determine the energy stored in the inductor, we first note that an increment of magnetic induction $dB$, caused by an increment of

---

[12]The units of magnetic induction are weber/meter$^2$ = volt second/meter$^3$. (See Table A.1.)

[13]The units of magnetic permeability are henry/meter = volt second/ampere meter. (See Table A.1.)

current $dI$ in a time interval $dt$, is related to the potential difference $\Delta\phi$ between the ends of the coil by Faraday's law of magnetic induction,

$$A\frac{dB}{dt} = \frac{\Delta\phi}{N}. \tag{5.12}$$

During this time interval, electric power of amount $\Delta\phi I$ is expended in increasing the magnetic induction, so that the free energy $F$ increase is

$$dF = \Delta\phi I\, dt = NAI\, dB = \frac{AL}{2\mu}\, dB^2 = \frac{\mu AL}{2}\, d\left(\frac{NI}{L}\right)^2$$

$$\frac{F}{\mathcal{V}} = \frac{B^2}{2\mu} = \frac{\mu(NI)^2}{2L^2}. \tag{5.13}$$

This relation may be expressed alternatively in terms of the *inductance* $\mathcal{L}$[14] of the coil, which is the ratio $\Delta\phi/(dI/dt) = \mu N^2 A/L$ by Equation (5.13), so that

$$F = \frac{\mathcal{L}I^2}{2}. \tag{5.14}$$

The storage of magnetic energy is complicated by two effects. Maintaining the coil current $I$ requires the expenditure of electric power to overcome its resistive losses, unless a superconducting coil is used. There is a practical upper limit to the magnetic induction $B$ related to the fact that the coil windings are subject to an outward force per unit length proportional to $IB$ that must be supported by a containing structure having a volume comparable to $AL$. This limits magnetic induction to values less than about 6 weber/m$^2$ and corresponding energy densities of 1.43E(7) J/m$^3$ and 1.8E(3) J/kg.

Superconducting magnetic energy storage systems have been proposed for use with electric utility systems. Designs project energy storage densities of 1 MJ/m$^3$, prospective costs of $180/kWh = $50/MJ, and an energy efficiency of 95%.

### 5.4.3  Electrochemical Energy Storage

We are familiar with the employment of batteries to supply electric power for a myriad of uses: flashlights, portable radios and telephones, watches, hearing aids, heart pacemakers, toys, and so on. Batteries that convert the energy of reactant chemicals to electrical work and are then discarded when discharged are called *primary batteries*. In contrast, *secondary batteries*, whose chemical constituents can be reconstituted by recharging with electrical power, are commonly termed storage batteries. The most common storage battery is that used in the automobile, primarily for supplying engine starting power.

A storage battery consists of two electrodes, a positive one and a negative one, each of different chemical composition, immersed in an electrolyte. The electrodes provide the store of chemical energy that is converted to electrical work as the battery is discharged. The electrolyte provides an internal electric current of negatively charged anions and/or positively charged cations closing the

---

[14]The unit of inductance is the henry = volt second/ampere. (See Table A.1.)

external electric circuit, which consumes or provides electric work as the battery is discharged or charged. In contrast with the fuel cell, which also converts chemical energy to electrical work but requires an external source of fuel and oxidant flow, the storage battery has only a limited amount of chemical energy stored within it that, once consumed, brings to an end its supply of electrical work. In this manner it resembles the other energy storage systems that we have been discussing in this chapter.

### 5.4.3.1  Lead-Acid Storage Battery

To illustrate the principle of operation of a storage battery, we will consider the example of the lead-acid storage battery. In its completely charged configuration, it consists of a positive electrode composed of lead dioxide ($PbO_2$), a negative one of pure lead (Pb) and an electrolyte that is a concentrated solution of sulfuric acid ($H_2SO_4$) in water. The sulfuric acid is dissociated into hydrogen cations ($H^+$) and hydrosulfate anions ($HSO_4^-$). The conversion of chemical energy to electrical work occurs partially at each electrode. At the negative electrode, a hydrosulfate ion at the electrolyte potential $\Phi_{el}$ reacts with the lead to form lead sulfate ($PbSO_4$), a hydrogen ion ($H^+$) at the electrolyte potential $\Phi_{el}$, and two conducting electrons in the negative electrode at its electric potential $\Phi_n$:

$$Pb + HSO_4^- \{\Phi_e\} \rightarrow PbSO_4 + H^+ \{\Phi_{el}\} + 2e^- \{\Phi_n\}. \tag{5.15}$$

At the positive electrode, two electrons at its potential $\Phi_p$ and a lead dioxide molecule combine with three hydrogen ions and a hydrosulfate ion at the electrolyte potential $\Phi_{el}$ to form a lead sulfate molecule and two water molecules,

$$PbO_2 + 2e^- \{\Phi_p\} + 3H^+ \{\Phi_{el}\} + HSO_4^- \{\Phi_e\} \rightarrow PbSO_4 + 2H_2O. \tag{5.16}$$

The net reaction in the storage battery is the sum of the cathode and anode reactions, Equations (5.15–5.16),

$$Pb + PbO_2 + 2(H^+ \{\Phi_{el}\} + HSO_4^- \{\Phi_{el}\}) \rightarrow 2PbSO_4 + 2H_2O + 2(e^- [\{\Phi_n\} - \{\Phi_p\}]). \tag{5.17}$$

In the discharge reaction of the lead acid battery, Equation (5.17), both electrode surfaces are each converted to lead sulfate while sulfuric acid is removed and water is added to the electrolyte, diluting it.[15] When both electrode surfaces are completely covered with lead sulfate, there is no difference between the electrodes and no further electric work may be drawn from it.

In the discharge reaction (5.17), the electrical work of moving the two electrons through the external circuit equals $2q_e(\Phi_p - \Phi_n)$, with $q_e$ being the magnitude of the electron charge. When this discharge occurs very slowly at fixed temperature and pressure, the electrical work is then equal to the reduction of free energy per unit mass $(\Delta f)_{la}$ when the reactants $Pb + PbO_2 + 2(H^+ + HSO_4^-)$ are converted to the products $2PbSO_4 + 2H_2O$. As a consequence, the battery potential difference $\Phi_p - \Phi_n$ can be written as

$$\Phi_p - \Phi_n = \frac{(\Delta f)_{la} \mathcal{M}_{la}}{2\mathcal{F}}, \tag{5.18}$$

---

[15]The degree of discharge of the lead-acid battery can be determined by measuring the specific gravity of the electrolyte, which decreases as the acid content is diluted.

where $\mathcal{M}_{la}$ is the molecular weight of the reactants and $\mathcal{F}$ is the Faraday constant.[16] At usual conditions, the lead-acid battery potential difference is 2.05 V. In the reaction of Equation (5.17), the free energy change per unit mass of reactants, $(\Delta f)_{la}$, is 171 Wh/kg = 0.6 MJ/kg.

The advantage of the storage battery is that the discharge reactions, such as those of Equations (5.15–5.16) for the lead-acid battery, can be reversed in direction by imposing an external potential difference between the positive and negative electrodes that exceeds the equilibrium value of Equation (5.18). In so doing, the products of the discharge reaction are converted back to reactants, renewing the energy stored in the battery, with that energy increment being supplied by electrical work from the external charging circuit. In the discharging and recharging of a storage battery there is usually a net loss of energy because more electrical work is required to charge the battery than is recovered in its discharge. This energy loss of the charge–discharge cycle is dependent upon the charging and discharging rates: When these rates are higher, the loss is greater. Some of the loss is caused by resistive heating resulting from the internal ion current of the battery while the rest is a consequence of electrochemical irreversibilities at the electrode surfaces.

The electrode reactions that release or absorb electrical work occur at the electrode surface. The electrodes of storage batteries have a porous, sponge-like structure to maximize the ratio of surface area to electrode volume. In this way the amount of energy stored per unit mass of electrode material may be maximized and the cost of energy storage minimized.

The lead-acid storage battery is a well-developed technology. There are about 200 million lead-acid batteries installed in U.S. road vehicles, each storing about 1 kWh of energy, for a total of about 7E(4) GJ of electric energy stored. This is about 1% of the daily electric energy produced in the United States. If storage batteries were used to level the daily electric utility supply, a very large increase in battery production, above that required for the automotive market, would be needed to satisfy this requirement.

The energy efficiency of lead-acid batteries is about 75% at low (multihour) charge and discharge rates. Manufacturing cost is about $50/kWh = $14/GJ. Battery life is usually about 1,000 full charge–discharge cycles, which makes storage batteries more expensive than pumped hydropower for electric utility storage systems. In addition, inverters and rectifiers would be needed to match the DC operations of storage batteries with the AC electric power system. However, electric storage batteries serve as an economical emergency power supply for computer systems.

If lead-acid batteries are overcharged, they emit hydrogen gas, which can be an explosion hazard.

### 5.4.3.2   Lithium-Ion Storage Battery

A recently developed storage battery, the lithium-ion battery, promises to be useful in automobiles employing hybrid electric or all-electric drives, storing more energy and power per unit weight than lead-acid batteries. It employs an anode (negative terminal), usually carbon, a cathode (positive terminal), usually lithium cobalt oxide ($LiCoO_2$), and an electrolyte of a lithium salt ($LiPF_6$) in in an inorganic solvent (ether). The charging or discharging internal current is a flow of lithium ions ($Li^+$) from anode to cathode (discharging) or vice versa when charging. When discharging, lithium in the cathode, in the form of $LiPF_6$, is converted to lithium in the anode, in the form of $LiC_6$, releasing 3.7 electron volts of energy for each electron passing through the external load. Unlike

---

[16] See Table A.3.

the lead-acid battery, there are no chemical changes in the electrolyte; it merely passes lithium ions from one electrode to the other, whether charging or discharging.

The electrode reactions during discharge are similar to those, (5.15) and (5.16), for the lead-acid battery. At the cathode, a $LiCoO_2$ molecule dissociates into a lithium ion at the electrolyte potential and an electron at the cathode potential, leaving a $CoO_2$ product molecule in the cathode:

$$LiCoO_2 \rightarrow Li^+\{\Phi_{el}\} + e^-\{\Phi_n\} + CoO_2. \tag{5.19}$$

At the anode, a lithium ion at the electrolyte potential enters the anode, combining with an electron at the anode potential and six carbon atoms to form the product $LiC_6$:

$$Li^+\{\Phi_{el}\} + e^-\{\Phi_p\} + 6C \rightarrow LiC_6. \tag{5.20}$$

The overall battery discharge reaction is the sum of (5.19) and (5.20) above, yielding

$$LiCoO_2 + 6C \rightarrow CoO_2 + LiC_6 + e^-\{\Phi_n\} - e^-\{\Phi_p\}, \tag{5.21}$$

in which the reactants ($LiCoO_2 + 6C$) change into the products ($CoO_2 + LiC_6$) by transfer of Li from the cathode to the anode and the passage of an electron through the potential increase ($\{\Phi_n\} - \{\Phi_p\}$), delivering electric work to the external battery load in an amount that does not exceed the decrease of free energy $\Delta f$ experienced when the reactants change into the products, as required by the second law of thermodynamics for this discharge process.

The charging process is the reverse of (5.21), with the external load becoming an input of electric power and the internal ion current and external electron current reversing their directions. Internal electrode and electrolyte losses inevitably require that the charging energy exceeds the discharging energy for any degree of charge.

For automotive use, lithium-ion batteries are lighter and less bulky than lead-acid batteries, for the same amount of energy storage, but need to be more carefully managed and charged to prevent internal damage.

### 5.4.3.3  Other Storage Batteries

Other well-developed storage batteries employ an alkaline electrolyte and a metal oxide-positive and a metal-negative electrode. The nickel–cadmium battery, commonly used in portable electronic equipment, consists of a NiOOH-positive and Cd-negative electrode, with a KOH electrolyte. Such batteries can provide more energy storage per unit weight than lead-acid batteries, but at greater economic cost. Table 5.1 lists the more prominent electric storage battery systems and their properties.

### 5.4.4  Mechanical Energy Storage

### 5.4.4.1  Pumped Hydropower

The common form of storing energy for use in electric utility systems is that of pumped hydropower. It consists of a normal hydroelectric plant that is supplied with water impounded behind a dam at high elevation and discharging to a body of water at a lower elevation through a turbine that drives an electric generator. But unlike the usual hydropower plant, the water flow may be reversed and pumped from the lower to the higher reservoir using electric power available from the utility

**TABLE 5.1**   Typical Properties of Storage Battery Systems

| Battery type | Energy/mass (Wh/kg) | Peak power/mass (W/kg) | Cost ($/kWh) | Efficiency (%) |
|---|---|---|---|---|
| Lead-acid | 40 | 250 | 130 | 80 |
| Lithium-ion | 150 | 250 | 200 | 85 |
| Nickel–cadmium | 50 | 110 | 300 | 75 |
| Nickel–metal hydride | 80 | 250 | 260 | 70 |
| Sodium–sulfur | 190 | 230 | 330 | 85 |

system during times of low demand. Operating on a diurnal cycle, the pumped storage system undergoes no net flow of water but does not deliver as much electrical energy as it uses during the pumping part of the cycle because its components (turbogenerator and pump motor) are less than 100% efficient.

Energy in a pumped hydropower system is stored by lifting a mass of water through a vertical distance $h$ in the earth's gravitational field. The gravitational energy stored in the reservoir water, per unit mass of water, is $gh$, while the energy per unit volume is $gh/\rho$, where g is the gravitational acceleration. For $h = 100$ m, $gh = 9.8E(2)$ J/m$^3$, and $gh/\rho = 0.98$ J/kg. These are extremely low-energy storage densities, requiring very large volumes of water to store desirable amounts of energy. Since water is essentially a free commodity, the cost of storage is related to the civil engineering costs of the reservoir and power house.

This system is by far the most commonly used energy storage in electric power utility systems. The largest pumped storage system in the United States is the Luddington, Michigan, plant that stores 15 GWh (= 5.4E(4) GJ) of energy, providing 2,000 MW of electrical power under an hydraulic head of 85 m. But the total U.S. pumped storage power is about 2% of the total U.S. electric power, implying an energy storage of 170 GWh (= 6E(5) GJ). Since normal hydropower provides about 11% of U.S. electrical power, pumped storage hydropower is not a negligible application of hydropower machinery.

The overall energy efficiency of hydropower storage is about 70% (i.e., the energy delivered to the electric power system is 70% of that withdrawn during the storage process). This creditably high value stems from the fixed speed and hydraulic head of the turbomachinery used to fill and withdraw water stored in the reservoir.

Hydroelectric machinery is safe, reliable, and cheap to operate. The capital cost of pumped storage installations is significantly related to the capital cost of the civil works (dam, reservoir, etc.). If the capital cost is 500 $/kW of power, then the capital cost of pumped storage energy is 23 $/GJ.

### 5.4.4.2   Flywheel Energy Storage

Flywheel energy storage systems are being developed for electric transmission system leveling and for emergency electric power supplies. In this device, an axially symmetric solid material is rotated about its axis of symmetry at a high angular speed $\Omega$. The material at the outer edge of the flywheel rim, whose radius is $R$, moves at the speed $\Omega R$ and has the kinetic energy per unit mass of $(\Omega R)^2/2$ and the kinetic energy per unit volume of $\rho(\Omega R)^2/2$, where $\rho$ is the flywheel mass density. If the flywheel rim has a radial thickness that is small compared with $R$, then it would experience a tangential stress $\sigma = \rho(\Omega R)^2$, which is twice the kinetic energy per unit volume. In other words, the maximum kinetic energy per unit volume equals $\sigma/2$ and the kinetic energy per unit mass is $\sigma/2\rho$, where $\sigma$ is the maximum allowable tangential stress in the flywheel. For high strength steel, where $\sigma = 9E(8)$ Pa and $\rho = 8E(3)$ kg/m$^3$, the kinetic energy per unit volume is 4.5E(8) J/m$^3$ and the kinetic energy per unit mass is 5.63E(4) J/kg.

The energy input and output are usually in the form of electric power. These systems have very high rotational speeds that decline as energy is withdrawn from them. Their overall energy efficiency is comparable to other forms of storage. Energy storage densities are about 50 Wh/kg = 0.18 MJ/kg. To reach these energy storage densities high strength-to-weight materials, such as carbon fiber, are used. In the event of a stress failure, the flywheel components become dangerous projectiles moving at the flywheel peripheral speed, so safety can be a problem.

## 5.4.5 Properties of Energy Storage Systems

For some energy storage systems the storage capacity per unit volume or mass (J/m$^3$, J/kg) is an important characteristic. For example, in electric-drive highway vehicles the mass or volume of a battery or flywheel system needed to supply enough traction energy for a desirable trip length may be too great for a practical design. Also, the capital cost of the energy storage system is relatable to its mass and affects the dollar cost per unit of stored energy ($/MJ). Values of these properties for the energy storage systems considered above are listed in Table 5.2.

Among the systems listed in Table 5.2, there is a large range in stored energy per unit volume and per unit mass. Yet even the best of them does not approach the high energy density of a hydrocarbon fuel, based upon its heating value, of 5E(7) J/kg and 4E(10) J/m$^3$. It is this great energy density of fossil fuel that makes highway vehicles, ships, and aircraft such productive transportation devices. On the other hand, the energy needs of a portable digital computer can be satisfied easily by high-grade storage batteries having a mass that is small compared with the computer mass.

The capital cost of energy storage is not greatly different among the systems listed in Table 5.2. If we value electric energy at 3 cents per kilowatt hour (= 2.8E(−2) $/MJ), then the capital cost of a typical storage system is about 3,000 times the value of the stored energy. If the stored energy is discharged each day, as it would be in a supply leveling system for an electric utility, the capital charge for storage would about equal the cost of electric energy, thus doubling the cost of the stored energy in this instance. The economic cost of storing energy is an important factor limiting its use in electric utility systems.

TABLE 5.2  Properties of Energy Storage Systems

| Type | Energy/volume (J/m$^3$) | Energy/mass (J/kg) | Cost ($/MJ) | Efficiency (%) |
|---|---|---|---|---|
| Capacitor | 4E(7) | 4E(4) | | 95 |
| Inductor | 1E(7) | 2E(3) | 50 | 95 |
| Pumped hydro | 1E(3) | 1 | 25 | 70 |
| Lead-acid storage battery | 3E(7) | 2E(5) | 15 | 75 |
| Flywheel | 2E(8) | 2E(5) | | 80 |

The efficiency of energy storage systems—the ratio of the output energy to the input energy—is exhibited in the last column of Table 5.2. These are relatively high values, reflecting the best that can be done with optimum energy management.

## 5.5  CONCLUSION

The generation and transmission of electric power is a necessary component of the energy supply of modern nations. Most such power is generated in thermal plants burning fossil or nuclear fuel or in hydropower plants. In either case, steam, gas, or hydro turbines supply mechanical power to electric generators that feed electric power via transmission and distribution lines to the end consumer. The electricity generation and distribution process is very efficient, with overall percentage losses being in single-digit numbers.

There is very little electric energy stored in the generation and transmission system, so electric power must be generated at the same rate as it is consumed if line voltages and frequencies are to be maintained. The diurnal pattern of electricity demand requires that the electric power network be capable of supplying the peak demand, which may be 25% or more above the average demand. This is usually accomplished by a mixture of base load and variable load electric generating plants. Alternatively, pumped hydroelectric plants are sometimes employed to store energy that may be used to supply daily peak electric power demand.

Electric energy may be stored for various purposes in storage batteries, capacitors, and inductors, but the only significant amount of energy storage of these types is that of lead-acid batteries in motor vehicles. The cost and weight of storage batteries has limited their use for primary power in vehicles, but improvements in storage batteries promise to enhance their use in the future.

# PROBLEMS

### Problem 5.1

From Figure 5.1, estimate the ratio of peak demand/average demand and that of minimum demand/average demand. If the electricity supply system must have a capacity 20% above the peak demand, estimate the system capacity factor; that is, the ratio of average demand to system capacity.

### Problem 5.2

A generator armature coil of length $L = 10$ cm and half width $r = 2$ cm, as shown in Figure 5.3b, is placed in a magnetic field of strength $B = 1$ weber/m$^2$. If the coil rotates at a frequency of 60 revolutions per second and a current $I$ is 10 amperes, calculate the peripheral velocity $V$, the force $F$ on each length $L$ of the coil and the corresponding electric field $E$, the electric potential difference $\Delta\Phi$ across the coil slip rings, and the external torque $T$ applied to the armature.

### Problem 5.3

A capacitor for storing 100 MJ of energy in an electric-drive vehicle utilizes a dielectric of thickness $h = 0.1$ mm and electric permittivity $\epsilon = 2E(-11)$ farad/meter. The maximum electric field $E$ is $3E(9)$ volts. For this capacitor, calculate the electric potential $\Delta\Phi$, the required capacitance $C$, the capacitor area $A$, and the volume of dielectric material.

## Problem 5.4

A typical automotive lead-acid battery stores 100 ampere-hours of charge. Calculate the charge in coulombs. Assuming that the battery can discharge its full charge at an electrode potential of 12 volts, calculate the energy stored in the fully charged battery.

## Problem 5.5

A pumped storage plant is being designed to produce 100 MW of electrical power over a 10-hour period during drawdown of the stored water. The mean head difference during this period is 30 m. Calculate the amount of electrical energy to be delivered and the required volume of water to be stored if the hydropower electric generator system is 85% efficient.

## Problem 5.6

The characteristics of a battery-powered electric vehicle, listed in column 1 of Table 9.3, give an electric energy storage of 18.7 kWh and a battery mass of 595 kg. If the batteries were replaced by an electric, magnetic, or flywheel energy storage system of characteristics given in Table 5.2, calculate the mass and volume of these alternative systems. Would any of these alternatives be practical?

# BIBLIOGRAPHY

Andreas, John C. *Energy-Efficient Electric Motors*, 2d. ed. New York: Marcel Dekker, 1992.

Dunn, P. D. *Renewable Energies: Sources, Conversion and Applications*. London: Peter Peregrinus, 1986.

Howes, Ruth, and Anthony Fainberg, eds. *The Energy Sourcebook. Guide to Technology, Resources, and Policy*. New York: American Institute of Physics, 1991.

Khartchenko, Nikolai, ed. *Advanced Energy Systems*. London: Taylor & Francis, 1998.

Panofsky, Wolfgang K. H., and Melba Phillips. *Classical Electricity and Magnetism*. Reading: Addison–Wesley, 1962.

Rand, D. A. J., R. Woods, and R. M. Dell. *Batteries for Electric Vehicles*. Somerset: Research Studies Press, 1998.

Ter-Gazarian, A. *Energy Storage for Power Systems*. Stevenage: Peter Peregrinus, 1994.

Twidell, John, and Tony Weir. *Renewable Energy Resources*, 2d ed. London: Taylor & Francis, 2006.

# Fossil-Fueled Power Plants

## 6.1 INTRODUCTION

Fossil-fueled power plants contribute 65% of the world's electricity supply. The rest is supplied by hydroelectric plants (16.6%), nuclear-fueled plants (15.7%), and renewables (geothermal, wind, solar, wood) (less than 3%).[1] Among the fossil-fueled plants, coal supplies 61%, oil 18%, and natural gas 21% of the fuel. Fossil-fueled power plants are major contributors to the anthropogenic emissions of $CO_2$ and other pollutants, such as $SO_2$, $NO_x$, products of incomplete combustion, and particulate matter (PM).

Power plants, for reasons of economy of scale, are usually built in large, centralized units, typically delivering power in the range of 500–1,000 MW. Therefore, much effort is being spent on their efficiency improvement and environmental control, as these efforts could result in significant reduction of pollutant emissions and conservation of fossil fuel reserves.

Almost all fossil-fueled power plants work on the principle of a heat engine, converting fossil fuel chemical energy first into mechanical energy (via steam or gas turbines) and then into electrical energy, as described in Chapters 3 and 5. Most large-scale power plants use the Rankine steam cycle,[2] in which steam is produced in a boiler heated by a combustion gas; the steam drives a steam turbine that drives a generator of electricity. Usually, coal-steam plants in combination with nuclear plants provide the base load for a regional grid. Peak loads are mostly supplied by gas turbine plants, which work on the principle of a Brayton cycle.[3] In those plants, natural gas is burned, and the combustion products directly drive a gas turbine, which in turn drives a generator.

---

[1] U.S. Energy Information Agency. *International Energy Annual*, 2007.

[2] See Sections 3.10.5 and 6.3.1.

[3] The exception would be a power plant that uses a fuel cell for the conversion of fossil fuel chemical energy directly into electrical energy. However, very few power plants utilizing the principle of a fuel cell exist today. The performance of a fuel cell electricity generator is described in Chapter 3.

The best steam cycle power plants can achieve a thermal efficiency above 40%; the United States average is 36%; the worldwide average is about 33%.[4] Gas turbine power plants achieve a thermal efficiency in the 25–30% range. More advanced power plants exist today that use a combination of Brayton and Rankine cycles. Combined cycle power plants can achieve an efficiency of about 45%. The relatively low thermal efficiency of power plants is the result of two factors. The first is a consequence of the second law of thermodynamics, whereby in a heat engine, after performing useful work, the residual of the fuel heat needs to be rejected to a cold reservoir, usually a surface water (river, lake, or ocean) or to the atmosphere via a cooling tower. The second factor is the result of parasitic heat losses through walls and pipes, frictional losses, and residual heat escaping with the flue gas into the atmosphere. Furthermore, some of the gross electricity generated is used for running the auxiliary equipment of the power plant, such as pumps, fans, coal conveyors, and pulverizers. It is an unfortunate fact of electricity generation that power plants use only about 25–50% of the input chemical energy of fossil fuels to generate electricity; the rest is wasted (i.e., it goes down the river and up in the air, so to speak).

In this chapter we describe the functioning of fossil-fueled power plants and their components: fuel storage and preparation, burner, boiler, turbine, condenser, and generator. We focus in particular on emission control technologies, which power plants in most countries are required by law to install to safeguard public health and the environment. We also review briefly advanced cycle power plants that have the promise of increasing thermal efficiency and reducing pollution.

## 6.2   FOSSIL-FUELED POWER PLANT COMPONENTS

In a fossil-fueled power plant the chemical energy inherent in the fossil fuel is converted first to raise the temperature and enthalpy of the combustion gases; that enthalpy is transferred by convection and radiation to a working fluid (usually water/steam); the enthalpy of the working fluid is converted to mechanical energy in a turbine; and finally the mechanical energy of the turbine shaft is converted to electrical energy in a generator.

The major components of a fossil-fueled power plant are as follows:

- Fuel storage and preparation
- Burner
- Boiler
- Steam turbine
- Gas turbine
- Condenser
- Cooling tower
- Generator
- Emission control

---

[4]Thermal efficiency is defined as the ratio of electrical energy output to fossil fuel energy input. An alternative definition is the heat rate. This is the amount of British thermal units or gigajoules of fuel energy spent per kWh of electricity produced.

## 6.2.1    Fuel Storage and Preparation

Coal is delivered to a power plant by rail or, in the case of a coastal or riverine plant, by ship or barge. Usually, power plant operators like to have several weeks of coal supply on site, in case of delivery problems or coal mine strikes. Since a 1,000-MW power plant, having a thermal efficiency of 35%, consumes on the order of 1E(4) metric tons of coal per day, the coal mounds near the plant may contain up to 3E(5) metric tons of coal. Some coal-fired power plants are situated right near coal mines (so-called mine-mouth plants). Even these plants store at least a month's supply of coal near the plant.

When coal arrives by rail, it is usually carried by a unit train, consisting of 100 wagons filled with coal, at 100 tons per wagon. The wagons are emptied by a rotary dump, and the coal is carried by conveyors to a stockpile or directly to the power plant.

Coal is delivered to a plant already sized to meet the feed size of the pulverizing mill (see below), in the order of a few to 10 centimeters per coal lump. In the United States and several other countries, coal is washed at the mine. Washing of coal removes much of the mineral content of the coal (including pyritic sulfur), thus reducing its ash and sulfur content and improving its heating value per unit mass. In preparation for washing, the coal is crushed at the mine mouth to less than centimeter size, called *nut* or *slack*.

Most modern steam power plants fire pulverized coal. The raw coal from the stockpile is delivered on a conveyor belt directly to pulverizing mills. Such mills are of the rotating ring, rotating hammer, or rotating ball type. The mills reduce the raw coal lumps to less than 1-millimeter particles. The pulverized coal is stored in large vertical silos from whence it is blown pneumatically into the burners at a rate demanded by the load of the plant.

At oil-fired power plants, oil is stored in large tanks (tank farm), to which oil is delivered by pipeline, railroad tankers, or tanker ship or barge if the plant is located near navigable waters. Power plants like to have at least a 30-day supply of oil in their tanks. For a 1,000-MW plant, 35% efficiency, this can amount to over 1E(5) metric tons of oil. The oil is purchased from refineries in the form it is combusted in the burners, with specified sulfur, nitrogen, and ash content, and other properties, such as viscosity and vapor pressure.

For natural gas-fired power plants, gas is delivered through transmission pipelines that are supplied directly from gas wells or from seasonal storage in underground caverns or import terminals receiving liquefied natural gas (LNG) by tanker transport from overseas suppliers.

## 6.2.2    Burner

The role of the burner is to provide a thorough mixing of the fuel and air so that the fuel is completely combusted. Ignition is accomplished by a spark-ignited light oil jet until the flame is self-sustaining. In the combustion chamber, a pulverized coal particle or atomized oil droplet burns in a fraction of a second. The coal particle or oil droplet burns from the outside to the core, leaving behind incombustible mineral matter. The mineral matter is called ash. In modern pulverized coal and atomized oil-fired power plants, more than 90% of the mineral matter forms the so-called fly ash, which is blown out of the boiler by forced or natural draft and later captured in particle collectors. About 10% of the mineral matter falls to the bottom of the boiler as bottom ash. The bottom of the boiler usually is filled with water. The settling bottom ash forms a wet sludge, which is sluiced away into an impoundment. Some of the fly ash, however, is deposited on the water pipes lining the boiler. This forms a scale, which hinders heat transfer. The scale needs to be removed from time to time by blowing steam jets against it or by mechanical scraping.

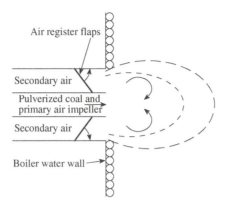

**FIGURE 6.1**    Pulverized coal burner.

Coal burns relatively slowly, oil faster, and gas the fastest. For complete combustion (carbon burnout), excess air is delivered, that is, more air than is required by a stoichiometric balance of fuel and the oxygen content of air. Pulverized coal requires 15–20% excess air; oil and gas require 5–10%.

A typical burner for pulverized coal is depicted in Figure 6.1. The central coal impeller carries the pulverized coal from the silo in a stream of primary air. Tangential doors (*registers*) built into the wind box allow secondary air to be admixed, generating a fast-burning turbulent flame. The impeller is prone to corrosion and degradation and has to be replaced once a year or so.

The burners are usually arranged to point nearly tangentially along the boiler walls. In such a fashion a single turbulent flame ensues from all four burners in a row, facilitating the rapid and complete burnout of the fuel. Depending on the power output of the boiler, as many as six rows of burners are employed, totaling 24 burners.

Some power plants employ cyclone furnaces, especially for poorer grades of coal with a high ash content. In a cyclone furnace the combustion of the pulverized coal is accomplished in a water-cooled horizontal cylinder located outside the main boiler wall. Combustion of the coal is completed inside the cylinder, and only the hot gases are conveyed to the main boiler. The advantage of a cyclone furnace is that the majority of the mineral matter forms a molten ash, called slag, which is drained into the bottom of the boiler, and only a smaller portion exits the boiler as fly ash. Thus, smaller and less expensive particle collectors are required. The disadvantage is that at the high temperatures experienced inside the cyclone furnace, copious quantities of $NO_x$ are formed. Nowadays, plants equipped with cyclone furnaces require the installation of flue gas denitrification devices for reducing $NO_x$ concentrations in the flue gas, largely vitiating the cost savings of cyclone furnaces in terms of coal quality and ash content.

Some older power plants and smaller industrial boilers employ stoker firing. In stoker-fired boilers, the crushed coal is introduced into the boiler on an inclined, traveling grate. Primary air is blown from beneath the grate, and secondary over-fire air is blown above the grate. By the time the grate traverses the boiler, the coal particles are burned out, and the ash left behind falls into a hopper. The carbon burnout efficiency is lower in stoker than in pulverized coal burners because of poorer mixing of coal and air that is achievable in stoker-fired boilers. Therefore, stoker-type boilers have a lower thermal efficiency compared with pulverized coal boilers.

### 6.2.3  Boiler

The boiler is the central component of a fossil-fueled steam power plant. Most modern boilers are of the water wall type, in which the boiler walls are almost entirely constructed of vertical tubes that carry the feed water into the boiler and steam out of the boiler. The first water wall boiler was developed by George Babcock and Stephen Wilcox in 1867. The early water wall boilers were used in conjunction with reciprocating piston steam engines. Only in the 20th century, with the advent of the steam turbine and its requirement for large steam pressures and flows, has the water wall boiler been fully developed. In a modern water wall boiler the furnace and the various compartments of the boiler are fully integrated.

Figure 6.2 shows a schematic flow diagram of a common water wall boiler. Water from the high-pressure *feed water heater* at a temperature of 230–260°C is further heated in the economizer section of the boiler to 315°C and then flows into the *steam drum*, which is mounted on top of the boiler. The steam drum measures typically 30 m in length and 5 m in diameter. In the steam drum liquid water is separated from the steam, usually by gravity. From the steam drum, liquid water flows down the *downcomer* tubes into the *header*. From there, the hot pressurized water flows upward

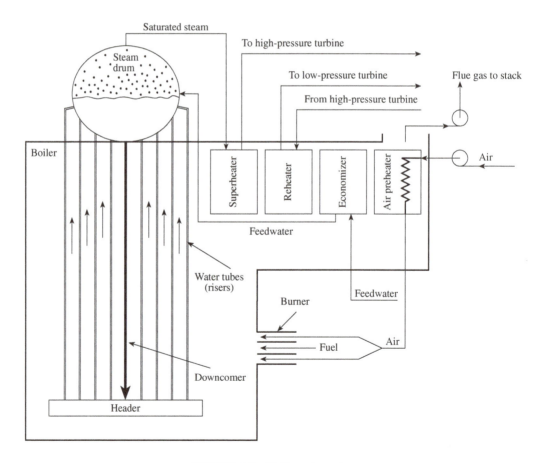

**FIGURE 6.2**  Boiler, schematic.

(because of a negative density gradient) through the *riser tubes*, where the actual boiling of water into steam occurs. The separated steam passes another section of the boiler, called the *superheater*, where its temperature is raised to 565°C at a pressure of 24 MPa. At this temperature and pressure the steam is in a supercritical state ($T_c = 374$°C, $p_c = 22$ MPa).[5] The supercritical steam drives the high-pressure turbine. The exhaust steam from the high-pressure turbine flows through the *reheater* section of the boiler, where the temperature is raised again to about 500°C at a pressure of 3.7 MPa. This steam drives the low-pressure turbine. The superheater and reheater sections of the boiler are usually situated past a bend in the boiler, called the *neck*. To optimize thermal efficiency, the combustion air is preheated to a temperature of 250–350°C in the *air preheater* section of the boiler.

Near the burners, heat is transferred from the combustion gases to the boiler tubes by radiation. Away from the burners, heat is transferred mainly by convection. Coal and oil flames are highly luminous in the visible portion of the spectrum because of the radiation from unburned carbon and ash particles. Gas flames are less visible because of fewer soot particles in the flame. However, most of the radiative transfer of all flames occurs in the nonvisible infrared portion of the spectrum.

The theoretical (Carnot) thermal efficiency of a heat engine that works between a temperature differential of 838 K (565°C) and 373 K (100°C), that is, between the temperature of the superheater and the condensation temperature of water, is $\eta = (T_h - T_l)/T_h = 55\%$. However, as mentioned in the introduction, typical efficiencies of steam power plants are in the 33–40% range, reckoned as electricity energy out per heat energy (enthalpy) in. Because heat is added to the water and steam at all temperatures between these limits, the Rankine cycle efficiency is necessarily lower than the Carnot value. Furthermore, parasitic efficiency degradation occurs because of heat losses through the walls of the boiler, ducts, turbine blades and housing, and frictional heat losses. Also, considerable internal energy has to be spent on running the auxiliary equipment, including coal transport and milling, fans, pumps, and emission control devices.

## 6.2.4 Steam Turbine

Steam turbines were first employed in power plants early in the 20th century. They can handle a much larger steam flow, pressure and temperature ratios, and rotational speed than reciprocating piston engines. Today, virtually all steam power plants in the world employ steam turbines. The steam turbine arguably is the most complex piece of machinery in the power plant, and perhaps in all of industry. There are only a score of manufacturers in the world that can produce steam (and gas) turbines.

The antecedent of the steam turbine is the water wheel. Just like water pushes the blades of a water wheel, steam pushes the blades of a steam turbine. Considering the high pressure and temperature of the steam, that the turbine must be leak-proof, the enormous centrifugal stresses on the shaft, and the fact that steam condenses to water while expanding in the turbine, thereby creating a two-phase fluid flow, one gets an idea of the technological problems facing the designer and builder of steam turbines.

The development of the modern steam turbine can be attributed to Gustav deLaval (1845–1913) of Sweden and Charles Parsons (1854–1931) of England. DeLaval concentrated on the

---

[5]At supercritical temperature and pressure the meniscus between the liquid and gaseous phase disappears; there is only one fluid phase.

development of an impulse turbine, which uses a converging–diverging nozzle to accelerate the flow speed of steam to supersonic velocities. That nozzle still bears his name. Parsons developed a multistage reaction turbine. The first commercial units were used for ship propulsion in the last decade of the 1800s. The first steam turbine for electricity generation was a 12-MW unit installed at the Fisk power plant in Chicago in 1909. A 208-MW unit was installed in a New York power plant in 1929.

### 6.2.4.1  Impulse Turbine

In an impulse turbine, a jet of steam impinges on the blades of a turbine. The blades are symmetrical and have equal entrance and exit angles, usually $20°$. Steam coming from the superheater, initially at $565°C$ and over 22 MPa, when expanded through a deLaval nozzle will have a linear velocity of about 1650 m s$^{-1}$. To utilize the full kinetic energy of the steam, the blade velocity should be about 820 m s$^{-1}$. Such a speed would generate unsustainable centrifugal stresses in the rotor. To reduce the rotor speed, turbines usually employ compounding or staging. In a staged turbine, two or more rows of moving blades (*rotors*) are separated by rows of stationary blades (*stators*, Figure 6.3a). Each pair of stator and rotor blades is called a stage. When the steam kinetic energy is divided among $n$ stages, the linear blade velocity of the rotors will be $1/2n$ that of a single rotor.

The force exerted on a rotor blade is $F = \dot{m}(v_s - v_b)$ newtons, where $\dot{m}$ is the mass flow rate of steam through the blade (kg s$^{-1}$), $v_s$ is the tangential velocity of the steam jet (m s$^{-1}$), and $v_b$ is the blade speed. The power generated by the blade is $P = Fv_b$ watts. It can be shown that maximum power obtained for $v_b = v_s/2$ is $P_{max} = \dot{m}v_s^2/4$.

### 6.2.4.2  Reaction Turbine

A reaction turbine consists of rows of fixed *stator* and moving *rotor* blades. The blades are shaped to form a converging nozzle (Figure 6.3b). Within the converging blades the steam pressure, density, and temperature decline while converting its enthalpy to kinetic energy. The steam pressure drops steadily through all rows of blades, stationary and moving, but the steam velocity oscillates depending on location within the blade formation. In a reaction turbine the optimum blade velocity

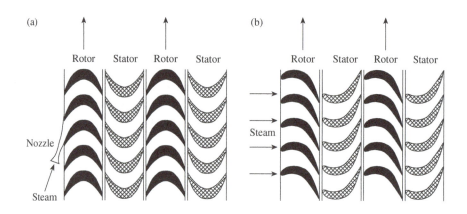

**FIGURE 6.3**    (a) Impulse turbine; (b) reaction turbine, schematic.

is $v_{b,opt} = v_s \cos\theta$, where $\theta$ is the leaving angle of the blades, and the maximum power obtained is $P_{max} = \dot{m}v_{b,opt}^2$.

In reaction turbines the pressure drops across both fixed and moving blades. This makes them less suitable for high-pressure steam because of leakage around the blade tips and consequent loss of efficiency. Therefore, impulse turbines are usually used for high-pressure steam and reaction turbines for intermediate- and low-pressure steam. In addition, the high-pressure impulse stage can reduce the steam flow to the turbine by closing off some of the first-stage nozzles, a flow control that is absent in reaction stages. In Figure 3.4 we have seen a photo of a 1,500-MW steam turbine complex for high-, intermediate-, and low-pressure stages.

In both the impulse and the reaction turbine efficiency losses are caused by supersaturation, fluid friction, leakage, and heat transfer losses. Supersaturation occurs primarily in reaction turbines when according to thermodynamic equilibrium in the expansion process steam ought to condense, releasing latent heat of condensation. Instead, steam remains for a while in a supercooled state before reverting to thermodynamic equilibrium. This results in a sudden release of energy, called the condensation shock. It is an irreversible process, causing loss in efficiency and energy availability.

Fluid friction occurs throughout the turbine, in the steam nozzles, along the blades, and along the rotor disks that carry the blades. In addition, the rotor and blade rotation impart a centrifugal action on the steam, causing it to be dragged along the blades. When the blades are not properly designed, flow separation may ensue, further increasing turbine losses.

Heat transfer losses are caused by conduction, convection, and radiation. Conduction is the result of heat transfer between metal parts of the turbine. Convection is the heat transfer between steam and the metal parts. Radiation is the heat given off by the turbine casings to the surroundings. Heat transfer losses are highest in the high-temperature, high-pressure sections of the turbines.

In addition, there are frictional losses in bearings, governor mechanism, and reduction gearing. Also, turbines must supply power for accessories, such as oil pumps. The combined efficiency losses and siphoning of auxiliary power amount to 10–20%, that is, turbines convert only 80–90% of the available steam enthalpy into mechanical energy that drives the generator.

## 6.2.5 Gas Turbine

In a gas turbine plant where oil, natural gas, or synthesis gas is used as a fuel, the hot combustion gases, are directly driving a gas turbine, rather than transferring heat to steam and driving a steam turbine. This requires a different turbine, appropriate for the much higher temperature but lower pressure of the combustion gases and their different thermodynamic properties compared with steam.

Gas turbines are easily brought on line and have flexible load match. But gas turbine (Brayton) cycle efficiencies are lower than steam turbine (Rankine) cycle efficiencies, and the fuel is more expensive, therefore, gas turbine power plants are mostly used for peak load production and for auxiliary power, such as during major plant outages. However, recently many natural gas-fueled gas turbine plants have been installed in the United States and other countries, but these usually employ a combined Brayton–Rankine cycle mode that has a higher efficiency than a single-cycle mode.

The Brayton cycle was described in Section 3.9.4. Compressed air enters a combustion chamber where it is mixed with liquid or gaseous fuel. The combustion of the fuel increases the temperature of the combustion gas, producing a net work output of the turbine–compressor system. The temperature of the combustion gases is in the order of 1100–1200°C, which is the maximum tolerable by present day steel alloys used for the gas turbine blades. Even at these temperatures, thermal stresses and corrosion problems are manifested, so that turbine blade cooling from the inside or outside of the blades by air or water is necessary.

Gas turbines are of the reaction type, where blades form a converging nozzle in which the combustion gases expand, thus converting enthalpy to kinetic energy. As in steam turbines, staged turbines are employed, consisting of several rows of moving and fixed blades.

The working fluid in gas turbines, composed of nitrogen, excess oxygen, water vapor, and carbon dioxide, is not recycled into the compressor and combustion chamber, but vented into the atmosphere. In some systems, a part of the energy still residing in the exhaust gas is recovered in heat exchangers to heat up the air entering the combustion chamber to enhance the overall thermal efficiency of the Brayton cycle, but eventually the exhaust gas is vented. This is in contrast to steam turbines where the working fluid—steam—is recycled into the boiler as condensed water.[6]

## 6.2.6  Condenser

In heat engine cycles, after performing useful work, the working fluid must reject heat to a cold reservoir, according to the second law of thermodynamics. Heat engines reject a significant amount of heat to the environment. Between 1.5 and 3 times as much heat is rejected as the plant produces work in the form of electricity. A 1,000-MW electric power plant working at 25% efficiency rejects 3,000 MW; that working at 40% efficiency rejects 1,500 MW. Some of that heat is rejected in a condenser and the rest by the emission of the hot flue gas into the atmosphere through a smoke stack.

In the Rankine steam cycle, after expansion in the turbine, the steam is condensed into water in a condenser. The condensed water is recycled into the boiler by means of a feed pump. The condenser serves not only the purpose of condensing the high quality feed water of the boiler, but also to lower the vapor pressure of the condensate water. By lowering the vapor pressure, a vacuum is created, which increases the power of the turbine. There are two types of condensers: *direct contact* and *surface contact* condensers.

A direct contact condenser is depicted in Figure 6.4a. The turbine exhaust passes an array of spray nozzles through which cooling water is sprayed, condensing the steam by direct contact. The warm condensate is pumped into a dry cooling tower where up-drafting air cools the condensate that flows in tubes. Part of the cooled condensate is recycled into the spray nozzles; the rest is sent as feed water to the boiler. Because of evaporative losses, make-up feed water needs to be added to the condensate. Throughout the condensing cycle the purity of the water must be strictly maintained to prevent fouling of the boiler water/steam cycle.

A surface contact condenser is depicted in Figure 6.4b. It is essentially a shell-and-tube type heat exchanger. The turbine exhaust passes an array of tubes in which the cooling water flows. The cooling water can be artesian or siphoned from a river, lake, or ocean. Because large power plants produce large volumes of steam that need to be condensed, the contact surface areas can

---

[6]Gas turbines have many applications other than for electricity generation. They are used for pipeline pumping of natural gas, ship propulsion, and, foremost, for airplane propulsion in turbojet aircraft. Here, air is compressed in the compressor, and jet fuel (kerosene) is added to the combustion chamber. The combustion gases drive the turbine that supplies power to the compressor and auxiliary systems (e.g., electricity generation), and the turbine exhaust gases pass a deLaval nozzle to provide forward thrust to the airplane.

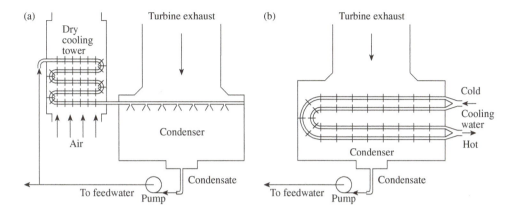

**FIGURE 6.4**   (a) Direct contact condenser; (b) surface condenser, schematic.

reach 100,000 m$^2$ for a 1,000-MW plant. The design of a properly functioning condenser involves complicated calculations of heat transfer. The tubes are surrounded by fins to increase the heat transfer area. The incoming steam needs to be deaerated of noncondensibles, mainly air that leaked into the system. The oxygen of air is corrosive. Also, noncondensibles in the condensate would increase its vapor pressure. We mentioned before that the lower the vapor pressure of the condensate, the higher the steam turbine efficiency.

### 6.2.7   Cooling Tower

The bulk of the rejected heat from the steam cycle occurs either to a surface water or to the atmosphere. In the past most power plants were located near a river, lake, or ocean. In those plants the hot water from the condenser is directly discharged to the surface water by means of diffusers or indirectly into a canal that leads to the surface water. The discharge of warm water into the surface water can cause thermal pollution and possible harm to aquatic organisms. Also, contaminants that leach into the discharge water from pipes and ducts may pollute the surface waters. As a consequence, environmental protection agencies in many countries mandate that heat rejection occur into the atmosphere via cooling towers, not to surface waters.

There are two types of cooling towers: *wet* and *dry*. There are also combinations of wet and dry towers, as well as combinations of cooling towers and surface water cooling.

### 6.2.7.1   Wet Cooling Tower

In a wet cooling tower the hot condensate water is cooled by rejection of sensible heat to the atmosphere and evaporation of part of the recirculated water itself. The evaporation causes water to cool because of the latent heat of evaporation. The cooling tower is usually visible as a gigantic spool-like structure constructed near a fossil or nuclear power plant. The spool configuration is advantageous from a structural standpoint—it requires less concrete for its size, and it is more resistant to strong winds.

A typical wet tower schematic with natural draft is depicted in Figure 6.5. Hot water from the condenser is sprayed over a latticework of closely spaced slats or bars, called *fill* or *packing*.

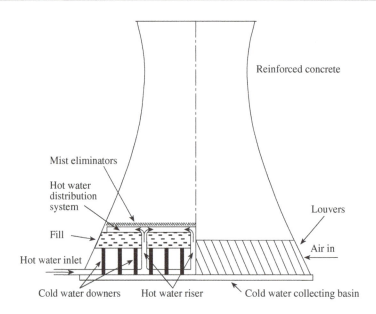

**FIGURE 6.5**  Wet cooling tower, schematic.

Outside air is drawn in through louvers in the bottom of the tower by natural draft. Heat is transferred from the cooling water to the air directly, and the water is further cooled by evaporation of part of the water. Cold water falls by gravity into the collecting basin from whence it is recirculated into the condenser. Because of evaporative losses, some make-up feed water is necessary. A mist or drift eliminator is placed above the fill. Nevertheless, a mist plume usually forms above the tower, especially in cold weather with a low dew point. This white plume is often mistaken as pollution; in fact, it is just composed of water droplets or ice crystals.

A 1,000-MW power plant working at 33% efficiency in a hot climate evaporates 0.63 $m^3s^{-1}$ water, which is 1.3% of the recirculating water. In a cold climate the same plant would evaporate only 1% of the recirculating water. This amounts to 2E(7) or 1.5E(7) $m^3y^{-1}$ water that the power plant must withdraw from a surface water, well, or municipal water, in hot or cold climate, respectively. In the United States it is estimated that power plants consume about 2% of all fresh water withdrawals.

### 6.2.7.2  Dry Cooling Tower

In a dry cooling tower, the recirculating water flows through finned tubes over which cold air is drawn. The advantages are that no water evaporates and dry towers are less expensive to maintain. The disadvantages are that they are more expensive to build and the back pressure on the turbines is higher, causing plant efficiency loss. Nevertheless, in arid areas where no make-up water is available, dry cooling towers need to be employed.

### 6.2.8  Generator

The generator is the heart of the power plant. That is where electricity is generated. Compared with the boiler, turbines, condenser, cooling tower, and other auxiliary equipment, the generator occupies a small fraction of the total plant outlay. Its noise level is also negligible compared with the hum and drum of the coal mills, burners, pumps, fans, and turbines.

The electromagnetic theory of the generator is described in Chapter 5. Briefly, the shaft of the turbine turns conducting coils within a magnetic field. This induces an electric current to flow in the coils. The electric power output of the generator equals the mechanical power input of the shaft, minus minor resistive losses in the coils and frictional losses. To prevent the overheating of the generator induced by these losses, generators are cooled by high thermal conductivity gases, such as hydrogen or helium.

The generator produces an AC, 60 Hz in the United States and Canada and 50 Hz in most other countries. The shaft of the turbine must rotate at a precise speed to produce the exact frequency of the AC. The voltage produced by the generator is enhanced in a step-up transformer and then transmitted into the grid. Because resistance losses are proportional to the square of the current, it is advantageous to transmit power at low current and high voltage. Long-distance electricity is usually transmitted in the hundreds of kilovolt range. At the user, the voltage is reduced in step-down transformers to 110 or 220 V.

### 6.2.9  Combustion Stoichiometry

The power of fossil-fueled power plants derives from the heat released during the combustion of the fossil fuel in air, that is, the oxygen content of air, which is approximately 21% by volume (nitrogen is 78%, other gases, mainly argon, is approximately 1%). We can approximate the chemical formula of coal as $CH$, oil as $CH_2$, and natural gas (mainly methane) as $CH_4$. Thus, the combustion stoichiometry (chemical formula balancing) of coal combustion in air is

$$CH + 1.25O_2 + 4.64N_2 \rightarrow CO_2 + 0.5H_2O + 4.64N_2. \tag{6.1}$$

The stoichiometric coefficient of $N_2$ arises because the volumetric ratio of $N_2/O_2 = 3.71$, times $1.25 = 4.64$. Similarly, the stoichiometry of oil combustion is

$$CH_2 + 1.5O_2 + 5.565N_2 \rightarrow CO_2 + 1.5H_2O + 5.565N_2. \tag{6.2}$$

That of natural gas combustion is

$$CH_4 + 2O_2 + 7.42N_2 \rightarrow CO_2 + 2H_2O + 7.42N_2. \tag{6.3}$$

In effect, coal is combusted with 20% excess air, oil with 10%, and gas with 5%. Thus, the stoichiometric coefficients of oxygen and nitrogen have to be multiplied by 1.2, 1.1, and 1.05, respectively.

Coal and oil usually also contain in their formula sulfur and nitrogen. A typical formula of coal is $C_1H_1S_{0.02}N_{0.01}$. Neglecting argon and other minor gases in air, the combustion of a typical coal in 20% excess air has the following stoichiometry:

$$C_1H_1S_{0.02}N_{0.01} + 1.536O_2 + 5.698N_2 \rightarrow$$
$$CO_2 + 0.5H_2O + 0.02SO_2 + 0.01NO_2 + 5.698N_2 + 0.307O_2. \tag{6.4}$$

The oxygen in the flue gas appears because excess oxygen is not consumed, it merely helps to complete the combustion process within the boiler confines.

Neglecting argon and the other minor gases in air, the volume (mole) fraction of the combustion products in the flue gas is calculated as follows. The total number of moles of the flue gas is 7.535. Thus, the volume (mole) fractions are

$$
\begin{aligned}
N_2 &= 5.698/7.535 = 0.756 & \text{(75.6\% by volume)} \\
O_2 &= 0.307/7.535 = 0.041 & \text{(4.1\% by volume)} \\
CO_2 &= 1/7.535 \quad\;\; = 0.133 & \text{(13.3\% by volume)} \\
SO_2 &= 0.02/7.535 \; = 0.002654 & \text{(2654 ppmv)} \\
NO_2 &= 0.01/7.535 \; = 0.001327 & \text{(1327 ppmv)}
\end{aligned}
$$

Similar calculations can be performed for oil and gas combustion. For a typical oil, the carbon/hydrogen ratio is 1:2, with a basic formula $C_1H_2$, when it is combusted in 10% excess air, the $CO_2$ volume (mole) fraction in the flue gas is 12%. For gas, with a basic formula $C_1H_4$, when it is combusted in 5% excess air, the $CO_2$ mole fraction is 9.2%. Generally, oil produces less $CO_2$ than coal, gas produces less than oil, and gas produces much less than coal. The reason is that for equal heat input the more hydrogen in the fuel molecular make-up, the less $CO_2$ is produced.

The volume flow rate of flue gas is calculated as follows. Neglecting argon and other minor gases in air, each mole of coal (13 g) produces 7.535 moles of flue gas. A 1,000-MW(el) coal-fired power plant working at 100% capacity, at 35% thermal efficiency, needs 2,857 MW thermal input. If the coal has a heating value of 30 MJ/kg and a molecular weight of 13 g (CH), neglecting the contribution of sulfur and nitrogen to the molecular weight, the power plant consumes 7,326 mol/s of coal. In the above example, that coal produces 55,200 mol/s of flue gas. Assuming the flue gas is an ideal gas, where the ideal gas law applies, $pV = nRT$ and considering the molecular volume of the flue gas at standard temperature and pressure $V_m = 0.02445$ m³, the volume flow rate of flue gas $Q = 1,350$ m³/s.

The volume air intake is calculated as follows. In the above example, each mole of coal (13 g) requires 7.065 mole of air. With the above molecular volume, the air intake amounts to 1,265 m³/s.

## 6.2.10  Emission Control

If left uncontrolled, power plants can emit quantities of air pollutants that exceed emission standards designed to protect human health and the environment. Suppose a 1,000-MW power plant burns coal with a 10% mineral content and 2% sulfur content (not unusual). It is a base-loaded plant that works at 100% capacity and 35% thermal efficiency. The coal has a heating value of 28 MJ kg⁻¹. All of the mineral content exits the smoke stack as particles (fly ash) and the sulfur as sulfur dioxide $SO_2$. This plant would emit 3.2 E(5) t y⁻¹ of particles and 1.3 E(5) t y⁻¹ of $SO_2$ (sulfur dioxide has twice the molecular weight of sulfur). In addition, the plant would emit copious quantities of nitrogen oxides, products of incomplete combustion (PIC), carbon monoxide, and volatile trace metals. Clearly, such an uncontrolled power plant would present a major risk to human health and the environment. Therefore, in most countries, environmental regulations require that the operator of the power plant install emission control devices for these pollutants. These devices contribute significantly to the capital and operating cost of the plant and reduce to some degree the thermal efficiency, because the devices siphon off some of the power output of the plant. These costs are passed on to the customers as added cost of the electricity. The control devices also produce multiple streams of waste, because what is not emitted into the atmosphere usually winds up as a solid or liquid waste stream.

### 6.2.10.1 Control of Products of Incomplete Combustion and Carbon Monoxide

The control of PIC and CO is relatively easy to accomplish. If the fuel and air are well mixed, as is the case in modern burners, and the fuel is burned in excess air, the flue gas will contain very little, if any, PIC and CO. It is in the interest of power plants to achieve a well-mixed, fuel-lean (air-rich) flame, not only for reducing the emission of these pollutants, but also for complete burnout of the fuel, which increases the thermal efficiency of the plant. PIC and CO emissions do occur occasionally, especially during start-ups and component breakdowns, when the flame temperature and fuel–air mixture is not optimal. Under those conditions a visible black smoke emanates from the smoke stack. These occurrences should be rare and should not contribute significantly to ambient concentrations of these pollutants.

### 6.2.10.2 Control of Particles

Particles, also called particulate matter (PM), would be the predominant pollutants emanating from power plants were they not controlled at the source. This stems from the fact that coal, and even oil, contain a significant fraction by weight of incombustible mineral matter. In older, stoker-fed and cyclone burner plants, the mineral matter accumulates in the bottom of the boiler as bottom ash and is discarded as solid waste or taken up in water and sluiced away. In modern pulverized coal-fired plants the majority ($\approx 90\%$) of the mineral matter is blown out from the boiler as fly ash. The fly ash contains a host of toxic metals, such as arsenic, selenium, cadmium, manganese, chromium, lead and mercury, and nonvolatile organic matter (soot), including polycyclic aromatic hydrocarbons (PAHs), which would pose a public health and environmental risk if emitted into the atmosphere. For that reason, most countries instituted strict regulations on particle emissions from power plants.

In the United States, power plants built between 1970 and 1978 had to meet a standard for PM emissions of a maximum of 0.013 kg per GJ heat input. For power plants built after 1978 there is no numerical standard, but the so-called Best Available Control Technology (BACT) standard applies. Currently, BACT for power plants is the electrostatic precipitator (ESP).

*Electrostatic Precipitator.* The ESP was invented in the early 1900s by F. G. Cottrell at the University of California, Berkeley, to collect acid mist in sulfuric acid manufacturing plants. It was soon applied for collecting dust in cement kilns, lead smelters, tar, paper and pulp mills, and other factories. Beginning in the 1930s and 1940s, ESP was applied to coal-fired power plants. The installation of early ESPs preceded environmental regulations; they were installed to protect the owners from possible liability suits because particle emissions could cause a health hazard.

The ESP works on the principle of charging particles negatively by a corona discharge and attracting the charged particles to a grounded plate. A schematic of an ESP is given in Figure 6.6. Several charging wires are suspended between two parallel plates. A high negative voltage, on the order of 20 to 100 kV, is applied to the wires. This causes an electric field to be established between the wires and the plates along which electrons travel from the wires toward the plates. This is called a corona discharge. The electrons collide with gas molecules, primarily with oxygen, creating negative ions. The molecular ions keep traveling along the field lines, colliding with particles, thus transferring the negative charge to the particles. Now the particles migrate to the plates, where their charge is neutralized. The neutral particles are shaken off the plates by rapping them periodically. The particles fall into a hopper from whence they are carted away.

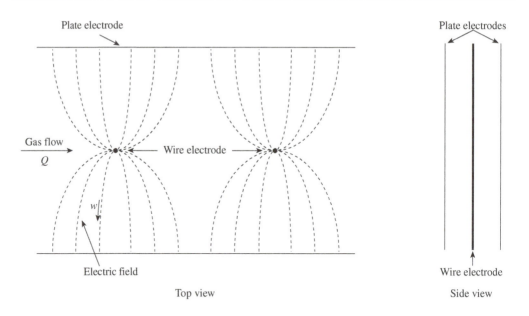

**FIGURE 6.6**  Electrostatic precipitator, schematic.

The collection efficiency (fraction of particles collected) of an ESP depends on many factors, primarily the particle diameter (the smaller the diameter the less the efficiency), the plate area, the volumetric flow rate of the flue gas passing between the plates, and the particle migration speed toward the plates. The efficiency is calculated by the Deutsch equation

$$\eta = 1 - \exp(-wA/Q), \tag{6.5}$$

where $w$ is the migration speed, $A$ is the total area of the plates, and $Q$ is the volumetric flow rate. The migration speed is approximated by $w \approx 0.05d_p$ m s$^{-1}$ where the particle diameter $d_p$ is given in micrometers. Actually, $w$ is a function of the voltage and the electrical resistivity of the particles, but the approximation works in most cases. Ironically, the higher the sulfur content of the coal the lower the resistivity of the fly ash and the higher the migration speed. However, the higher the sulfur content of the coal the higher acid gas emissions, so there is a trade-off.

The ESP collection efficiency plotted as a function of plate dimension and particle diameter is given in Figure 6.7. It is seen that with large plate dimensions, particles larger than 2 μm in diameter are collected at close to 100% efficiency, 1-μm particles at only 90–95%, and submicrometer particles are not collected efficiently at all.

A cutaway drawing of a large ESP is shown in Figure 6.8. Typically, the overall width is about 20–30 m and the overall length 18–20 m. Typical plate dimensions are 10 × 10 m, plate spacing is 0.25 m, and the number of plates is in the hundreds. The gas volumetric flow rate is up to 20,000 m$^3$ s$^{-1}$ for a 1,000-MW plant. After exiting from the boiler, the flue gas is transferred to the ESP in large ducts. From the ESP, the cleansed flue gas enters the smoke stack. The ESP, together with its housing, transformers, hoppers, and truck bays, may occupy a structure half the size of the boiler house of a power plant.

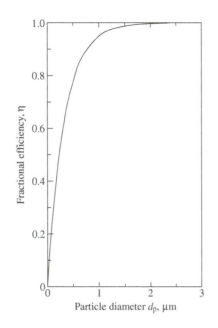

**FIGURE 6.7** Particle collection efficiency of the electrostatic precipitator as a function of particle diameter calculated from the Deutsch Equation (6.5). Flue gas volumetric flow rate $Q = 1260$ $m^3 s^{-1}$, plate area $A = 75,500$ $m^3$, and drift velocity $w = 0.05 d_p$ m $s^{-1}$, where $d_p$ is given in micrometers.

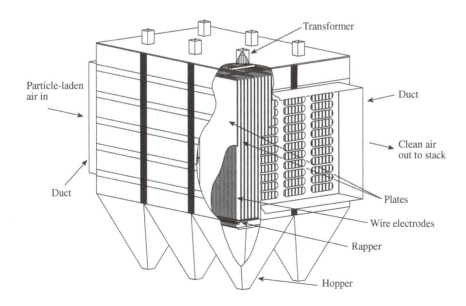

**FIGURE 6.8**    Electrostatic precipitator, cutaway.

The power drawn by the ESP, including the fans that drive the flue gas through it, and the mechanical energy for rapping the plates consume less than 0.1% of the plant's power output. The levelized cost (amortized capital and operating cost) of running an ESP may add 6–10% to the generating cost of electricity.[7]

The relatively low efficiency of small particle collection by ESPs poses a problem to the operators of existing power plants because environmental protection agencies in several countries, notably the United States, introduced new ambient standards for particles less than 2.5 μm in diameter. The new standard may require retrofitting power plants with devices that more efficiently collect small particles, such as the fabric filter.

*Fabric Filter.* A fabric filter, also called baghouse, works on the principle of a vacuum cleaner. Particle laden gas is sucked into a fabric bag, the particles are filtered out, and the clean gas is vented into the atmosphere. The pore size of the fabric can be chosen to filter out any size of particles, even submicrometer particles, albeit at the expense of power that is required to drive the flue gas through the pressure drop represented by the fabric pores.

A typical fabric filter schematic is shown in Figure 6.9. Long, cylindrical tubes (bags) made of selected fabric are sealed at one end and open at the other end. The sealed end of the tubes are hung upside down from a rack that can be shaken mechanically. The particle laden flue gas enters through the bottom, open end of the tubes. The clean flue gas is sucked through the fabric of the tubes by fan-induced draft or special pumps and vented through the smoke stack. A baghouse for a large power plant may contain several thousand tubes, each up to 4 meters high and 12.5 to 35 cm in diameter. It is necessary to distribute the incoming flue gas equally to all tubes, which is done in the plenum. The tubes provide a large surface area per unit of gas volumetric flow rate. The inverse is called the air-to-cloth or filtering ratio, which is equal to the superficial gas velocity; it typically ranges between 0.5 and 4 cm s$^{-1}$.

The collected particles can be removed by mechanical shaking of the tubes or by the reverse jet method. Mechanical shaking is induced by a cam-driven moving rack on which the tubes hang. In the reverse jet method, a strong air flow is blown from the outside of the tubes toward the inside, dislocating the particle cake that has built up. The removal of particles is not always complete, as particles cling to the fabric and are lodged in the pores. This causes frequent breakdowns and requires replacement of the fabric tubes about once a year. Ironically, the more clogged the tubes the better the removal efficiency, but at the expense of pumping power, which has to be supplied to maintain the superficial gas velocity. The removed particles fall into the bottom hopper from whence they are carted away by trucks.

The flue gas at the entrance to the baghouse has a relatively high temperature of 300–350°C. Also, the flue gas may contain corrosive gases and moisture. These conditions require a heat- and corrosion-resistant fabric. Usually, fiberglass is chosen for coal-fired power plants, whereas other fabrics, natural or synthetic, are applicable for other industries, like cement kilns, ferrous and nonferrous smelters, and paper and saw mills.

Fabric filters with a high collection efficiency for small particles are expected to have a higher capital and operating cost than ESPs.

---

[7]Data on emission control efficiency and cost are obtained from Takashita, M. *Air Pollution Control Costs for Coal-Fired Power Stations.* Coal Research Publication IEAPER/17. London: International Energy Association, 1995.

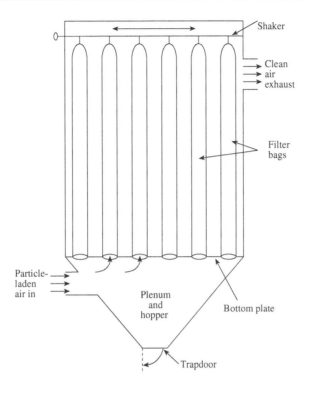

**FIGURE 6.9**  Baghouse with mechanical shakers, schematic.

### 6.2.10.3  Sulfur Control

Because living organisms contain sulfur in their cellular make-up, this sulfur is mostly retained in the fossilized remnants of these organisms. Coal can contain up to 6% by weight of sulfur and oil up to 3%. However, most coals and crude oils contain a much lower percentage of sulfur. Generally, bituminous, subbituminous, and lignite coals are used in power plants. They contain 0.7–3% by weight of sulfur. Residual oil used in power plants contains 0.7–2% sulfur. Without sulfur emission control devices, the oxidized sulfur, mainly sulfur dioxide $SO_2$, and minor quantities of $SO_3$ and sulfuric acid $H_2SO_4$ would be emitted through the smokestack into the environment. The oxides of sulfur are precursors to acid deposition and visibility impairing haze (see Chapter 10). Because coal-fired power plants emit the majority of all sulfur oxide emissions worldwide (industrial boilers, nonferrous smelters, diesel, and home heating oil make up the rest of sulfur emissions), operators of these plants are required to limit the emissions by either switching to fuels containing low sulfur concentrations or installing sulfur emission control devices.

Sulfur emissions can be reduced by treating the fuel before combustion or by treating the flue gas after combustion.

**Before Combustion.**

(a) *Coal Washing.* When coal is removed from the coal seams in underground or surface mines, there is always some mineral matter included in the coal. The majority of mineral matter is composed of silicates, oxides, and carbonates of common crustal elements, such as calcium, magnesium, aluminum, and iron, but some of it contains pyrites, which are sulfides of iron, nickel, copper, zinc, lead, and other metals. Because the specific gravity of mineral matter, including the pyrites, is greater than that of the carbonaceous coal, a part of the mineral matter can be removed by "washing" the coal. Coal washing not only reduces the sulfur content of coal, but also reduces its ash content, thereby increasing its heating value (J/kg) and putting a lesser load on the particle-removing systems.

Coal washing is usually performed at the mine mouth. Typically, the crushed raw coal is floated in a stream of water. The lighter coal particles float on top, and the heavier minerals sink to the bottom. The wet coal particles are transferred to a dewatering device, generally a vacuum filter, centrifuge, or cyclone. The coal can be further dried in a hot air stream.

One problem with coal washing is that the stream containing mineral matter may contain dissolved toxic metals, and it can be acidic. Strict regulations have been introduced in the United States and elsewhere to prevent dumping of these toxic acidic streams into the environment without prior treatment.

In the United States, about 50% of all coals delivered to power plants are washed. These are mainly coals from eastern and midwestern shaft and strip mines. Western coals generally have a low mineral and sulfur content, so washing is not necessary. Coal washing can remove up to 50% of the pyritic sulfur, which is equivalent to 10–25% removal of the total sulfur content of the coal.

*(b)* **Coal Gasification.** Coal can be converted by a chemical process into a gas, called synthesis gas, or syngas for short. In the process of coal gasification, most of the sulfur can be eliminated before the final stage of gasification. The clean, desulfurized syngas can be used to fuel a gas turbine or combined cycle power plant. The process of coal gasification will be described in Section 6.3.2.

*(c)* **Oil Desulfurization.** Refineries can reduce the sulfur content of crude oil almost to any desired degree. This is usually done in a catalytic reduction/oxidation process called the Claus process. First, sulfur compounds in the oil are reduced to hydrogen sulfide by blowing gaseous hydrogen through the crude oil in presence of a catalyst,

$$RS + H_2 \rightarrow H_2S + R, \qquad (6.6)$$

where R is an organic radical. Next, $H_2S$ is oxidized by atmospheric oxygen to elemental sulfur, also in the presence of a catalyst:

$$H_2S + 0.5O_2 \rightarrow H_2O + S. \qquad (6.7)$$

The elemental sulfur is an important by-product of oil refining and can be often seen as a yellow mound within the refinery complex. It is a major raw material for sulfuric acid production.

Even though the sulfur in crude oil is a saleable by-product of oil refining, refineries charge more for low sulfur oil products. Thus, utilities pay a premium price for low sulfur oil, and we pay higher rates for electricity generated by the oil-burning power plant. However, it is usually cheaper for a utility to buy low sulfur oil than to remove $SO_2$ from the stack gas.

**After Combustion.** The removal of sulfur oxides from the flue gas after combustion of the fuel in a furnace or boiler is called *flue gas desulfurization* (FGD). There are several methods of FGD: sorbent injection and wet or dry scrubbers.

(a) **Sorbent Injection.** In sorbent injection (SI), a sorbent, usually dry sintered $CaCO_3$ or $CaO$, or a slurry thereof, is injected into the flue gas in the upper reaches of the boiler, past the neck. The sorption of $SO_2$ proceeds by forming a mixture of calcium sulfite and sulfate. The capture efficiency is dependent on many factors: the temperature, oxygen and moisture content of the flue gas, time of contact between the sorbent and $SO_2$, and the characteristics of the sorbent (e.g., sintered sorbent, porosity, admixture of other sorbing agents). The resulting particles, consisting of hydrated calcium sulfite and sulfate, and unreacted sorbent, in addition to the fly ash, need to be captured in an electrostatic precipitator or baghouse.

Sorbent injection can be retrofitted to existing coal-fired power plants, albeit the particle removal system may have to be upgraded to collect the considerable larger load of particles. The sulfur capture efficiency is on the average 50%, which may be adequate for meeting emission reduction quotas for existing power plants, but not enough for emission standards for new power plants.

(b) **Wet Scrubber.** In a wet scrubber the flue gas is treated with an aqueous slurry of the sorbent, usually limestone ($CaCO_3$) or calcined lime ($CaO$), in a separate tower. A schematic of a wet scrubber is shown in Figure 6.10. After exiting the electrostatic precipitator, the flue gas enters an absorption tower where it is sprayed through an array of nozzles with a slurry of the sorbent. The following sequence of reactions takes place between $SO_2$ and the sorbent slurry:

$$CaCO_3 + SO_2 + 0.5H_2O \rightarrow CaSO_3 \cdot 0.5H_2O + CO_2 \tag{6.8}$$

$$CaSO_3 \cdot 0.5H_2O + 1.5H_2O + 0.5O_2 \rightarrow CaSO_4 \cdot 2H_2O \tag{6.9}$$

The water molecules that are attached to calcium sulfite and sulfate are called water of crystallization. Hydrated calcium sulfate is similar to natural gypsum.

The formed mixture of hydrated calcium sulfite and sulfate, together with some unreacted limestone, falls to the bottom of the wet scrubber in the form of a wet sludge, where it is transferred into a funnel-shape thickener. The top liquor of the thickener is decanted into an overflow tank, where it is recycled to make up a fresh slurry of limestone. The settled thick sludge is pumped into a vacuum filtering system where it is dewatered as much as possible. Both hydrated $CaSO_3$ and $CaSO_4$ are difficult to dewater; they form a gelatinous sludge. This sludge may be thickened further with fly ash coming from the ESP and then disposed into a landfill.

Above the spray nozzles, a mist eliminator condenses water. The clean flue gas enters a reheater (to add buoyancy) and then exits through the smokestack.

While perfected over the past decades, the wet scrubber still poses many operational problems. The spray nozzles tend to clog; sludge often clings to the bottom and side of the absorption tower, where it has to be removed mechanically; the slurry of lime and limestone is highly corrosive; the dewatering system is prone to breakdowns; the dewatered sludge is difficult to transport to the disposal site. Frequent outages may still be experienced, and

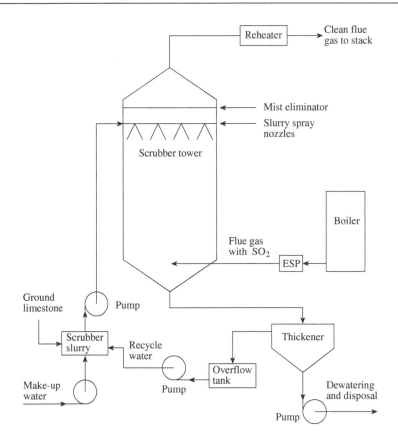

**FIGURE 6.10**   Wet limestone scrubber, schematic.

since power plants cannot afford to install dual systems, during scrubber outages the $SO_2$ containing flue gas is simply bypassed directly into the smokestack.

A well-designed wet scrubber can remove as much as 90–99% of the sulfur in the flue gas. Its power requirements (pumps, filters, reheater, etc.) may siphon off 2–3% of the power plant's electrical output, thereby reducing the overall thermal efficiency by the same amount. Its amortized capital and operating cost may add 10–15% to the electricity generating cost.

(c) **Dry Scrubber.** The chemical reaction mechanism in the dry scrubber is similar to that of the wet scrubber, namely that $CaCO_3$ or $CaO$, or both, are used to absorb the $SO_2$ from the flue gas, forming a mixture of calcium sulfite and sulfate. The difference is that in the dry scrubber the sorbent is introduced as a very fine spray of an aqueous slurry. The hot flue gas is blown countercurrent against the slurry spray. The proportions of slurry and flue gas are carefully metered, so that the slurry completely evaporates within the scrubber. In such a fashion, a dry powder of calcium sulfite, sulfate, and unreacted sorbent is created. Here, the particle removal system, usually a fabric filter, is installed downstream of the scrubber in contrast to the wet scrubber, where the particle removal system, usually an ESP, is upstream of the scrubber. Thus, the dry scrubber does not create a gelatinous sludge, which is difficult to transport and dispose of, but it does impose a larger load on the particle removal system.

The dry scrubber $SO_2$ removal efficiency is not as high as that of the wet scrubber, amounting to 70–90%. Its capital and operating costs are somewhat lower than those of a wet scrubber.

In the United States, all new coal-fired power plants must install a scrubber, either a wet or a dry scrubber, depending on the sulfur content of the coal. At present, FGD units were being used in 27 countries and there were 678 FGD units operating on a total power plant capacity of about 229 GW. About 45% of that FGD capacity was in the United States, 24% in Germany, 11% in Japan, and 20% in various other countries. Approximately 79% of the units, representing about 199 GW of capacity, were using lime or limestone wet scrubbing. About 18% (or 25 GW) utilized spray-dry scrubbers or sorbent injection systems.

### 6.2.10.4  Nitrogen Oxide Control

The other major gaseous pollutants that are emitted from fossil fuel combustion are nitrogen oxides, called $NO_x$, which include nitric oxide NO, nitrogen dioxide $NO_2$ (and its dimer $N_2O_4$), nitrogen trioxide $NO_3$, pentoxide $N_2O_5$, and nitrous oxide $N_2O$. Other than NO and $NO_2$, the other oxides are emitted in minuscule quantities, so that $NO_x$ usually implies the sum of NO and $NO_2$.

$NO_x$ are pernicious pollutants because they are respiratory tract irritants and they are precursors to photo-oxidants, including ozone, and acid deposition (see Chapter 10). Because of the introduction in 1997 of a new U.S. ambient ozone standard of 0.08 parts per million by volume (8-hour average concentration), the implication for fossil-fueled power plants is that $NO_x$ emissions need to be minimized as much as possible.

Coal and oil contain organic nitrogen in their molecular structure. When burned in air, the intrinsic (organic) nitrogen of the fuel oxidizes to $NO_x$. This is called *fuel* $NO_x$. In addition, fossil fuel combustion produces *thermal* $NO_x$. This results from the recombination of atmospheric nitrogen and oxygen under conditions of the high temperatures prevailing in the flame of fossil fuel combustion:

$$N_2 + O_2 \rightarrow 2NO. \tag{6.10}$$

The recombination involves intermediate radicals, such as atomic oxygen and nitrogen, and organic radicals, which are formed at the high flame temperatures. During transit from the boiler to the exhaust stack, a part of the NO oxidizes into $NO_2$ and other nitrogen oxides.

Coal and oil combustion produce both fuel and thermal $NO_x$. Because natural gas does not contain intrinsic nitrogen, it produces only thermal $NO_x$ at the high flame temperature of combustion. As a rule of thumb, coal and oil combustion produce about equal amounts of fuel and thermal $NO_x$. The flue gas of uncontrolled coal and oil combustion contains thousands of parts per million by volume of $NO_x$; that of natural gas combustion contains half as much.

Because intrinsic nitrogen cannot be removed prior to combustion of the fuel, $NO_x$ emission control can only be achieved during and after combustion.

**During Combustion.**

***Low-$NO_x$ burner.***  A low-$NO_x$ burner (LNB) employs a process called *staged combustion*. LNB exploits the fact that $NO_x$ formation is a function of air-to-fuel ratio (by weight) and temperature of the flame. This ratio and the flame temperature affect the availability of free radicals that participate in the $NO_x$ formation process. A plot of $NO_x$ concentration in the flame versus air-to-fuel ratio is presented in Figure 6.11. It is seen that under stoichiometric conditions (air-to-fuel ratio $\approx 15$)—when exactly as much air oxygen is present as necessary for complete combustion of the fuel—maximum $NO_x$ is formed. Less $NO_x$ is formed both under fuel-rich and fuel-lean combustion conditions. Fuel-rich conditions are to the left of Figure 6.11; fuel-lean conditions are to the right.

A schematic of an LNB is presented in Figure 6.12. Fuel (e.g., pulverized coal) and air are injected through the central annulus of the burner. The ratio of air-to-fuel is less than stoichiometric,

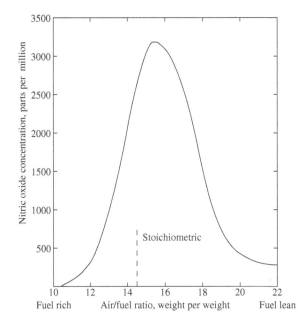

FIGURE 6.11    Nitric oxide concentrations in flue gas versus air-to-fuel mass ratio.

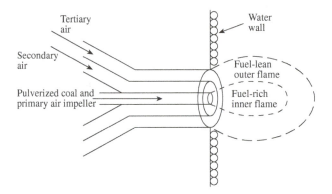

FIGURE 6.12    Low-$NO_x$ burner, schematic.

that is, fuel rich. This produces a luminous flame, with some of the pulverized coal left unburned, but also with low $NO_x$ formation, according to the left side of Figure 6.11. Secondary and tertiary air arrives through outer annuli, creating an outer flame envelope that is fuel lean. Here, all the unburned carbon burns up and less $NO_x$ is formed, according to the right side of Figure 6.11. The net result is complete burnout of the fuel and less $NO_x$ formation, circumventing the peak $NO_x$ formation indicated in Figure 6.11.

The retrofitting of boilers with LNB is relatively easy and inexpensive to accomplish. The incremental cost of electricity production using LNB is only 2–3%. In fact, most fossil-fueled power plants and industrial boilers in the United States and elsewhere in the world have installed LNBs. The problem is that LNB can reduce $NO_x$ formation only by 30–55% compared with regular burners. Because of the prevailing problems of acid deposition and high ozone concentrations in urban–industrial regions of continents, the pressure is mounting on operators of fossil-fueled power plants and industrial boilers to further reduce $NO_x$ emissions by more effective means than LNB.

### After Combustion.

*Selective catalytic reduction.*  In the selective catalytic reduction (SCR) process either ammonia or urea is injected into a catalytic reactor through which the flue gas flows. The following reaction takes place when ammonia is injected:

$$4NO + 4NH_3 + O_2 \rightarrow 4N_2 + 6H_2O. \qquad (6.11)$$

Thus, NO is reduced by ammonia, and ammonia is oxidized by NO and $O_2$ to form elemental nitrogen and $H_2O$. The catalyst is a mixture of titanium and vanadium oxides dispersed on a honeycomb structure. The SCR reactor is placed between the economizer and air preheater sections of the boiler, where the flue gas temperature is 300–400°C. The reaction is 80–90% complete; thus, 10–20% of the $NO_x$ escapes through the smokestack, and 10–20% of the unreacted ammonia also escapes. This is called *ammonia slip*. While ammonia is a toxic gas, by the time the flue gas plume disperses to the ground, its concentration is not considered harmful.

The catalyst is easily "poisoned," especially if the flue gas contains fly ash and sulfur oxides. (Note that the catalytic reactor is upstream from the ESP and scrubbers.) Therefore, the catalyst requires frequent replacement, which is one of the major cost elements in operating an SCR. The incremental cost of electricity production in a coal-fired power plant using SCR is about 5–10%.

*Selective noncatalytic reduction.*  A reduction of $NO_x$ can be accomplished at a higher temperature without a catalyst using the process of selective noncatalytic reduction (SNCR). In this process, urea is used instead of ammonia. An aqueous solution of urea is injected into the superheater section of the boiler, where the temperature is about 900–1000°C, which is sufficiently high for the reaction to proceed to near completion:

$$4NO + 4CO(NH_2)_2 + O_2 \rightarrow 4N_2 + 4CO_2 + 2H_2O. \qquad (6.12)$$

Many operators of electric power plants prefer this option because it needs no catalyst. However, urea is more expensive than ammonia. This option is also more amenable to retrofitting of existing plants because the urea injector can be mounted directly onto the wall of the boiler. For SCR a major reconstruction is necessary, as the catalytic reactor must be inserted between the economizer and air preheater sections of the boiler.

SNCR can be used in conjunction with low-$NO_x$ burners. The final reduction of $NO_x$ is in the 75–90% range, and the incremental cost of electricity production is 3–4.5%.

## 6.2.10.5  Mercury Control

In recent years mercury emissions from coal-fired power plants—as well from other industrial activities, such as municipal incinerators, medical and hazardous waste incinerators, and chlorine manufacturing plants—became a major health concern. High levels of mercury can have a toxic effect on the nervous systems of humans.[8] A part of the mercury emissions is deposited on land and surface waters. Fish and other aquatic organisms may bioaccumulate mercury and pass it on through the food chain to humans. Therefore, frequent advisories are posted by health organizations to limit the intake of fish and other seafood, especially by pregnant mothers and infants whose brain cells are still in development.

Coal-fired power plants are estimated to emit about one third of all mercury emissions. The U.S. Department of Energy estimated that in 1999, coal-fired power plants in the United States emitted about 48 t $y^{-1}$ of mercury. The U.S. EPA Clean Air Mercury Rule anticipates that mercury emissions from U.S. coal-fired power plants will gradually be reduced to 15 t $y^{-1}$.

**Mercury Control at Power Plants.**  One problem with mercury emission control at coal-fired power plants is that mercury can be emitted in two forms: elementary, gaseous mercury $Hg^0$ and oxidized mercury $Hg^{2+}$ (e.g., $HgO$ or $HgCl_2$). Some studies indicate that the emissions are equally divided between the two forms.[9] Different technologies have to be applied to capture the mercury in the two forms.

**Capture of Gaseous Mercury.**  Elemental, gaseous mercury $Hg^0$ can be captured by activated carbon injection (ACI). Dry, powdered activated carbon is injected into the flue gas duct between the air preheater and the electrostatic precipitator or baghouse, typically in the range of 120 to 180°C. The flue gas may have to be atempered by spray cooling before ACI injection. Gaseous mercury is adsorbed by the activated carbon, and the carbon particles are subsequently removed by the electrostatic precipitator or baghouse. Depending on the activated carbon quality and the characteristics of the electrostatic precipitator or baghouse, between 60 and 95% of the gaseous mercury can be removed with this technology.[10]

**Capture of Oxidized Mercury.**  Because oxidized mercury appears in the solid form, it usually is incorporated into fly ash particles. Therefore, commonly used particle control devices, such as electrostatic precipitators and baghouses can remove the majority of oxidized mercury. Baghouses appear to be superior than ESPs in removing particulate mercury. Between 74 and 86% removal of oxidized mercury has been achieved with baghouses.

---

[8]The term "mad hatter" is associated with the fact that mercury was used in leather tanning. The leather was used in hat manufacturing, and the workers often developed nervous disorders.

[9]Galbreath, K. C. and C. J. Zygarlicke. *Environ. Sci. Technology*, **30**, 2421–26, (1996).

[10]Durham, M. *Tools for Planning and Implementing Mercury Control Technology*, 43–46. Washington, D.C.: American Coal Council, 2003.

### 6.2.10.6   Toxic Metals

In addition to mercury, mineral matter (ash) of coal may contain several toxic metals, such as arsenic, selenium, cadmium, manganese, nickel, vanadium, and chromium. Arsenic and selenium are semivolatile and may escape from the smokestack in the vapor phase. The others are mostly found in the oxidized form adhering to the fly ash particles. Insofar as an efficient particle control system is used, especially a baghouse, the toxic metal emissions will be greatly reduced. However, from older power plants with less efficient particle control the oxides and salts of these metals are found in dry and wet deposition on land and water.[11]

### 6.2.11   Waste Disposal

Coal-fired power plants produce a significant amount of solid waste. Oil-fired plants produce much less waste and gas-fired plants practically none. We calculated before that a 1,000-MW plant fired with coal that contains 10% by weight mineral matter produces about $3.2E(5)$ t $y^{-1}$ of fly ash. If the coal contains 2% by weight of sulfur, and if that sulfur is removed by flue gas desulfurization using a wet limestone scrubber, another $3–4E(5)$ t $y^{-1}$ of wet sludge is created containing hydrated calcium sulfite, calcium sulfate, and unreacted limestone. While some plants succeed in selling, or at least giving away the fly ash for possible use as aggregate in concrete and asphalt or as road fill material, the scrubber sludge has practically no use. For some coals the fly ash and the scrubber sludge may contain toxic organic and inorganic compounds. In that case, the waste needs to be disposed of in a secure landfill. The landfill must be lined with impenetrable material, so that leaching into the soil and groundwater is to be prevented. Typical liner materials are natural clays or synthetic fabrics. Because transport costs can be substantial, the landfill area should be in the vicinity of the power plant. The mixed fly ash and sludge from a 1,000-MW coal fired power plant would add about 0.3 m $y^{-1}$ of sludge to a 4- to 8-hectare landfill. Thus, a new coal-fired power plant must be situated where a suitable landfill area is available.

## 6.3   ADVANCED CYCLES

The pulverized coal-fired single steam cycle power plant is the workhorse of the power industry worldwide. These plants normally provide the base load of the power grid (a) because they are cheaper to run than other fossil-fueled plants and (b) because it is not easy to adjust the firing rate to the variable diurnal load. As noted above, the worldwide average thermal efficiency of these plants is 33%, which means that two thirds of the thermal energy inherent in coal is not available for power production. Furthermore, these plants emit copious quantities of conventional pollutants—PM, CO, $SO_x$, $NO_x$, Hg, PAHs—and the global warming gas, $CO_2$. Therefore, the quest is to employ coal- and other fossil-fueled power plants that have a higher thermal efficiency and that emit fewer conventional pollutants and $CO_2$.

### 6.3.1   Combined Cycle

We described the thermodynamic principles of a combined cycle in Section 3.9.5. A power plant schematic using a GTCC is shown in Figure 6.13. In the first cycle—the *topping* cycle—a suitable

---

[11]Golomb, D., et al. *Atmospheric Environ.* 31, 1349–59, (1997).

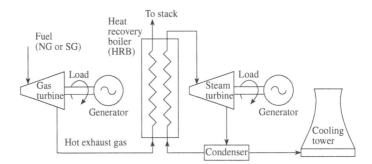

**FIGURE 6.13**   Gas turbine combined cycle power plant, schematic.

fluid fuel, usually natural gas, powers a gas turbine. The still-hot exhaust gas of the gas turbine passes through a heat exchanger, called a heat recovery boiler (HRB), and then to the power plant stack. In the HRB, feed water is boiled into steam, which powers a steam turbine—the *bottoming* cycle. Sometimes, more fuel (heat) is added to the gas turbine exhaust gas in a combustion chamber before the hot gases enter the HRB. The combination of the two cycles can achieve a thermal efficiency of 45–50%, an improvement over either a single cycle steam turbine power plant or a gas turbine plant.

The problem with combined cycle is that the primary fuel, natural gas, is more expensive per unit of heating value than coal. Also, gas reserves may not last as long as coal reserves (see Chapter 2). Combined cycle power plants are suitable where gas supplies are plentiful and cheap and where environmental regulations impose a heavy technical and financial burden on coal-fired power plants. They are especially attractive in urban environments because they require practically no fuel storage facilities (compressed gas arrives in pipes to the power plant), no particle removal system, and no scrubber for $SO_2$ removal. There is also no solid waste to dispose of. However, combined cycle power plants may require a $NO_x$ control system because in the gas turbine copious quantities of thermal $NO_x$ may be created. Of course, a condenser and cooling tower are also required for the steam portion of the combined cycle.

Combined cycle power plants are able to supply variable load demand and can complement the intermittent supply of most renewable energy systems such as wind and solar.

## 6.3.2   Coal Gasification Combined Cycle

The combined Brayton–Rankine cycle power plants described in the previous section run on natural gas. As natural gas resources may become scarcer and the price becomes higher, natural gas may be replaced by synthetic gas—syngas for short—which is a product of coal gasification. Coal gasification has also the advantage that most impurities can be removed in the gasification process, rather than postcombustion removal of impurities from the flue gas. Integrated coal gasification combined cycle power plants (IGCC) have been demonstrated to achieve a higher thermal efficiency, reckoned as electricity output per coal thermal input, than single-cycle pulverized coal-fired power plants. However, the increased capital and operating cost of IGCC may not compete economically with single-cycle pulverized coal-fired plants, even if the cost of postcombustion pollutant control systems is included. On the other hand, IGCC can be designed to capture $CO_2$ in the gasification process rather than emit it into the atmosphere as conventional pulverized coal-fired power plants do. IGCC with $CO_2$ capture will be described in Section 12.5.1.5.

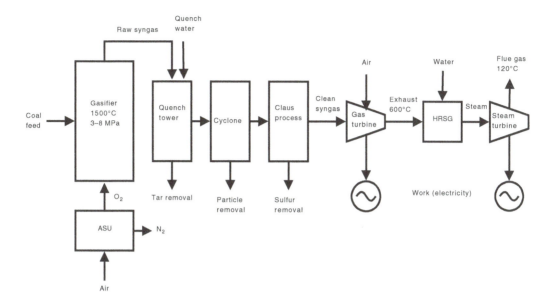

**FIGURE 6.14** Integrated coal gasification combined cycle process without $CO_2$ capture, schematic.

Various coal gasification methods were developed already in the 19th century for providing piped gas for home heating, cooking, and lighting. This was called *city gas* and was used before natural gas became widely available in the second half of the 20th century. In the world wars, Germany used gasified coal as a fuel for automobiles, trucks, and military vehicles.

A schematic of an IGCC is presented in Figure 6.14. The coal is supplied to the gasifier, where it is partially oxidized under pressure (3 to 8 MPa). The gasifier needs relatively pure oxygen supplied from an air separation unit (ASU). A 1,000-MW IGCC may need up to 20,000 t d$^{-1}$ of oxygen! The temperature in the gasifier may exceed 1,500°C. At these temperatures most of the ash is converted into molten slag, which is poured out from the gasifier vessel. The resulting syngas is quenched with water that removes a tar-like substance. The purified syngas enters a particle removal device, usually a cyclone. Because the syngas may contain a significant quantity of hydrogen sulfide, it needs to be removed from the syngas by a Claus or similar process (see Section 6.2.10.3).

The purified syngas is burned in a gas turbine in a Brayton (topping) cycle. The exhaust gas from the gas turbine is hot, about 600°C. The heat from the exhaust gas is used to boil water into steam in a heat recovery steam generator (HRSG). This steam is used to run a steam turbine in a Rankine (bottoming) cycle.

In the United States there is one operating IGCC—the Wabash River Power Station in Indiana. Construction started in 1993 for demonstration purposes with subsidies from the U.S. Department of Energy. Since 2000, it has been operating commercially. The gasifier uses technology initially developed by Dow Engineering Co. and now offered commercially by Global Energy Inc. as the E-GASTM technology. The combustion turbine generates 192 MWel and the steam turbine generates 104 MWel. With the system's parasitic load of 34 MWel (including oxygen production in the air separation unit), gross power production is 330 MWel, but average power delivery to the grid amounts to 262 MWel. Its thermal efficiency reckoned as net electricity output per coal heat input is 39.7%. Acid gas treatment removes over 99% of the sulfur after the gasification process.

Nitrogen oxide emissions are estimated at 64.5 g per GJ heat input compared with the U.S. $NO_x$ emission standard of 130 g per GJ heat input. Particle emissions are not detectable.

### 6.3.3  Cogeneration

Cogeneration is the term applied to systems that provide both electrical power and useful heat from the burning of fuel. In industrial or commercial installations the heat may be used for space heating or material processing. The incentive for cogeneration is primarily financial in that the cost of supplying electricity and heat via a cogeneration scheme might be less than supplying them separately, such as purchasing electric power from a supplier while generating heat from an in-plant furnace or boiler. Whether a cogeneration system reduces the amount of fuel needed to supply the electricity and heat depends upon the details of the cogeneration system. Also, emission controls may be more efficiently and cheaply affected on a large-scale electric power plant than on a smaller scale cogeneration plant unless the latter is fueled by natural gas, which inherently produces less pollutant emissions.

When a heat engine drives an electric generator to produce electricity, it also provides a stream of hot exhaust gas. When the exhaust gas is warm enough to be used for process or space heat, some of the exhaust gas enthalpy may be extracted to satisfy the heat requirement in a cogeneration plant. If $Q_{fuel}$ is the rate of fuel heat consumption needed to generate electric power $P_{el}$ and process heat $Q_{proc}$ in a cogeneration plant, then

$$Q_{fuel} = P_{el} + Q_{ex} \tag{6.13}$$

$$P_{el} = \eta_{th} Q_{fuel} \tag{6.14}$$

$$Q_{proc} = \eta_{xch} Q_{ex} = \eta_{xch}(1 - \eta_{th})Q_{fuel}, \tag{6.15}$$

where $Q_{ex}$ is the enthalpy flux of the exhaust gas, $\eta_{th}$ is the thermal efficiency of the electrical generation process, and $\eta_{xch} \leq 1$ is the fraction of the exhaust stream enthalpy that is delivered as heat for processing. The split of fuel heat $Q_{fuel}$ between electric power, $\eta_{th}Q_{fuel}$, and process heat, $\eta_{xch}(1 - \eta_{th})Q_{fuel}$, depends upon the thermal efficiency $\eta_{th}$ of the heat engine and the heat exchanger effectiveness $\eta_{xch}$. The latter depends principally on the temperature at which process heat is delivered compared with the temperature of the exhaust gas, being greatest when the difference is large and least when it is small. When process heat is needed at high temperatures, the usable process heat from a cogeneration system may be too small compared with the electric power generated to justify this complex system, and it may be more economical and fuel efficient to buy electricity from an efficient central power plant and generate process heat in an efficient in-plant boiler or furnace.

Cogeneration is most useful when process or space heat is required at a low temperature. Then $Q_{ex}$, instead of being dumped in a condenser and cooling tower, is heat exchanged with the device that needs the relatively low-temperature heat, say for drying or space heating. When a central power station (or incinerator) is located in a densely populated or commercial downtown area, $Q_{ex}$ may be piped directly into the buildings for space and water heating. This is called district heating.

### 6.3.4  Fuel Cell

The functioning of a fuel cell was described in Chapter 4. A fuel cell is not a heat engine. In a fuel cell some of the chemical energy of the fuel is directly converted into electrical energy, with the rest appearing as heat rejected to the environment. Its theoretical thermal efficiency in terms of electrical energy generated versus fuel chemical energy input can be very close to 100% when producing a low output of power. However, because of parasitic heat losses (e.g., ohmic resistance) and because of the power requirements of auxiliary equipment (e.g., pumps, fans), current fuel cells using natural gas or hydrogen and air (instead of oxygen) have a much lower thermal efficiency, in the 45–50% range. Furthermore, if hydrogen is used as a fuel, it has to be generated separately in some fashion, which requires energy. For example, hydrogen can be generated from water by electrolysis. But the splitting of water into hydrogen and oxygen requires about 18 MJ of electric energy per kilogram of water, more than can be generated in the fuel cell using the hydrogen from the electrolysis of water. Clearly, the electrolysis of water must use alternative energy sources, such as solar, wind, geothermal, biomass, or nuclear. At present, most, if not all hydrogen is derived from natural gas reforming or coal gasification.

## 6.4  CONCLUSION

Fossil-fueled power plants consume about 26% of the world's primary energy. Worldwide, about two thirds of the electrical energy is generated by fossil energy, 61% of which is coal, 21% natural gas, and 18% oil. Almost all fossil-fueled power plants work on the principle of a heat engine where 25–50% of the fossil energy input is converted into electrical energy, and 50–75% of the input energy is wasted in the form of heat rejection to a cold reservoir and parasitic heat losses. Almost all coal-fired power plants use pulverized coal that is injected pneumatically into the boiler. Coal-fired power plants usually provide the base load. Coal-fired plants utilize the Rankine single steam cycle. Oil- and gas-fired power plants usually provide the peak load, as they can be brought on-line more easily than coal-fired plants. Many of the oil- and gas-fired plants utilize the combined Brayton–Rankine cycles.

Fossil-fueled power plants, especially coal-fired plants, produce a host of pollutants, including particulate matter, sulfur and nitrogen oxides, and toxic organic and inorganic by-products of combustion. The emissions of these pollutants must be controlled to safeguard public health and the environment. Particulate matter with particle size greater than 1 to 2 $\mu$m is effectively (99+%) and relatively economically controlled by the ESP. For smaller particle control, a fabric filter, also called a baghouse, would be necessary but adds to the cost and power requirement to run it. Sulfur oxide emissions can be reduced by sorbent injection (about 50% emission reduction), by a dry scrubber (about 70% reduction), or by a wet scrubber (95+% reduction). Most of these methods rely on a chemical reaction between $SO_2$ and lime (CaO) or limestone ($CaCO_3$). Nitrogen oxide emission can be controlled by a low-$NO_x$ burner (about 30% emission reduction), selective noncatalytic reduction (together with low-$NO_x$ burner about 75% reduction), or selective catalytic reduction (about 90% reduction). Emission control devices subtract from the thermal efficiency of power plants and add to the electricity generating cost. Power plants also produce large amounts of solid and liquid waste, consume significant amounts of fresh water, and last, but not least, emit substantial quantities of $CO_2$, a greenhouse gas that contributes to global warming.

However, life in an urban-industrial society cannot be imagined without electricity. We only can strive toward increasing the thermal efficiency of fossil-fueled power plants, improving the emission control technologies, and gradual substitution of fossil-fueled power plants with solar, wind, and other renewable-energy power plants. With the looming crisis of global warming and the increased demand of electricity, nuclear-fueled power plants may also have to be utilized in greater proportions than is now the case.

# PROBLEMS

### Problem 6.1

A coal-fired power plant has a rated power of 1,000 MW(el). It works at 100% capacity (base loaded) at a thermal efficiency of 35%. It burns coal with a heating value of 30 MJ/kg and a sulfur content of 2% by weight, a nitrogen content of 1% by weight, and a mineral (ash) content of 10% by weight.

(a) How much electricity does the power plant produce per year (kWh/y)?

(b) How much coal does the power plant consume per year (t/y)?

(c) How much $SO_2$ does the power plant emit per year (t/y)? (Note that $SO_2$ has twice the molecular weight of S.)

(d) How much $NO_2$ does the power plant emit per year (t/y)? (Note that $NO_2$ has 3.28 times the molecular weight of N.)

(e) How much fly ash does the power plant emit per year (t/y)? Assume all mineral content is emitted as fly ash.

Tabulate the results of (a), (b), (c), (d), and (e).

### Problem 6.2

Using the sulfur and nitrogen content of coal as given in Problem 6.1, calculate the emission rate of $SO_2$ and $NO_2$ per unit heat input (g $SO_2$/GJ and g $NO_2$/GJ). Compare with U.S. emission standards for coal-fired large boilers: 516 g $SO_2$/GJ and 260 g $NO_2$/GJ.

### Problem 6.3

A 1,000-MW(el) power plant of 35% thermal efficiency and 90% average capacity factor uses coal with a heating value of 30 MJ/kg. The combustible part of coal has the chemical formula CH. The coal is burned with 20% excess air. Calculate the volumetric flow rate of the flue gas $Q$ in $m^3 s^{-1}$ at STP.

### Problem 6.4

How much air ($m^3$/s at 1 atm) needs to be pumped through the boiler of the plant in Problem 6.3 to burn the coal (chemical formula CH) with 20% excess air?

### Problem 6.5

The coal in the power plant of Problem 6.3 also contains 10% by weight mineral matter that is converted in the boiler to fly ash. Calculate the mass concentration of the fly ash in the flue gas ($g/m^3$).

## Problem 6.6

The collection efficiency $\eta$ of particles in an electrostatic precipitator is given by the Deutsch equation

$$\eta = 1 - \exp(-wA/Q),$$

where $A$ is the area of collection plates, $w$ is the drift velocity ($= 0.05\, d_p$ m/s, where $d_p$ is the particle diameter in micrometers), and $Q$ is the volumetric flow rate. (a) Calculate the collection efficiency for $d_p = 0.1, 0.3, 1$, and 3 $\mu$m, using $A = 75,500$ m$^2$ and $Q$ from Problem 6.3. Tabulate and plot these results. (b) Calculate the plate area necessary for 99% removal efficiency of 1-$\mu$m-diameter particles.

## Problem 6.7

The plant with the parameters given in Problem 6.1 has an electrostatic precipitator with the removal efficiencies and weight distribution in the given size ranges (shown in Table P6.7). Determine the rate of fly-ash emission (g/s). Plot collection efficiency vs. $d_p$.

TABLE P6.7

| Particle size ($\mu$m) | 0–5 | 5–10 | 10–20 | 20–40 | >40 |
|---|---|---|---|---|---|
| Removal efficiency (%) | 70 | 92.5 | 96 | 99 | 100 |
| Weight (%) | 14 | 17 | 21 | 23 | 25 |

## Problem 6.8

A coal has a heating value of 30 MJ/kg and the molecular composition $C_{100}H_{100}S_1N_{0.5}$. It is burned in air, without excess oxygen. (a) Calculate the emission rate of $SO_2$ and $NO_2$ in g/GJ. Compare this with the 1970 U.S. emission standards for large coal-fired boilers. (b) Calculate the mole fraction and volume fraction (ppmv) of $SO_2$ and $NO_2$ in the flue gas.

## Problem 6.9

The flue gas of the plant in Problem 6.8 is to be treated with flue gas desulfurization (FGD) using a limestone ($CaCO_3$) wet scrubber to remove the $SO_2$. Assume that 2 mol of $CaCO_3$ are necessary for every mol of $SO_2$. How much limestone is consumed per ton of coal?

## Problem 6.10

The flue gas of the plant in Problem 6.8 is to be treated with flue gas denitrification (FGN) using urea ($CO(NH_2)_2$) injection directly into the boiler. Assume that a molar ratio $NO_2:CO(NH_2)_2 = 1:1$ is necessary. How much urea is consumed per ton of coal?

## Problem 6.11

A 1,000-MWe1 pulverized coal steam plant operating at 35% thermal efficiency rejects one third of the coal heating value (30 MJ/kg) to the once-through seawater cooling system. At the discharge point the seawater can only be 5°C warmer than at the intake point. How much seawater needs to be pumped through the cooling system (m$^3$/s)?

# BIBLIOGRAPHY

Cooper, C. D., and F. C. Alley. *Air Pollution Control*, 3rd ed. Prospect Heights, IL: Waveland Press, 2002.

Decher, R. *Energy Conversion*. New York and Oxford: Oxford University Press, 1994.

El-Wakil, M. M. *Powerplant Technology*. New York: McGraw–Hill, 1984.

Wark, K., C. F. Warner, and W. T. Davis. *Air Pollution: Its Origin and Control*, 3rd ed. Menlo Park, CA: Addison–Wesley–Longman, 1998.

Weston, K. C. *Energy Conversion*. St. Paul, MN: West Publishing Co., 1992.

# Nuclear-Fueled Power Plants

## 7.1 INTRODUCTION

Nuclear-fueled power plants (NFPP) contribute a significant fraction to the world's electricity supply. In the United States, about 20% of the electricity is generated by NFPP; in France, about 80%; and in Japan, about 30%. In absolute terms, the United States produces the largest quantity of electricity from NFPP. In the United States there are more than 100 NFPP; worldwide there are more than 400.

The first commercial scale nuclear power plant of 180-MW capacity went into operation in 1956 at Calder Hall, England. In the United States, a 60-MW station started operating in 1957 at Shippingport, Pennsylvania. Previously, an experimental breeder reactor that produced electricity was demonstrated in 1951 near Detroit, Michigan. The first nuclear-powered submarine, the Nautilus, was launched in 1954. Submarine reactors produce steam that drives a turbine, which in turn propels the submarine.

The great advantage of NFPP is that they do not use fossil fuels. As described in Chapter 2, the supply of petroleum and natural gas is rapidly declining. The supply of coal may last longer, but deep-shaft mining carries considerable risk to the miners, and surface ("strip") mining leaves deep scars on the environment. All fossil-fueled power plants emit a host of pollutants into the atmosphere, including carbon dioxide—the principal greenhouse gas that causes global climate change.

However, NFPP are very controversial because of the real and perceived risks to human health, the environment, and nuclear weapons proliferation. The principal risk stems from radioactivity associated with the nuclear fuel cycle. Radiation already starts at the uranium mines and accompanies uranium fuel processing. The greatest radiation risk comes from nuclear waste disposal. The very source of nuclear energy is the fission of uranium, thorium, and plutonium isotopes. Almost all fission products of these isotopes are highly radioactive. No country has yet devised a safe and permanent method of radioactive waste disposal. While NFPP have a proven, impressive safety record over the 50 or so years of operation, we all remember the accidents at the Three Mile Island power plant in Pennsylvania in 1979 and at the Chernobyl, Ukraine, power plant in 1986. In response to the perceived risks, some countries are phasing out their NFPP and/or banning construction of new ones; however, other countries are introducing or expanding the use of nuclear energy for power production.

Currently, the capital investment for NFPP is much greater than that for fossil-fueled plants of equal electric power, even when the investment for pollution control equipment of these plants is included. The economic calculus may change, however, if fossil-fueled plants will be subjected to a carbon tax or are mandated to install carbon capture and sequestration equipment.

In this chapter we describe the fundamentals of NFPP, from fuel mining and processing to nuclear reactor operations and spent fuel reprocessing and disposal. The basic physical processes that are involved in liberating energy from the nucleus will be discussed.

## 7.2   NUCLEAR ENERGY

Nuclear energy is derived from the binding force (the "strong" force) that holds the nucleons[1] of the atomic nucleus together. The binding force per nucleon is greatest for elements in the middle of the periodic table and smallest for the lighter and heavier elements (see Figure 7.1). When lighter nuclei *fuse* together, heavier elements are created; therefore, the binding force per nucleon increases. Likewise, when heavier elements *fission*, lighter elements are created; therefore, the binding force per nucleon also increases. We shall see that in the process of fission and fusion there is a conversion of mass to energy. In a fusion or fission reaction, the mass of the products is not the same as the mass of the reactants, with the mass of the products being smaller than the mass of the reactants. This is called *mass deficit*. The mass deficit can be related to energy via the famous Einstein equation $E = mc^2$. The greater the mass deficit in a nuclear reaction, the greater the energy evolved in the reaction. Since all commercial nuclear power plants are based on fission, we shall discuss fission reactions first and relegate fusion reactions to the end of the chapter.

### 7.2.1   Nuclear Energy from Fission

When a nucleus of $^{235}U$ (an isotope of uranium) is bombarded with a *neutron*, it usually splits into two lighter elements. The mass of one of the elements centers around atomic mass number 90 and the other around 140. The exact split into the two elements (or their isotopes) is unpredictable; it is a probabilistic process. The fission products (i.e., isotopes) have an excess or deficiency of neutrons in comparison with naturally occurring elements (see Figure 7.2). These isotopes are highly unstable and transmute via several intermediates to the final stable elements with the emission of $\beta$ and $\gamma$ rays. Furthermore, the fission of the heavy element by bombardment with a neutron usually results in the release of two to three neutrons. The released neutrons can propagate further fissions of the heavy element in an exponential fashion. This is called a *chain reaction*.[2]

---

[1]For our purposes we shall consider as nucleons only positively charged protons and chargeless neutrons. Other particles have been observed in nuclear disintegration experiments, but they are not germane to our discussion.

[2]A chain reaction propagates exponentially because of the release of more than one reactant (in this case a neutron) per step than the one that initiated it.

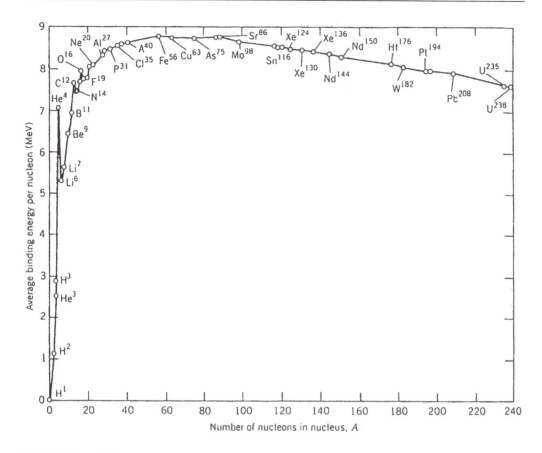

**FIGURE 7.1**  Binding energy per nucleon of the isotopes. Binding energies are maximum in the middle of the periodic table and less for the lighter and heavier isotopes.

For example, one of the fission reactions is the splitting of $^{235}$U into $^{144}$Ba and $^{89}$Kr, with the release of three neutrons plus 177 MeV of energy.[3]

$$^{235}\text{U} + {}^1\text{n} \rightarrow {}^{144}\text{Ba} + {}^{89}\text{Kr} + 3{}^1\text{n} + 177 \text{ MeV}, \qquad (7.1)$$

where $^1$n stands for a neutron. The large evolvement of energy in the fission reaction (7.1) is on account of the mass deficit. The sum of the masses of the products in reaction (7.1), $^{144}$Ba and $^{89}$Kr plus $3\,{}^1$n, does not equal the sum of the masses of the reactants, $^{235}$U plus $^1$n. The difference of the masses is the mass deficit, which is converted into energy. For the calculation of the mass deficit and the conversion into energy see Section 7.7.

---

[3]Electron volt (eV) = energy gained by an electron when accelerated through an electrical potential difference of one volt. 1 eV = 1.602E(−19) J; 1 MeV = 1.602E(−13) J.

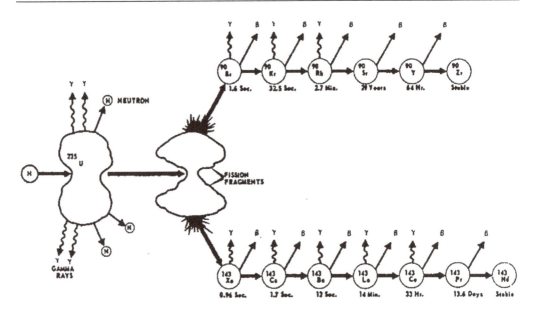

**FIGURE 7.2**    Uranium fission and possible fission products.

Reaction (7.1) is but one of the many fissions that can result when $^{235}$U is bombarded with a neutron. Other fission products include $^{131}$I, $^{90}$Sr, $^{137}$Cs, $^{90}$Rb, $^{90}$Y, $^{135}$Xe, and several isotopes of these elements.

Most of the fission products are radioactive (see Section 7.3 below). Because more than one neutron is released in the fission reaction, a chain reaction develops, with an increasing amount of energy released. The greater portion (about 80%) of the released energy is contained in the kinetic energy of the fission products, which manifests itself as sensible heat. A part of the remaining energy is immediately released in the form of $\gamma$ and $\beta$ rays. The rest of the fission energy is contained in delayed radioactivity of the fission products.

At the same time that $^{235}$U splits into fission products with the release of two to three neutrons per fission, a part of the neutrons can be absorbed by the more abundant $^{238}$U in the fuel, converting it in a series of reactions to an isotope of plutonium, $^{239}$Pu:

$$^{238}\text{U} + {}^{1}\text{n} \rightarrow {}^{239}\text{U} + \gamma \rightarrow {}^{239}\text{Np} + \beta \rightarrow {}^{239}\text{Pu} + \beta. \tag{7.2}$$

This reaction series is accompanied by $\gamma$ and $\beta$ radiation. $^{239}$Pu is a fissile element that can sustain a chain reaction. A *fissile* nucleus is one that can split after absorption of a thermal neutron. Examples are $^{235}$U, $^{239}$Pu, and $^{233}$U. A *fertile* nucleus is one that can convert into a fissile nucleus after absorption of a fast neutron. Examples of fertile nuclei are $^{238}$U and $^{232}$Th.

Because plutonium is a different element than uranium, it can be extracted chemically from the spent fuel and used as fresh fuel for reloading a reactor. The extracted plutonium can also be used in atomic bombs. Thus, reprocessing of spent fuel carries the risk of nuclear weapons proliferation. Reprocessing of spent fuel for the purpose of extracting plutonium is currently banned in the United States.

It must be emphasized that while the chain reaction in a nuclear power plant should be carefully controlled, a nuclear reactor cannot explode like an atomic bomb. Atomic bombs contain highly enriched $^{235}$U (95%+) or $^{239}$Pu, in contrast to nuclear reactors that contain a maximum of 3–4% enriched $^{235}$U or $^{239}$Pu. In an atomic bomb, the chain reaction proceeds so fast that all fissile isotopes are fissioned almost instantaneously, resulting in a powerful explosion. In a reactor, not all of the fissile isotopes can fission before there is a complete disintegration ("meltdown") of the reactor, as has occurred in the accidents at the Three Mile Island, Pennsylvania, power plant in 1979 and the Chernobyl, Ukraine, power plant in 1986.

## 7.3   RADIOACTIVITY

Radioactivity is the spontaneous decay of certain nuclei, usually the less stable isotopes of an element, both natural and manmade, which is accompanied by the release of very energetic radiation. After the emission of radiation, an isotope of the element is formed, or even a new element, which is usually more stable than the original element. We note that radiation emanates directly from the nucleus, not the atom as a whole. This is an important distinction, because X-ray radiation, although equally as damaging as radioactivity, emanates from the inner electronic shells of the atom, not the nucleus.

In radioactive decay, there are three types of radiation: $\alpha$, $\beta$, and $\gamma$. The latter is a form of electromagnetic radiation; the former are emissions of very high-energy particles. All three are called ionizing radiation because they create ions as their energy is absorbed by matter through which they travel. Ionizing radiation can travel only so far into matter, depending upon its energy and character.

**Alpha radiation.** In $\alpha$ radiation, a whole nucleus of a helium atom ($^{4}$He), containing two protons and two neutrons, is emitted. Because two protons of the original nucleus are lost (and consequently, two electrons must be lost from the electronic orbitals to maintain electric neutrality), the daughter isotope moves two elements backward from the parent element in the periodic table. For example, the isotope $^{239}$Pu disintegrates into the isotope $^{235}$U with the emission of an $\alpha$ particle:

$$^{239}\text{Pu} \rightarrow {}^{235}\text{U} + \alpha. \tag{7.3}$$

Another example is the disintegration of an isotope of radon $^{222}$Rn (which is a gas at normal temperatures) into polonium $^{218}$Po. The latter emits another $\alpha$ particle with the formation of a stable isotope of lead $^{214}$Pb:

$$^{222}\text{Rn} \rightarrow {}^{218}\text{Po} + \alpha \tag{7.4}$$

$$^{218}\text{Po} \rightarrow {}^{214}\text{Pb} + \alpha. \tag{7.5}$$

Because $\alpha$ particles are relatively heavy, their penetration depth into matter is very short, on the order of a millimeter. A sheet of paper or a layer of dead skin on a person can stop $\alpha$ radiation.

**Beta Radiation.** In $\beta$ radiation, an electron is emitted. This electron does not stem from the electronic orbitals surrounding the nucleus, but from the nucleus itself—a neutron converts into a proton with the emission of an electron. In that conversion, a parent isotope converts into a daughter isotope, which is an element forward in the periodic table because a proton has been added to the nucleus. An example is the decay of strontium $^{90}$Sr into yttrium $^{90}$Y:

$$^{90}\text{Sr} \rightarrow {}^{90}\text{Y} + \beta. \tag{7.6}$$

Because $^{90}$Sr is one of the products of uranium fission, this isotope is a major source of radiation from the spent fuel of a nuclear reactor. The emitted electron is relatively light; therefore, it can penetrate deeper into matter, on the order of centimeters. To shield against $\beta$ radiation, a plate of metal, such as lead, is necessary.

**Gamma Radiation.** Generally, the emission of $\beta$ radiation is followed by the emission of $\gamma$ radiation. Because the number of protons or neutrons does not change in that radiation, the radiating isotope does not change its position in the periodic table. Gamma radiation is essentially the emission of very shortwave, and hence energetic, electromagnetic radiation. An example is the emission of $\gamma$ radiation from an isotope of cobalt $^{60}$Co:

$$^{60}\text{Co} \rightarrow {}^{60}\text{Co} + \gamma. \tag{7.7}$$

The $\gamma$ radiation from $^{60}$Co finds uses in medicine, such as destroying cancer cells, and hence provides a therapy for certain types of cancer. Because $\gamma$ rays carry no mass, its penetration into matter is very deep, on the order of meters, and very heavy shielding is necessary to protect against it.

In addition to the above three types of radiation, there is the emission of neutrons. Neutrons are emitted as a consequence of nuclear fission. We already mentioned the fission of $^{235}$U with the emission of two to three neutrons per fission. These neutrons can enter other nuclei of $^{235}$U, which leads to a chain reaction. Also, neutron irradiation of other nuclei can lead to their destabilization, with the successive emission of $\beta$ or $\gamma$ radiation.

Radioactivity accompanies the whole nuclear power plant fuel cycle, from mining of uranium ore through uranium extraction, isotope enrichment, fuel preparation, fuel loading, reactor operations, accidents, plant decommissioning, and last but not least, spent fuel disposal.

## 7.3.1    Decay Rates and Half-Lives

The decay rate of an ensemble of radioactive nuclei, to which the diminishing intensity of radioactivity is proportional, is governed by the law of exponential decline,

$$-dN/dt = kN, \tag{7.8}$$

where $N$ is the number of decaying nuclei, or their mass, present at time $t$, and $k$ is the radioactive decay constant in units of $t^{-1}$. Integration yields

$$N = N_0 \exp(-kt), \tag{7.9}$$

where $N_0$ is the number of nuclei, or their mass, at the start of counting time. The time after which the number of decaying nuclei is halved is called the half-life $t_{1/2}$.

$$t_{1/2} = \frac{\ln 2}{k}. \tag{7.10}$$

**TABLE 7.1**   Some Isotopes in the Nuclear Fuel Cycle, with Half-Lives and Radiation

| Isotope | $t_{1/2}$ | Activity |
|---------|-----------|----------|
| Krypton-87 | 76 min | $\beta$ |
| Tritium ($^3$H) | 12.3 y | $\beta$ |
| Cesium-137 | 30.2 y | $\beta$ |
| Xenon-135 | 9.2 h | $\beta$ and $\gamma$ |
| Barium-139 | 82.9 min | $\beta$ and $\gamma$ |
| Radium-223 | 11.4 d | $\alpha$ and $\gamma$ |
| Radium-226 | 1600 y | $\alpha$ and $\gamma$ |
| Thorium-232 | 1.4E(10) y | $\alpha$ and $\gamma$ |
| Thorium-233 | 22.1 min | $\beta$ |
| Uranium-233 | 1.65E(5) y | $\alpha$ and $\gamma$ |
| Uranium-235 | 7.1E(8) y | $\alpha$ and $\gamma$ |
| Uranium-238 | 4.5E(9) y | $\alpha$ and $\gamma$ |
| Neptunium-239 | 2.35 d | $\beta$ and $\gamma$ |
| Plutonium-239 | 2.44E(4) y | $\alpha$ and $\gamma$ |

Some radioactive nuclei decay very fast—their half-lives are measured in seconds; others decay slowly: their half-lives can be days, years, or even centuries. The radioactive decay constant has great importance in regard to radioactive waste disposal. For example, spent fuel of a nuclear power plant contains many radioactive isotopes, such as $^{90}$Sr (half-life 28.1 y), $^{137}$Cs (half-life 30.2 y), and $^{129}$I (half-life 1.7E(7) y). $^{90}$Sr and $^{137}$Cs will decay to small amounts in hundreds of years, but $^{129}$I will stay around practically forever.[4] Table 7.1 lists some radioactive isotopes which play a role in the fuel cycle of nuclear power plants, their radiation type and half-life.

## 7.3.2   Units and Dosage

The level of radioactivity of a sample of substance is measured by the number of disintegrations per second. The SI unit of radioactivity is the becquerel (Bq), which is one disintegration per second (see Table A.1). A more practical unit of measurement of radioactivity is the curie (Ci).

$$1 \text{ Ci} = 3.7\text{E}(10) \text{ Bq}. \tag{7.11}$$

The radioactivity of 1 gram of radium-226 is 1 Ci and that of cobalt-60 is 1 kCi.[5] But for mixtures of radioactive isotopes, such as is found in ore samples or spent reactor fuel, the radioactive level measured in Ci cannot tell us the amount or composition of the radioactive components, only their total disintegration rate.

---

[4]With a half-life of 28.1 years, $^{90}$Sr will decay to 1% of its initial value in 187 years; $^{137}$Cs with a half-life of 30.2 years will decay to 1% in 201 years.

[5]The specific radioactivity level of a sample is obtained by dividing the radioactivity level by the mass or volume of the sample (e.g., Ci/kg or Ci/L).

Because exposure of humans to $\alpha$, $\beta$, and $\gamma$ radiation can be harmful, we need practical units of measurement of exposure to them. The SI unit of absorbed dose of radiation is the gray (Gy), which equals 1 joule of absorbed energy per kilogram of matter penetrated by the radiation. Another commonly used unit of absorbed dose is the rad, which equals 1E(-2) Gy.[6]

The absorbed energy is not entirely satisfactory as a measure of the harmfulness to human populations of ionizing radiation, as other qualities of the radiation are also important. To take these into account, a different unit, the sievert (Sv), is used to measure what is called the dose equivalent. Like the Gy, it has the dimensions of J/kg. The Sv takes into account the quality of the absorbed radiation. An equivalent dose of 1 Sv is received when the actual dose of radiation (measured in grays), after being multiplied by the dimensionless factors $Q$ (the so-called quality factor) and $N$ (the product of any other multiplying factors), is 1 J/kg. $Q$ depends on the nature of radiation; it has a value of 1 for X-rays, $\gamma$-rays, and $\beta$ particles, a value of 10 for neutrons, and 20 for $\alpha$ particles. $N$ is a factor that takes into account the distribution of energy throughout the dose. An alternative unit of dose equivalent is the rem, defined as 1E(-2) Sv. An accumulation of absorbed $\alpha$, $\beta$, and $\gamma$ radiation over time is called a radiation dosage. For example, the average person in the United States receives in 1 year a dosage of about 360 millirems (3.6 mSv), of which 200 is from radon-86, 27 from cosmic rays, 28 from rocks and soil, 40 from radioactive isotopes in the body, 39 from X-rays, 14 from nuclear medicine, and 10 from consumer products and other minor sources.

The average person on earth receives about 2.2 mSv y$^{-1}$. A short-term dose of 1 Sv causes temporary radiation sickness; 10 Sv is fatal. After the Chernobyl accident in the Ukraine, the average dose received by people living in the affected areas surrounding the plant over a 10-year period, 1986–1995, was 6–60 mSv. The 31 radiation fatalities at Chernobyl appear to have received more than 5 Sv in a few days; those suffering acute radiation sickness averaged 3–4 Sv.

### 7.3.2.1  Health Effects of Radiation

The greatest risk to humans of nuclear power plant operations is associated with radioactivity, also called ionizing radiation, because of the creation of ions left by the passage of $\alpha$-, $\beta$-, and $\gamma$-rays and neutrons. Radioactivity affects humans and animals, causing somatic and genetic effects.[7] Acute effects ensue when an organism is subjected to large doses of radiation over a short period; chronic effects ensue when the exposure is at low levels, but over a protracted period. Acute effects include vomiting, hemorrhage, increased susceptibility to infection, burns, hair loss, blood changes, and ultimately, death. Chronic effects, which usually manifest themselves over many years, include eye cataracts and the induction of various types of cancer, such as leukemia, thyroid cancer, skin cancer, and breast cancer. Genetic effects may become apparent in later generations but not in the exposed person. These effects are the result of mutations in the genetic material (e.g., chromosome abnormalities or changes in the individual's genes that make up the chromosomes).

### 7.3.2.2  Radiation Protection Standards

The prescription of radiation protection standards is an onerous and controversial task. In the United States, this task was vested in the Committee on the Biological Effects of Ionizing Radiation of the National Academy of Sciences and internationally in the United Nations Scientific Committee on

---

[6]In the treatment of cancer with radiation, the absorbed dose needed to kill cancer cells is of the order of 100 Gy.

[7]Somatic effects pertain to all cells in the body; genetic effects pertain to egg and sperm cells.

the Effects of Atomic Radiation. The task is difficult because direct evidence on biological effects of radiation comes—unfortunately—from high-level exposures, such as received by the population of Hiroshima and Nagasaki during atomic bombing and by the workers at the Chernobyl accident. Lower-level exposure data can only be obtained from animal studies extrapolated to humans. However, even in animals, the effects of low-level exposure can only be established statistically, observing a large cohort of animals over lengthy periods and over many generations.

In setting radiation standards, the following three assumptions are made:

(a) There is no threshold dose below which radiation has no effect.

(b) The incidence of any delayed somatic effect is directly proportional to the total dose received.

(c) There is no dose–rate effect.

Basically, these assumptions constitute the linear-no-threshold (LNT) hypothesis. The assumptions mean that even the slightest radiation could lead to delayed somatic or genetic effects, and the occurrence of delayed effects does not depend on the stochastic nature of radiation or whether the given dose is received over a short or extended time.

Based on the LNT hypothesis, the Nuclear Regulatory Commission in the United States set a standard of exposure for workers in nuclear power plants to 50 mSv y$^{-1}$ ($= 5$ rem y$^{-1}$). For the general population, that is, any person in the region outside the plant boundary, the standard is 1 mSv y$^{-1}$ $= 100$ mrem y$^{-1}$.

## 7.4  NUCLEAR REACTORS

The first controlled nuclear reactor was built and demonstrated by Enrico Fermi in 1942. It was constructed under the bleachers of the stadium at the University of Chicago. The reactor had dimensions 9 m wide, 9.5 m long, and 6 m high. It contained about 52 tons of natural uranium and about 1,350 tons of graphite as a moderator. Cadmium rods were used as a control device. The reactor produced an output of only 200 W and lasted only a few minutes. Fermi's "pile" ushered in the nuclear age.

A nuclear reactor in a nuclear-fueled power plant is a pressure vessel enclosing the nuclear fuel that undergoes a chain reaction, generating heat that is transferred to a fluid, usually water that is pumped through the vessel. The heated fluid can be steam, which then flows through a turbine generating electric power, or it can be hot water, a gas, or liquid metal that generates steam in a heat exchanger. The steam drives a turbine. We have discussed before that the energy of a nuclear reactor is derived from splitting a fissile heavy nucleus, such as $^{235}$U or $^{239}$Pu. In a nuclear reactor of a power plant, the splitting of the nucleus and sustaining of the ensuing chain reaction has to proceed in a controlled fashion.

The basic ingredients of a nuclear reactor are *fuel rods*, a *moderator*, *control rods*, and a *coolant*. A schematic of a reactor is depicted in Figure 7.3.

**Fuel Rods.** The fuel rods contain the fissile isotopes $^{235}$U and/or $^{239}$Pu. Natural uranium contains about 99.3% $^{238}$U and 0.7% $^{235}$U. In most power plant reactors, the concentration of the fissile isotope $^{235}$U in natural uranium is not enough to sustain a chain reaction; therefore, this isotope

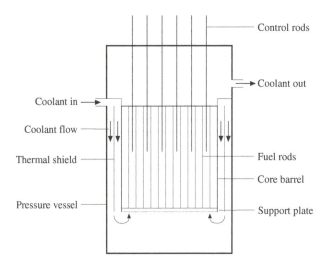

**FIGURE 7.3**    Schematic of nuclear reactor with water as coolant/moderator.

needs to be "enriched" to 3–4% (for enrichment processes, see Section 7.5).[8] The fuel rods contain metallic uranium, solid uranium dioxide ($UO_2$), or a mix of uranium dioxide and plutonium oxide, called MOX, fabricated into ceramic pellets. The pellets are loaded into zircalloy or stainless steel tubes, about 1 cm diameter and up to 4 m long.

**Moderator.**  Moderators are used to slow the energetic neutrons that evolve from the fission reaction, yielding low-energy neutrons, also called thermal neutrons. This increases the probability for the neutrons to be absorbed in another fissile nucleus, so that the chain reaction can be propagated. Moderators contain atoms or molecules whose nuclei have high neutron scattering and low neutron absorption characteristics. Typical moderators are light water ($H_2O$), heavy water ($D_2O$), graphite (C), and beryllium (Be). The light- or heavy-water moderators circulate around the fuel rods. Graphite or beryllium moderators constitute a block into which fuel rods are inserted. For example, the original Fermi "pile" consisted of a graphite block into which metallic uranium fuel was inserted. The Chernobyl-type reactors also use graphite as a moderator.

**Control Rods.**  Control rods contain elements whose nuclei have a high probability of absorbing thermal neutrons, so that they are not available for further splitting of fissile nuclei. In the presence of control rods, the chain reaction is controlled or stopped altogether. Typical control rods are made of boron (B) or cadmium (Cd).

The chain reaction inside the reactor is governed by the *neutron economy coefficient k*. Under a steady state, the number of thermal neutrons is invariant with time, $dn/dt = 0$, and $k = 1$. The reactor is then in a critical condition. When $k < 1$, the reactor is subcritical; when $k > 1$, it is supercritical. A nuclear reactor becomes critical when control rods are lifted out of the core of the

---

[8]Some power plant reactors use natural uranium containing 0.7% $^{235}$U as the fuel with different moderator and coolant combinations.

reactor to a degree when more than one neutron released by the fission of a fissile nucleus survives without being absorbed by the control rods. The position of the control rods determines the power output of the reactor. Monitoring the critical condition in a nuclear reactor while varying the output is quite complicated. Generally, nuclear power plants are run at full load, providing the base load of a grid. Running the plant at full load is also more economical.

Once the remaining fuel in the rods cannot sustain the rated capacity of the plant, even with complete withdrawal of the control rods, the fuel rods need to be replaced. This occurs every 2 to 3 years.

**Coolant.** Heat must be constantly removed from the reactor. Heat is generated not only by the fission reaction, but also by the radioactive decay of the fission products. Heat is removed by a coolant, which can be boiling water, pressurized water, a molten metal (e.g., liquid sodium), or a gas (e.g., He or $CO_2$). The accident at the Three Mile Island power plant near Harrisburg, Pennsylvania, in 1979 occurred because after shutdown (full insertion of the control rods), the reactor was completely drained of its coolant, so that the residual radioactivity in the fuel rods caused a meltdown of the reactor. The heat removed by the coolant, in the form of steam or pressurized hot water, is used in conventional thermodynamic cycles to produce mechanical and electrical energy. In addition to the control rods position, the mass flow of the coolant also determines the plant's power output.

## 7.4.1 Boiling Water Reactor

About one quarter of U.S. nuclear power plants are of the *boiling water reactor* (BWR) kind. The schematic of a BWR is depicted in Figure 7.4; the reactor assembly itself is depicted in Figure 7.5. Most BWRs use 3–4% enriched $^{235}U$ as fuel. Light water serves both as a coolant and moderator.

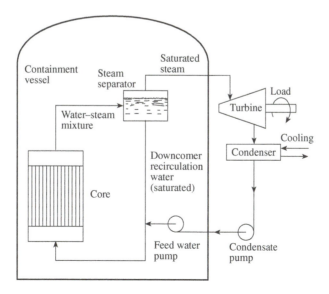

**FIGURE 7.4**   Schematic of a boiling water reactor power plant.

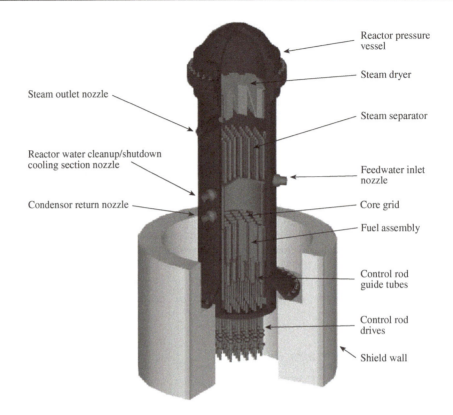

**FIGURE 7.5**   Cutaway of a boiling water reactor.

As the control rods are withdrawn, the chain reaction starts, and the coolant-moderator water boils. The saturated steam has a temperature of about 300°C and pressure of 7 MPa. After separation of condensate in the steam separator, the steam drives the turbine that drives the generator. After expansion in the turbine, the steam is condensed in the condenser and returned via the feed-water pump to the core.

The direct cycle has the advantage of simplicity and relatively high thermal efficiency, as the steam generated in the core directly drives a turbine without further heat exchangers. The thermal efficiency of a BWR is on the order of 33%, reckoned on the basis of the inherent energy of the nuclear fuel that has been consumed in power production.

In a BWR there is no need for an extra moderator, as the coolant light water slows down the fast neutrons to thermal velocities, which can engage in further fission reactions. Another advantage of a BWR is the fact that it is self-controlling. When the chain reaction becomes too intense, the coolant water boils faster. Because of its low density, steam has little or no moderating propensity. Thus, the reduction of the liquid content of the coolant water automatically slows the chain reaction.

In a BWR, the core is enclosed in the primary containment vessel made of steel and surrounded by reinforced concrete. A secondary containment vessel made of prestressed concrete (the dome-shape building visible at nuclear power plants) contains the steam separator and the spent fuel storage pool. The steam turbine, condenser, and electric generator are located outside of the secondary containment vessel. Even though the coolant water is demineralized, some radioactive material may leach from the core into the coolant water and be transferred by the steam to the steam turbines. Furthermore, the coolant water may contain traces of mildly radioactive isotopes of hydrogen (tritium, $^3H$) and nitrogen ($^{16}N$ and $^{17}N$). The steam remaining in contact with the turbine and other equipment loses its radioactivity quite fast, and with proper precaution no significant exposure is presented to workers in the power plant or the population outside the plant.

## 7.4.2 Pressurized Water Reactor

The majority of nuclear power plants in the United States and worldwide are of the *pressurized water reactor* (PWR) type. A schematic of a PWR is depicted in Figure 7.6. In the primary loop surrounding the reactor core, the coolant water is kept at a high pressure of about 15 MPa, so that the water does not boil into steam even though the temperature is in the 340–350°C range. At that temperature, the pressure of the water exceeds the vapor pressure, so there is only liquid phase. The hot, nonboiling water is pumped into a heat exchanger, which is located together with the reactor core inside a heavy steel and reinforced concrete containment vessel. In the heat exchanger, feedwater is boiling into steam at about 7 MPa in a secondary loop. The steam drives a turbine that drives the electric generator in a conventional Rankine cycle. The steam turbine and condenser are located outside of the containment vessel.

One advantage of the PWR is that because of the single phase of the coolant water, the moderating capacity of the (light) water can be precisely adjusted, unlike in a BWR, where the coolant is in two phases, liquid and vapor. Usually, boron in the form of boric acid is added to the coolant to increase the moderating capacity. In such a fashion, fewer control rods are necessary to maintain the reactor at the design capacity. The other advantage is that the steam that is generated in the heat exchanger never comes into direct contact with the coolant water. Thus, any radioactivity

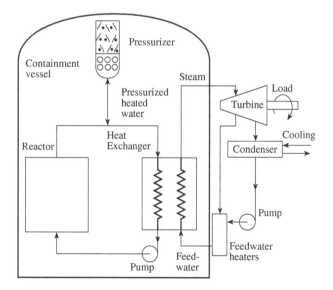

**FIGURE 7.6**   Schematic of a pressurized water reactor power plant.

that may be present in the coolant is confined to the primary loop inside the containment vessel. Because of the heat exchanger, the overall thermal efficiency of a PWR is somewhat lower than that of a BWR, on the order of 30%.

A special type of PWR is the Canadian Deuterium Uranium (CANDU) reactor. This type of reactor is also used in Argentina, India, Pakistan, and Korea. CANDU uses natural uranium fuel (without enrichment) and heavy water ($D_2O$) as the moderator. Heavy water absorbs practically no neutrons; thus, its neutron economy is superior to that of light water. On the other hand, its moderating capacity is less than that of light water; therefore, neutrons have to travel twice as far to be slowed down (become thermal) as in light water. The fissile material is $^{235}U$ as in enriched uranium reactors, but some of the fertile $^{238}U$ is converted into fissile $^{239}Pu$, which can participate in the chain reaction or be extracted for fuel recycling or for weapons production. In the CANDU-type reactor the fuel rods are contained in pressure tubes with circulating pressurized water. A schematic of the CANDU-type reactor is depicted in Fig. 7.7.

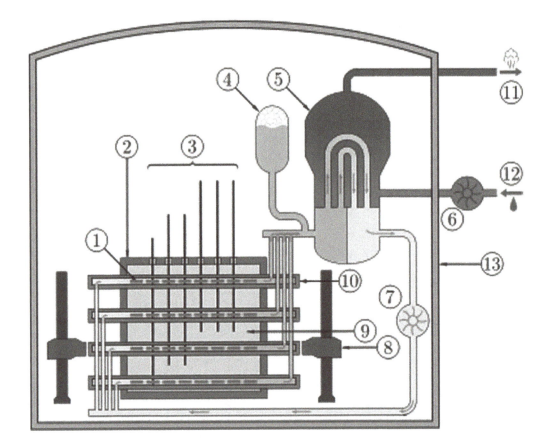

**FIGURE 7.7** Schematic of a CANDU reactor: (1) fuel rods, (2) reactor core, (3) control rods, (4) heavy-water pressure reservoir, (5) steam generator, (6) light-water pump, (7) heavy-water pump, (8) refueling machines (9) heavy-water moderator, (10) pressure tube, (11) steam to turbine, (12) cold water return from turbine, and (13) containment vessel. (Source: Atomic Energy of Canada, Ltd.)

Instead of uranium enrichment facilities, the CANDU-type reactor requires heavy-water production stills. Depending on which facilities a country possesses, the economic incentives of a particular government, and the wish of a country to produce weapons-grade plutonium, it may be more advantageous to operate a natural uranium/heavy-water reactor than to operate an enriched uranium/light-water reactor. In the United States, because of the availability of enriched uranium, no heavy-water power plants are operating.

### 7.4.3  Gas-Cooled Reactor

A reactor type that was developed in particular in Great Britain is the *gas-cooled reactor* (GCR). In fact, the first commercial power plant put into operation in 1956 at Calder Hall was a GCR. Generally, these reactors are fueled by natural or enriched uranium, either metallic or ceramic uranium oxide. The moderator is graphite, and as the name implies, the coolant is a gas, normally $CO_2$, but helium can also be used. Because of the lower heat transfer capacity of gases compared with liquids, the contact surfaces and flow passages in the reactor must be larger than in liquid-cooled reactors. To obtain a reasonable thermal efficiency, GCRs are run at higher temperatures than PWRs or BWRs. This necessitates cladding and piping materials that can withstand the higher temperatures. Some GCRs are using enriched uranium to boost the thermal efficiency.

An example of a high-temperature gas-cooled ($CO_2$) reactor is operating at Hinkley Point in the United Kingdom. Its net output is 1,250 MW. The fuel is $UO_2$ with 2.6% $^{235}U$. The $CO_2$ leaves the reactor at 655°C and a pressure of 4.3 MPa. The $CO_2$ is pumped to a heat exchanger where steam is generated at 540°C and 17 MPa. The plant achieves a thermal efficiency of close to 42%.

In the United States, a 40-MW high-temperature gas-cooled power plant was constructed and operated near Philadelphia, Pennsylvania. It is now decommissioned. Another 330-MW plant was operated near Platteville, Colorado. While providing interesting experience, the plant had many engineering problems and is by now also decommissioned. Research and development on GCRs is continuing, and their revival may occur in the future.

### 7.4.4  Breeder Reactor

In a *breeder reactor*, fissile nuclei are produced from fertile nuclei. The principal breeder mechanism is the conversion of $^{238}U$ to $^{239}Pu$ as was shown in Equation (7.2). The intermediary $^{239}U$ has a half-life of 23 minutes, which converts to neptunium $^{239}Np$ with a half-life 2.4 days, which in turn decays to $^{239}Pu$, with a half-life of 24,000 years. The $^{239}Pu$ formed, while a fissile nucleus, does not participate to a significant extent in the chain reaction, but accumulates in the spent fuel from which it is later extracted and reused.

Unlike $^{235}U$, which efficiently fissions with slow thermal electrons whose energy is in the tenths of eV range, $^{238}U$ captures efficiently fast neutrons in the MeV range. To obtain this wide spectrum of neutron energies, a different coolant/moderator combination is required than light or heavy water. The preferred coolant is liquid sodium, and such a reactor is called the *liquid-metal fast-breeder reactor* (LMFBR). The sodium nucleus has a larger mass than hydrogen or deuterium; therefore, a neutron colliding with a sodium nucleus bounces off with nearly its original momentum, whereas a neutron colliding with a hydrogen nucleus (e.g., a proton) imparts nearly half of its momentum to hydrogen.

The efficiency of fuel utilization in a breeder reactor is expressed as the ratio of the number of fissile nuclei formed to the number destroyed. This ratio is called *breeding ratio* (BR):

$$BR = \frac{\text{Number of fissile nuclei produced}}{\text{Number of fissile nuclei destroyed}}. \tag{7.12}$$

When BR is greater than 1, breeding occurs. Note that both in the numerator and in the denominator of (7.12) fissile nuclei include not only $^{235}U$, but also $^{233}U$ and initially present $^{239}Pu$. Breeder reactors could also use thorium as a fuel. $^{232}Th$ is a fertile nucleus that can be converted into fissile $^{233}U$ by the reaction sequence

$$^{232}Th + n + \gamma \rightarrow {}^{233}Th + \beta \rightarrow {}^{233}Pa + \beta \rightarrow {}^{233}U. \tag{7.13}$$

Here $^{233}Pa$ is an isotope of element 91, protactinium. Since worldwide thorium ores are about as abundant as uranium ores, the use of thorium would extend the nuclear fuel resources by about a factor of 2. So far, thorium-based breeder reactors are only in the planning and development phase.

A major problem with breeder reactors is the need to use liquid sodium as a coolant. (A helium-cooled breeder reactor was designed by Gulf General Atomic Corporation in the 1960s, but was not constructed.) In addition to the appropriate moderating characteristics of sodium, it also has excellent heat transfer capacity. Sodium melts at 90°C and boils at 882°C. This allows the reactor to run hotter and, consequently, a higher thermal efficiency is obtained. But sodium is a nasty chemical. It burns spontaneously in air and reacts violently with water. Furthermore, $^{23}Na$ (the natural isotope) can absorb a neutron to convert first into $^{24}Na$ and then into stable $^{24}Mg$. The intermediary $^{24}Na$ emits very energetic $\beta$ and $\gamma$ radiation. Therefore, more than one sodium loop is required, a separate loop for the sodium coolant circulation and another loop for the water/steam cycle. A schematic of a LMFBR with two sodium loops is presented in Figure 7.8. The reactor

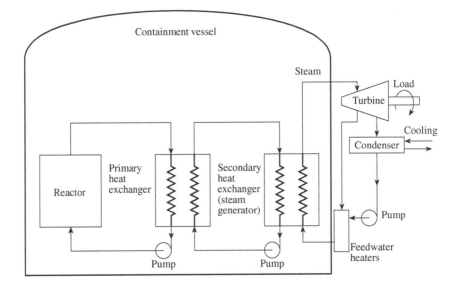

**FIGURE 7.8**   Schematic of a liquid-metal fast-breeder reactor power plant.

core, sodium loops, heat exchangers, and steam loops are all located in a containment vessel; the steam turbine, condenser, and the rest of the generating plant are outside of the vessel.

Several fast breeder reactors were operated in the United States, the United Kingdom, France, Germany, India, Japan, and Russia. The largest (400-MW) breeder reactor for the production of electricity in the United States was to be constructed at Clinch River, Tennessee, but construction was terminated in 1983 when the U.S. Senate voted 56 against 40 to deny any further funding for the project because of escalating costs and concern about nuclear proliferation. Breeder reactors still in operation are the 250-MW Phenix in France (the 1,240-MW Superphenix was shut down in 1998), the 40-MW FBTR in India, the 100-MW Joyo in Japan, the 135-MW BN 350 formerly in the USSR but now in Kazakhstan, and the 10-MW BR 10, 12-MW BOR 60, and the 600-MW BN 600 in Russia.

Most breeder reactors, with the exception of those in Japan, were operated for the primary purpose of producing weapons-grade plutonium. This is certainly one of the drawbacks of breeder-type power plants. While they produce more fissile fuel than they consume in the process of power production, the fissile fuel, plutonium, can be fairly easily extracted to make atomic bombs. Strict international supervision will be necessary if in the future breeder-type power plants will commonly be used for electric power production.

## 7.5   NUCLEAR FUEL CYCLE

The nuclear fuel cycle starts from mining uranium (or thorium) ore, through extraction of the useful uranium concentrate, gasification to $UF_6$, enrichment of $^{235}U$, conversion to metallic uranium or oxide of uranium, fuel rod fabrication, loading of a reactor, retrieval of spent fuel, reprocessing of spent fuel, and finally fuel waste disposal. A block diagram of the nuclear fuel cycle is presented in Figure 7.9.

### 7.5.1   Mining and Refining

Uranium ores containing variable concentrations of uranium are found in many parts of the world. Rich ores may contain up to 2% uranium by weight, medium-grade ores 0.5–1%, and low-grade ores less than 0.5%. In the United States, ores are found in Wyoming, Texas, Colorado, New Mexico, and Utah. Large deposits are found in Australia, Kazakhstan, Canada, South Africa, Namibia, Brazil, and Russia.

The most economic way of extracting the ore is from open surface mines. Since uranium deposits are always associated with decay products (daughters) of uranium, such as radium and radon, these ores can be radioactive, and workers' protection must start at the mining phase. Open pit mines are well ventilated, so most radiation, especially that associated with radon, which is a gas, escapes into the atmosphere. However, masks must be worn to prevent inhalation of mining dust, which can contain radioactive elements.

The ore is crushed and ground. The ground ore is leached with sulfuric acid. Uranium, together with some other metals, dissolves. Uranium oxide with the approximate composition $U_3O_8$, called *yellow cake*, is precipitated, dried, and packed into 200-liter drums for shipment. The radiation from these drums is negligible. However, the solids remaining after leaching with acid may contain radioactive isotopes of radium, bismuth, and lead. These solids are pumped as a slurry to the tailings heap, also called tailings dam. The tailings must be covered with clay or other impenetrable material to protect humans and animals from radiation exposure.

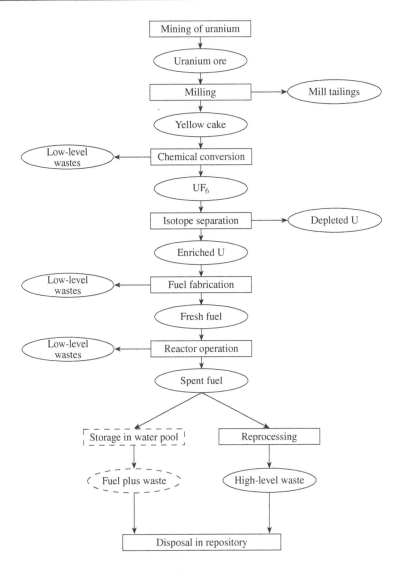

**FIGURE 7.9**   Nuclear fuel cycle.

## 7.5.2  Gasification and Enrichment

The $U_3O_8$ concentrate is shipped from the mines to the enrichment facilities. The concentrate has the normal isotope distribution, about 99.3% $^{238}$U and 0.7% $^{235}$U. With the exception of CANDU-type heavy-water moderated reactors, all other reactors need uranium enriched with $^{235}$U.

For enrichment, a gaseous uranium compound is necessary. The concentrate is treated with hydrogen fluoride gas. In such a fashion, uranium oxide is converted to uranium hexafluoride $UF_6$. This is a white solid that sublimates to a vapor at a pressure of 1 atm at 56°C. The gaseous $UF_6$ is enriched by one of the following processes:

**Gaseous Diffusion.** Membranes with small pores let gases diffuse through them. The rate of diffusion is a function of pressure, temperature, and the molecular mass of the diffusing gas. $^{235}UF_6$ has a slightly smaller mass than $^{238}UF_6$. Because of the smaller mass, $^{235}UF_6$ diffuses somewhat faster through the pores than $^{238}UF_6$. In one passage, the enrichment factor is very small, but by forcing the gases to pass many membranes ("stages"), almost any degree of enrichment can be achieved. For light-water moderated nuclear reactors, the desired enrichment is 3–4%. If the enrichment factor is a fraction of a percent per stage, hundreds to thousands of stages are required for the desired enrichment level. This requires large and expensive facilities with huge electricity consumption for the vacuum pumps and the blowers that force the gases through the membranes. Therefore, gas diffusion facilities are usually built near a supply of abundant and cheap electricity. In the United States, gas diffusion plants are located in Oak Ridge, Tennessee, and Hanford, Washington, where large hydroelectric facilities are operating.

In the gaseous diffusion process about 85% of the uranium feed is rejected as depleted uranium. Currently, the "tailings" are stockpiled for eventual future use in fast breeder reactors.

The gas diffusion facilities were established during the second world war for the production of weapons-grade highly enriched uranium. The relatively low enriched uranium used for power plant reactors is a by-product of the wartime effort. Critics of nuclear power plants often claim that if enriched uranium were not available from diffusion plants constructed and operated by governments, nuclear electricity would be far more expensive or not available at all.

**Gas Centrifuge.** Recently, in the United States, Europe, and elsewhere, enrichment plants have been built that work on the principle of a gas centrifuge. When uranium gas ($UF_6$) is spun very fast in a centrifuge, the heavier $^{238}UF_6$ spreads toward the periphery of the centrifuge; the lighter $^{235}UF_6$ concentrates toward the center. Here, the enrichment is dependent on the rate of revolutions and the time spent in the centrifuge. Gas centrifuges are less energy consuming than diffusion plants and require lower capital investment.

**Laser Enrichment.** Development is in progress in various countries on laser enrichment of uranium. In this process, metallic uranium is vaporized in an oven. A stream of vaporized uranium atoms exits the oven port. A laser beam with a very narrow wavelength band is shined onto the atomic stream to differentially excite only $^{235}U$ to a higher electronic state, but not the heavier isotope. The excited atom is then ionized with ultraviolet light. The ionized $^{235}U$ is collected on a negatively charged plate.

After enrichment, $UF_6$ gas or metallic uranium is converted to uranium dioxide $UO_2$. This is a ceramic-like material that is fabricated into pellets. The pellets are loaded into fuel rods. A 1,000-MW power plant needs about 75 metric tons of uranium dioxide pellets per load.

## 7.5.3  Spent Fuel Reprocessing

In boiling- and pressurized-water reactors the fuel stays in the reactor for 2–3 years, generating electricity. After that period, the level of fission products and other neutron absorbers has built up, and the fission reaction has slowed down, with a concomitant decline in steam and electricity production. At that time, the fuel rods have to be replaced with fresh ones. In CANDU-type reactors, fuel rods have to be replaced every 18 months. The retrieved fuel rods emit high levels of radiation because of the radioactive fission products and other neutron activated isotopes that have accumulated in the spent fuel rods. The extracted fuel rods are stored in the containment vessel of the power plant, in steel- and concrete-walled water pools or dry casks.

After the decline of radioactivity, the spent fuel is reprocessed. About 96% of the original uranium in the fuel is still present, although it contains less than 1% $^{235}$U. Another 1% of the uranium has been converted to $^{239}$Pu. The spent rods are chopped up and leached in acid. Uranium and plutonium dissolve and are separated chemically from the rest of the dissolved elements. The recovered uranium is sent back to the enrichment facilities. The recovered plutonium is mixed with natural uranium and made into fresh fuel, called mixed oxide (MOX). France reprocesses about 2,000 tons per year of spent fuel, the United Kingdom 2,700 tons, Russia 400 tons, India 200 tons, and Japan 90 tons, but Japan sends spent fuel abroad for reprocessing. These amounts do not include plutonium reprocessing from defense establishment reactors for weapons fabrication. Currently, the United States is not reprocessing nuclear power plant spent fuel for fear of weapons proliferation.

### 7.5.4   Temporary Waste Storage

The liquid wastes generated in reprocessing plants are stored temporarily in cooled stainless steel tanks surrounded by reinforced concrete. After a cooling period, the liquid wastes are calcined (evaporated to dry powder) and vitrified (encased in molten glass). The molten glass is poured into stainless steel canisters. In the United Kingdom, France, Belgium, and Sweden the canisters are stored in deep silos, pending permanent disposal.

### 7.5.5   Permanent Waste Storage

Perhaps the largest problem facing nuclear power plants is the permanent disposal of spent fuel, or the waste remaining after extracting the still useful fuel from the spent fuel. The level of radioactivity of the spent fuel declines about 10-fold every 100 years. After about 1,000 years, the radiation level reaches that of the original ore from whence the fuel (uranium) was extracted.

The only practical way of disposing the waste would be in stable geologic formations known not to suffer from periodic earthquakes and where the water table is either absent or very deep beneath the formation. For example, deep salt formations have these characteristics; otherwise, the salt would have leached out long ago.

In the United States, a formation at Yucca Mountain, Nevada, was selected for permanent disposal. The repository facility was excavated from volcanic tuff at a depth of about 300 meters beneath Yucca Crest, which is 300 meters above the local water table. The repository was projected to store 70,000 tons of spent fuel and 8,000 tons of high-level military waste. Many studies have been undertaken to model the fate of the disposed waste. Most studies concluded that the waste would be undisturbed for periods of 10 thousands to millions of years. Yet some uncertainties remained in these assessments, and there is strong political opposition to using Yucca Mountain for permanent spent fuel disposal in the United States. No other country has yet solved the permanent disposal problem of radioactive waste from nuclear power plants.

## 7.6   FUSION

As noted in Section 7.2, a large amount of energy is evolved when light nuclei fuse together. For example, the following fusion processes are accompanied by energy evolvement:

$$^2D + {}^3T \rightarrow {}^4He + {}^1n + 17.5 \, \text{MeV} \tag{7.14}$$

$$^2D + {}^2D \rightarrow {}^3He + {}^1n + 4 \text{ MeV} \tag{7.15}$$

$$^2D + {}^3He \rightarrow {}^4He + {}^1p + 18.3 \text{ MeV}. \tag{7.16}$$

The sum of the masses of the nuclei that fuse (the left side of the equations) is not exactly the sum of the masses of the fused nucleus plus the mass of the ejected neutron or proton (the right side of the equations). The *mass deficit* appears as the evolved energy (see Section 7.7). The ejected neutrons or protons collide with surrounding matter so that their kinetic energy is converted into sensible heat. Such fusion reactions power the sun and other self-luminous stars, as well as thermonuclear bombs, also called hydrogen bombs.

It would be desirable to perform fusion reactions under controlled conditions, so that the evolved heat energy can be transferred to a coolant working fluid, which in turn would drive a turbomachinery. The advantages of fusion-based power plants are threefold:

(a) The "raw" material or fuel available for fusion reactors is almost unlimited, because deuterium is a natural isotope of hydrogen to the extent of 1 deuterium atom per 6,500 hydrogen atoms. Tritium is not found in nature, but can be manufactured from an isotope of lithium in the following reaction:

$$^6Li + {}^1n \rightarrow {}^3T + {}^4He + 4.8 \text{ MeV}. \tag{7.17}$$

(b) The fusion reactions would produce a minimal amount of radioactivity. Some radioactive isotopes may be created as a result of the absorption of neutrons in materials surrounding the fusion reactor. Also, tritium is mildly radioactive, emitting low-energy $\beta$-rays with a half-life of about 12 years.

(c) There is no spent fuel waste from which ingredients could be extracted for fabricating fission-type nuclear weapons.

The difficulty of achieving a controlled fusion is in overcoming the electrical repulsion force of the positively charged nuclei. To overcome the repulsive force, the colliding nuclei must have a kinetic energy commensurate to a temperature of tens of million degrees. At such temperatures, atoms are completely dissociated into positively charged nuclei and free electrons, the so-called plasma state. For the release of significant amounts of energy, many nuclei must collide. Hence, the plasma needs to be confined to a small volume at a high pressure.

Attempts of controlled fusion processes have been conducted since the 1950s. So far only limited success has been achieved, with success meaning that an equal or greater amount of energy is released as was consumed in the fusion experiment, the so-called "break-even" point. Optimistic estimates predict that commercial power plants based on fusion will be operative in the next 40–50 years. Pessimistic estimates claim that fusion-type power plants will never be practical or they will be too expensive compared with power plants based on fission reactors or renewable energy, let alone fossil energy.

## 7.6.1 Magnetic Confinement

Most approaches to plasma confinement and inducement to fusion rely on magnetic field confinement. The magnetic field is created inside cylindrical coils in which a current flows. The cylindrical coils form a circle, so that the magnetic field has the shape of a toroid (doughnut). The plasma particles travel in helical revolutions along the magnetic field lines.

The first toroidal magnetic field reactor was constructed in the former USSR, hence the acronym Tokamak, short in Russian for toroidal magnetic chamber. Further confinement of the plasma is provided by an additional current flowing in the plasma itself. The plasma current is induced by transformer action from external coils. The plasma is heated by a combination of the resistive dissipation from the current flowing in the plasma and from external sources, such as radiofrequency waves. Also, energetic particles may be injected into the plasma, such as high-velocity ions from an accelerator.

Tokamak-type fusion machines are operating in Russia, Europe, Japan, and the United States. In 1993, the Tokamak reactor at the Princeton Plasma Physics Laboratory, using the deuterium–tritium fusion, achieved a nominal temperature of 100 million degrees and a power of 5 million watts for about 4 seconds. Plans are underway to build a $1.2 billion International Thermonuclear Experimental Reactor, jointly funded by the United States, Japan, Russia, and several European countries. This facility will use deuterium–tritium as a fuel and superconducting magnets for confinement.

### 7.6.2  Laser Fusion

Laser-induced fusion apparatuses are operating at the Lawrence Livermore Laboratory, California, Los Alamos Laboratories, New Mexico, and the University of Rochester, New York. A fuel (e.g., a mixture of deuterium and tritium) is contained in a small (1-mm) sphere made of glass or steel. The mixture is irradiated with several high-intensity laser beams. The sphere is compressed and heated by implosion. The glass or steel outer layer evaporates. The temperature of the content gases increases to tens of millions degrees, and the pressure rises to hundreds of MPa. Laser beams at MJ energy are employed in very short pulses, a billionth of a second long. The power at the center of the sphere reaches 1E(15) W. The initial beam is from a neodymium–glass laser radiating at 1 μm. A crystal is used to generate higher harmonics in the ultraviolet range. Unfortunately, the laser system operates at about 1% efficiency, so that 99% of the input energy does not result in light emission. Higher-efficiency lasers pulsing at a higher rate will be necessary for break-even power production, let alone for a commercial power plant.

Plans are underway to build the National Ignition Facility at the Lawrence Livermore Laboratory, California, employing a deuterium–tritium fuel and 192 laser beams.

### 7.7  ENERGY EVOLVEMENT IN NUCLEAR FISSION AND FUSION REACTIONS

The energy evolved in a fission reaction can be calculated by means of the famous Einstein equation that ties energy to mass, $E = mc^2$. The masses on the left side of Equation (7.1) are $^{235}$U = 235.04394 amu (1 atomic mass unit = 1.66E(−27) kg), n = 1.00867 amu; on the right side $^{144}$Ba = 143.92 amu, $^{89}$Kr = 88.9166 amu, and 3n = 3.026 amu. If we subtract the sum of the masses on the right side from the sum of the masses on the left side, there is a *mass deficit* $\Delta m = 0.19$ amu. This mass deficit is converted into energy $E = mc^2 = 0.19$ amu = 1.66E(−27) kg × 3E(8) m s$^{-2}$ = 2.84E(−11) J = 177 MeV. (1 amu deficit is equivalent to 931.5 MeV.) The fission of $^{235}$U produces 2.84E(−11) × 6.023E(23) atoms mole$^{-1}$ = 0.235 kg mol$^{-1}$ = 7.3E(13) J per kg of $^{235}$U. Similarly, in the fusion reaction (7.13), the masses on the left side of the equation ($^2$H+$^3$H) total 5.03005 amu. The masses on the right side of the equation ($^4$He + n) total 5.0113 amu. The mass deficit $\Delta m = 0.01875$ amu. Multiplying by the square of the velocity of light and using energy conversion units, the energy evolved in the fusion reaction (7.14) is 17.5 MeV , or 0.84E(15)

J per kg of $^2$H (deuterium). In comparison, the combustion of carbon produces about 3.3E(7) J per kg of carbon. Thus, a fission reaction produces about 2.5 million times more energy per unit weight than a combustion reaction, and a fusion reaction produces about 25 million times more energy per unit weight than a combustion reaction.

## 7.8  CONCLUSIONS

More than 400 nuclear power plants are currently operating in the world, supplying about 17% of the global electricity demand of more than 2E(15) kWh per year. While complicated, the technology is well developed, and the power plants operate relatively reliably.

Most of the current power plants use a reactor of the PWR type in which the fuel is uranium enriched to 3–4% uranium-235; the pressurized light water serves both as the coolant and as the moderator, and cadmium or boron control rods serve to control the power output or to shut down the reactor. The BWR uses the same fuel and control rods, but the coolant/moderator is in two phases, liquid water and steam. In some BWR the moderator is graphite. Another type of reactor is used in the CANDU-type power plants, where the fuel is natural or slightly enriched uranium, the coolant/moderator is heavy water, and the control rods are similar to the other reactors. Breeder reactors have been primarily used to produce weapons-grade plutonium. They use natural uranium and a coolant/moderator, which usually is liquid sodium. Most of the breeder reactors have been phased out because sufficient plutonium for weapons production has been stockpiled, and there is no immediate scarcity of natural uranium.

Further advances in developing novel nuclear power plants that will increase their thermal efficiency, use less expensive fuels, such as natural uranium and thorium, and foremost, ensure their absolute safety are possible.

Despite persisting concerns about nuclear power plant safety, only one acknowledged serious accident occurred—that at Chernobyl in 1986—involving 31 mortalities and a yet unknown number of latent morbidities. The Three Mile Island accident in 1979 is not known to have caused excess mortality or morbidity.

Nevertheless, the future of nuclear power plants is very much uncertain. In part the uncertainty stems from the public's perception that nuclear power plants are inherently unsafe, notwithstanding the contrary records and statistics.

# PROBLEMS

**Problem 7.1**

    (a) Calculate the mass deficit $\Delta m$ in atomic mass units (amu) of the following fission reaction. (Use literature values for the exact masses of the isotopes and neutrons.)

$$^{235}U + {}^1n \rightarrow {}^{139}Xe + {}^{95}Sr + 2{}^1n$$

    (b) Calculate the energy (MeV) released per one fission.

    (c) Calculate the energy released per kilogram of $^{235}U$, and compare with the energy released in the combustion of 1 kg of carbon.

## Problem 7.2

(a) Calculate the mass deficit $\Delta m$ in atomic mass units (amu) of the following fusion reaction.

$$^2D + {}^3T \rightarrow {}^4He + {}^1n$$

(b) Calculate the energy released (MeV) per one fusion.

(c) Calculate the energy released per kilogram of deuterium.

## Problem 7.3

In a nuclear accident there is a release of $^{90}$Sr that emits $\gamma$-rays with a half-life of 28.1 y. Suppose 1 μg was absorbed by a newly born child. How much would remain in the person's body after 18 and 70 years if none is lost metabolically?

## Problem 7.4

The isotope $^{223}$Ra has a half-life of 11.4 days. $^{223}$Ra decays at a rate of 1 Ci per gram of radium isotope. (Ci=curie=3.7E(10) disintegrations/sec.) What will be the decay rate (Ci) of the 1 gram sample after 10, 100, and 1,000 days?

## Problem 7.5

Prove that the radioactivity of 1 gram of fresh radium-226 is 1 Ci.

## Problem 7.6

The isotope $^{129}$I has a half-life of 15.7E(6) years. In a nuclear power plant accident, 1 kg of the iodine isotope is dispersed into the surroundings of the plant. How much of the iodine isotope will remain in the surroundings after 1E(3), 1E(4), 1E(5), and 1E(6) years?

## Problem 7.7

Define the following terms and give examples as appropriate:

- fissile nucleus
- fertile nucleus
- chain reaction
- neutron economy, sub- and supercriticality
- fuel rod
- moderator
- control rod
- coolant

## Problem 7.8

For each of the following, create a schematic diagram and describe and list its advantages and disadvantages.

(a) Boiling water reactor

(b)  Pressurized water reactor

(c)  Breeder reactor

## Problem 7.9

Create a block diagram of and describe the nuclear fuel cycle. List the risks at each step.

## Problem 7.10

Why is there no danger of a nuclear power plant reactor exploding like an atomic bomb?

# BIBLIOGRAPHY

Bodansky, D. *Nuclear Energy—Principles, Practices and Prospects*. New York: American Institute of Physics, 1996.

El-Wakil, M. M. *Nuclear Energy Conversion*. La Grange Park: American Nuclear Society, 1982.

El-Wakil, M. M. *Powerplant Technology*. New York: McGraw–Hill, 1984.

Glasstone, S., and Sesonske, A. *Nuclear Reactor Engineering*. New York: Chapman and Hall, 1994.

Murray, R. L. *Understanding Radioactive Waste*. Columbus: Battelle Press, 1989.

Murray, R. L. *Nuclear Energy*, 5th ed. Boston: Butterworth–Heinemann, 2001.

Turner, J. E. *Atoms, Radiation, and Radiation Protection*. New York: Pergamon Press, 1986.

# Renewable Energy

## 8.1 INTRODUCTION

Renewable energy systems draw energy from the ambient environment rather than from the consumption of mineral fuels (coal, oil, gas, and nuclear). The ultimate source of most renewable energy production is the sun, whose total radiant energy flux intercepted by the earth provides a much greater source of power than can be captured by practical renewable energy schemes.[1] Despite the vast amounts of energy that are potentially available from renewable sources, collecting and utilizing that energy in an economical and effective manner is far from an easy task.

There are many reasons for the growth of interest in renewable energy systems. Such systems are independent of fuel supply and price variabilities and thereby economically less risky. Renewable energy resources are more uniformly distributed geographically than are fossil fuels, providing indigenous energy resources for most fuel-poor nations. Some forms of renewable energy operate efficiently in small units and may be located close to consumers, reducing energy transmission costs. In the United States in recent years, governmental regulation of the electric utility system has provided some economic incentive for the construction of renewable energy systems as a small part of the mix of electric power sources. Since renewable energy sources have lesser environmental effects than conventional energy sources, especially in regard to the emission of toxic and greenhouse gases to the atmosphere, they are expected to become an important source of energy in future years as environmental controls become more stringent.

The major obstacles to greater use of renewable energy are its generally higher cost, compared with conventional sources, and the intermittent availability of renewable energy collection on a diurnal, seasonal, and annual basis.

The renewable technologies that currently are employed or show promise of becoming practical are: (a) hydropower, (b) biomass power, (c) geothermal power, (d) wind power, (e) solar heating

---

[1]Exceptions are tidal power, which derives its energy from the gravitational energy of the earth–moon–sun system, and geothermal power, whose source is nuclear isotope decay in the earth's interior.

and thermal electric power, (f) photovoltaic (PV) power, (g) ocean tidal and tidal current power, (h) ocean wave power, and (i) ocean thermal electric power. Renewable energy amounted to 7% of the total U.S. 2006 energy production and 7.7% of the world energy production in 2004, mostly in the form of hydropower and biomass, but the small remainder is growing rapidly.

Some renewable-energy systems produce only electrical power, which has a higher economic value than heat. Among these are hydro, wind, photovoltaic, tidal, and ocean wave power. Nevertheless, biomass, geothermal, and solar systems, which can deliver both electric power and heat, are equally important renewable energy sources. In 2006, U.S. renewable energy production, by type of source, was 42% hydro, 48% biomass, 5% geothermal, 4% wind, and 1% solar thermal/PV. Of the electrical power component, hydropower accounted for 75%, biomass 11%, geothermal 8%, wind 7%, and solar 0.1%. Together, renewable electricity accounted for 10% of U.S. 2006 electricity production. (On a worldwide basis, renewable electricity is 19% of the total 2004 electricity production.)

Renewable energy is characterized by several important traits:

• Ubiquity of renewable energy sources;
• Low intensity of energy fluxes being captured compared with conventional systems;
• Random, intermittent nature of renewable energy fluxes;
• High capital cost per unit of power output compared with conventional sources.

Some form of renewable energy is available everywhere on earth. Averaged over a year, the entire earth's surface, including the polar regions, intercepts some sunlight, making solar insolation universally available, although low-latitude, dry (clear) climates have the most and the polar regions the least solar flux at ground level. Solar insolation induces circulation in the earth's atmosphere, giving rise to wind and ocean waves and precipitation into elevated drainage basins, providing secondary sources of solar energy. Biomass crops and forests grow in tropical and temperate lands having sufficient precipitation. Ocean tides are noticeable along continental margins. Temperature gradients beneath the earth's surface exist everywhere, providing the possibility of geothermal energy. But it is generally true that there is considerable geographic variability of each form of energy, so that some locations are more favorable for its development than others.

Fossil- and nuclear-fueled power systems operate under conditions of intense energy transfer within their components, typically of the order of $1E(5)$ W/m$^2$, so that they produce large amounts of power per unit volume. In contrast, renewable energy fluxes are much smaller, although quite different among the various sources, as listed in Table 8.1, requiring larger structures and areas to reap the same amount of power. As a consequence, the capital cost per unit of average power output is higher for renewable than for conventional power plants, but the cost of fuel for the conventional plant may more than offset its lower capital cost, making renewable energy cheaper.

Energy from sunlight is only available about half of the time and reaches its higher intensities in the several hours on either side of local noon. On cloudy and partly cloudy days, there is a considerable reduction in solar insolation compared with clear days. Electric power generated from sunlight must be stored or integrated into a distribution network with other power sources to be useful in an industrial society. Similar considerations apply to wind, wave, and tidal power systems, which are also intermittent sources. On the other hand, hydropower, biomass, and geothermal systems have energy storage capabilities that permit them to deliver power when it is needed, as do conventional systems.

**TABLE 8.1**  Annual Average Energy Flux in
Renewable-Energy Systems

| Source | Area | Heat (W/m$^2$) | El. power (W/m$^2$) |
|---|---|---|---|
| Solar | Collector | 100 | 20 |
| Photovoltaic | Cell | | 40 |
| Hydropower | Drainage basin | | 0.01 |
| Wind turbine | Turbine disk | | 40 |
| Wind farm | Farm | | 2 |
| Geothermal | Field | 0.1 | 0.02 |
| Biomass | Field | 0.5 | 0.1 |
| Ocean tidal | Tidal pond | | 1 |
| Ocean thermal | Source area | | 0.1 |
| Ocean wave | Farm | | 10 |

The economic cost of generating renewable energy is primarily associated with its high capital cost per unit of average power output when compared with conventional sources, such as fossil-fueled electric power plants. These higher costs reflect the expense of collecting renewable energy from the low-intensity, intermittent renewable source while converting it to the useful compact form of electricity for distribution to consumers. On the other hand, renewable energy is free of air and water pollutant emissions, especially carbon emissions, and will become less costly compared with conventional sources as the latter are required to reduce such emissions, usually at increased capital and operating cost. Nevertheless, it will be seen that renewable energy projects are designed to minimize the unit cost of production, just as is the case for conventional plants.

There are two figures of merit for optimizing renewable-energy systems. One we label the *effectiveness*, or the fraction of the renewable resource that is collectable by the system. The second is the *capacity factor*, the ratio of the long-term average power output to the rated power output of the plant. We would like both of these parameters to be as close to unity as possible, so as to make the most efficient and economical use of the renewable resource. But it will be seen that these objectives are incompatible and that the most economical system involves a compromise between cost and performance.

In this chapter we explain the physical basis for each of the renewable-energy sources listed above and the technology used to collect and utilize that energy on a large commercial scale. In some cases it is possible to identify costs and performance for systems in use. We also discuss environmental effects that, while greatly reduced compared with traditional energy sources, can still be significant.

## 8.2  HYDROPOWER

Before steam engines were developed, mechanical power generated by river water flowing past water wheels was the major source of power for industrial mills. These mills had to be located near river falls so that water could be diverted from an impoundment upstream of the falls and fed to a water wheel or turbine discharging to a lower level downstream of the falls. The need for mechanical

**TABLE 8.2**  Hydropower Development in the United States in 1980[a]

| Region | Number of plants | Installed capacity (GW) |
|---|---|---|
| Pacific Northwest | 160 | 28.1 |
| California | 173 | 7.5 |
| South Atlantic and Gulf | 119 | 5.6 |
| Great Lakes | 203 | 3.9 |
| Tennessee | 47 | 3.7 |
| Other | 682 | 14.5 |
| Total | 1,384 | 63.3 |

[a] Data from Gulliver, John S., and Roger E. A. Arndt. *Hydropower Engineering Handbook*. New York: McGraw–Hill, 1980.

power and sites for industrial facilities soon outgrew the availability of hydropower, leading to the introduction of steam engines and, eventually, steam electric power plants that distribute their power via electric lines to consumers far removed from the power plant site. Nevertheless, today hydropower continues to be an important source of electric power generation, supplying 7.5% of U.S. and 17% of world electricity production.

In the United States in 1980, the installed hydropower capacity totaled about 63 GW generated in 1,384 plants distributed throughout the country, as listed in Table 8.2, operating at an average capacity factor of 51%. It is estimated that the total potential capacity in the United States is about three times this amount. Most of the U.S. hydropower plants are small, with the average power being 46 MW. The largest, Grand Coulee, located on the Columbia River, can generate 7.5 GW. Worldwide there are eight hydropower installations that generate more than 5 GW, located in China (18.2), Venezuela (10.6), the United States (7.5), Brazil/Paraguay (7.4), the former Soviet Union, (6.4 and 6.0), and Canada (5.3 and 5.2). Such giant hydroelectric plants supply only a small fraction of total hydropower capacity, both worldwide and in the United States. Global hydropower installations are increasing with time, especially in developing nations with significant undeveloped hydropower resources.

An economical hydropower plant requires a reliable supply of water flow from an elevated source that can be discharged at a lower elevation nearby. Most sites require the construction of a dam at a location along a river where a large volume of water can be impounded behind the dam at a level higher than the river bed. A pipe connected to the upstream reservoir (called a *penstock*) conducts high-pressure water to the inlet of a hydroturbine located at a level at or below the downstream riverbed, into which the turbine discharge water is released, generating mechanical power to turn an electric generator. If the difference in height of the water between upstream and downstream of the dam is $h$ (called the *head*) and the volume flow rate of water through the turbine is $Q$, then the maximum power $P$ the turbine can generate is

$$P = \rho g h Q, \tag{8.1}$$

where $\rho = 1\text{E}(3)$ kg/m$^3$ is the mass density of water and g $= 9.806$ m/s$^2$ is the acceleration of gravity. For a 1,000-MW hydropower plant utilizing a head of 10 m, a flow of more than $1\text{E}(4)$ m$^3$/s would be required. Only the largest of rivers can supply flows of this magnitude.

River water flow depends upon precipitation in the drainage basin upstream of a dam site. Only a fraction of the precipitation reaches the river course, the remainder being lost to evaporation into the atmosphere and recharging of the underground water aquifer. On a seasonal or annual basis, precipitation and river flow are variable, so that it is desirable to impound a volume behind the hydropower plant dam that can be used in times of low river inflow. In locations where a substantial portion of annual precipitation is in the form of snow, its sudden melting in the spring will produce a surge in flow that may exceed the capacity of the hydroplant to utilize or store. It is usually not possible or economical to utilize the entire annual river flow to produce power.

In mountainous terrain it may be possible to locate sites having a large head, of the order of several hundreds of meters, but with only moderate or low flow rates. Such sites may prove economical because of the large power output per unit of water flow. On the other hand, low head "run of the river" plants that store no water behind their dams but can utilize whatever flow is available during most of the year have also proved economical.

The technology of hydropower is well developed. Hydroturbines are rotating machines that convert the flow of high-pressure water to mechanical power. For low-head, high-volume rate flows the Kaplan turbine is an axial flow device that somewhat resembles a ship's propeller in shape (see Figure 8.1). For higher heads, the Francis turbine is a radial inflow reaction turbine in which the flow is approximately the reverse of a centrifugal pump. For high-head, low-volume rate flow, the Pelton wheel utilizes a nozzle to accelerate the high-pressure water stream to a high-velocity jet that subsequently impinges on the peripheral blades of the turbine wheel. Powering a synchronous electric generator, hydroturbine installations can convert more than 80 percent of the ideal water power available to electric power.

Most hydropower plants require the construction of a dam and spillway, with the latter to bypass excess water flow around the dam when the reservoir is full and the turbines are operating at full capacity. The civil works required for the dam, spillway, and power house can be a considerable fraction of the cost of the installation, so that the overall cost per unit of electrical power generation is higher than that for the hydroturbine and generator alone. A favorable site for a hydropower plant is one for which the cost of the civil works is not large compared with the power-producing equipment.

**FIGURE 8.1**    The rotating component of a hydroturbine: low-head, high-volume rate flow Kaplan turbine (left), medium-head Francis turbine (center), and high-head, low-volume rate flow Pelton turbine (right). (By permission of VA Tech Hydro.)

**FIGURE 8.2**    The Noxon Rapids dam on the Clark Fork River in Montana generates 466 MW. (By permission of George Perks/Avista Corp.)

An example of a hydropower installation is shown in Figure 8.2. On the far left and right is an earthen dam that backs up the river flow, forming a higher-level pool extending miles upstream. In the center, on the left, is a spillway that passes excess river flow and, on the right, is a power house with five turbine-driven generators discharging water to a lower level downstream.

Given the unevenness of river flow and the finiteness of reservoir volume, hydropower plants do not operate at their rated capacity year round. A typical capacity factor is 50%, which represents a compromise between the desire to utilize all the hydroenergy available in the river flow and the necessity to limit the capital costs of doing so.

Hydroplants have indefinitely long useful lives. The civil works and power-generating machinery are very robust and operating expenses are very low. Many large hydropower plants have been constructed by national governments and their cost paid in part from government revenues as well as electricity consumers. The construction of the dam may provide other benefits than electric power, such as water for irrigation or flood control, that reduce the costs allocated to electricity production. Nevertheless, there are hydropower sites that can be developed that are economically competitive with fossil fuel power plants.

## 8.2.1  Environmental Effects

Hydropower plants can have severe environmental effects. Where a reservoir is formed behind a dam, aquatic and terrestrial ecosystems are greatly altered. Downstream of the dam, the riverine flow is altered, interfering with the ecosystems that have adjusted to the natural variable river flow pattern. The reservoir interrupts the natural siltation flow in the river and its contribution to alluvial deposits downstream. The flooding of land that previously served for agriculture and human habitation may be a significant social and economic loss. The construction of dams interferes with the migration of anadramous fish and adversely affects their populations, even where fish ladders are employed. Currently, the removal of U.S. hydropower dams in the Pacific northwest and Maine has been recommended to revive endangered salmon fisheries.

## 8.3  BIOMASS ENERGY

Until the onset of the industrial revolution led to the exploitation of fossil fuel deposits, wood was the principal fuel available to humans, being used for cooking, space heating, and materials manufacture. In many developing nations today, wood supplies up to half of the total national energy consumption, which is very low on a per capita basis compared with that of industrial nations. A renewable resource, wood crops replace themselves every 50 to 100 years. On a global basis, the amount of non-food-producing land available for energy crops like wood is far from sufficient to supplant current fossil fuel consumption. Nevertheless, a variety of agricultural and silvicultural crops or their by-products can contribute marginally to energy supplies, thereby replacing some fossil fuel consumption and its attendant carbon dioxide emissions.

Another form of biomass that can be converted to a useful fuel is animal waste. Digesters can generate methane from farm animal or human wastes, with the residue of this process being suitable for crop fertilizing. Organic matter in municipal waste landfills generates methane by bacterial action that can supply low heating value gas.

In the United States in 2006, biomass provided 48% of renewable energy and 3.4% of total energy. It also contributed 11% of renewable electric power and 1.1% of all electric power.

Why is annual biomass energy use so small compared with fossil fuel sources? First, it is renewable on an annual basis; that is, it is a chemical form of solar energy stored in plants during the annual growing season, whereas fossil fuel is a residue of millions of years of plant growth, encapsulated in underground sedimentary reservoirs that can be recovered at annual rates greatly exceeding that at which they were formed eons ago. Second the fraction of annual solar insolation falling on a square meter of ground that is stored annually in plants is very small. Finally, most of the earth's surface that can grow food or other plant-based crops is already used to supply food for human and food animal use year round; there is little unused or underused land that can be devoted to energy crop production.

Scientists have estimated the net primary production of plant species on land in terms of the annual average rate of plant carbon mass fixed by photosynthesis, expressible in units of kilograms carbon per square meter per year. For example, for North America this rate is 0.4 kg $C/m^2y$.[2] If we convert this to an energy capture rate by multiplying by the heating value of carbon, the corresponding primary plant energy annual fixation rate is 0.42 $W/m^2$. Comparing this with the U.S. annual average solar insolation to the ground of 230 $W/m^2$, we find that primary plant productivity is only 0.2% of solar insolation.[3]

The food energy stored in edible agricultural crops like grains is only a fraction of that in the entire crop mass, so that agricultural crop residues are potential sources of biomass energy. It has been estimated that crop and forest residues in the United States have a heating value equal to 12% of fossil fuel consumption, but only about one fifth of this is readily usable.[4] The energy required to collect, store, and utilize this residue further reduces the amount of energy available from it to

---

[2]Haberl, H., et al. *Proceedings of the National Academy of Sciences* 104 (2007): 11206.

[3]The total North America net primary productivity (in energy units) is 9.8 GW, of which 2.2 GW is used for human purposes, mostly food production. U.S. primary energy consumption is 2.8 GW. It is not possible to replace a significant fraction of current energy consumption by bioenergy sources without serious consequences for food production and ecological integrity.

[4]D. Pimentel et al. *Science* 212 (1981): 1110.

replace fossil fuels. Furthermore, crop residues are often used to build soil mass and fertility by composting, during which some, but not all of the residue carbon is released to the atmosphere as $CO_2$. Using crop residues as fuel reduces the amount of carbon uptake in soil.

### 8.3.1  Photosynthesis

Organic matter in terrestrial plants and soil is, among other things, a temporary storage system for solar energy. The conversion of atmospheric $CO_2$ and $H_2O$ to organic matter by photosynthetic reactions in plants stores the energy of visible light from the sun in the form of chemical energy of the organic matter. The latter may be utilized in the same manner as fossil fuels, releasing $CO_2$ back to the atmosphere. Thus no net emissions of $CO_2$ to the atmosphere result from this cycle.

The overall photosynthetic process by which water and carbon dioxide in plant leaves are combined to form carbohydrate molecules in plants may be represented by a global chemical reaction as

$$nCO_2 + mH_2O \rightarrow C_n(H_2O)_m + nO_2, \tag{8.2}$$

where $C_n(H_2O)_m$ is a generic form of various carbohydrate molecules such as sugar, starch, or cellulose that constitute most of the plant mass.[5] The first step of this overall process is a photo-chemical one, where solar radiation provides the high energy needed to start the process in which the small reactant molecules, $CO_2$ and $H_2O$, are split and rearranged to form $CH_2O$ and $O_2$. This is followed by a low-energy biochemical process in which the $CH_2O$ complexes are stitched together to form the much larger carbohydrate molecules that constitute plant mass. The chlorophyll in plant leaves provides a catalytic template for these reactions.

The chemical energy required for the overall reaction (8.2) is 4.07 eV per carbon atom.[6] The photons active in photosynthesis are about 700 nm in wave length, or 1.9 eV of energy. Thus at least two photons must be captured to supply the reaction energy. As pointed out in Section (8.3) above, the solar energy flux available for plant synthesis is typically about 500 times that stored in plant photochemical reaction (8.2). Much of this surplus energy is required for growth of the whole plant. Photosynthesis provides the necessary primary energy that supports all biological organisms on the earth.[7]

### 8.3.2  Biofuels

Agricultural crops supply humans with food and materials (e.g., lumber, paper, textiles), food animals with food, and wood fuel energy for human use. On an energy basis, the price to consumers of food commodities exceeds that of fossil fuels, often by a considerable amount. To compete economically with traditional agricultural commodities for scarce agricultural resources, biofuel

---

[5]For most carbohydrates of interest, $n \simeq m$ and $5 < n < 10^4$.

[6]The electron volt, the energy needed to move an electron through an electric potential of one volt, is a suitable energy unit for measuring the energy changes in rearranging the atoms in molecules. Its value is 1.6030 E($-$19) J (see Table A.3).

[7]This description is necessarily approximate. For a full discussion, see Smil, Vaclav. *Energy in Nature and Society*. Cambridge, Mass: MIT Press, 2008.

energy supplies must use otherwise undervalued by-products and waste streams and efficient, low-capitalization facilities that convert primary biomass to higher-value fuels.

The principal processes that utilize the energy content of primary biomass are as follows.

- Combustion
- Gasification
- Pyrolysis
- Fermentation
- Anaerobic digestion

Woody plants and grasses can be burned directly in stoves, furnaces, or boilers to provide cooking, space, or process heat and electric power. Alternatively, biomass may be converted to a synthetic fuel, such as a gaseous fuel composed of $H_2$ and CO, via a thermal process that conserves most of its heating value, with the fuel being combustible in boilers and furnaces. Pyrolysis, the thermal decomposition of biomass, produces a combination of solid, liquid, and gaseous products that can be fashioned into useful fuels. Fermentation and distillation of food grains produces ethanol ($C_2H_5OH$), a valuable liquid fuel that is commonly blended with gasoline for motor vehicle use. Anaerobic digestion produces a gaseous mixture of $CO_2$ and $CH_4$ that is a combustible gas of low heating value. The gaseous fuels can be upgraded to more desirable forms, albeit at some loss of energy.

The processing of biomass to a form that is more readily used as a replacement for fossil fuel inevitably results in a loss of some of its heating value and generates a production cost of the conversion. To compete economically with fossil fuels, which have small processing costs, biomass-derived synthetic fuels must utilize low-cost forms of primary biomass.

Except for wood, which may be harvested year round, biomass energy crops, like food crops, must be stored after harvesting to provide a steady supply of feed stock for the remainder of the year. The economic cost of storage and maintaining a year's inventory adds to the price differential between biomass and fossil fuels, which have much smaller inventories. In addition, the cost of transporting the feed stock from farm to refinery increases with refinery (and farm area) size, a diseconomy of scale.

During the 1970s, when international oil prices rose abruptly because of supply restrictions, many agricultural-based, biomass-derived liquid fuel production schemes were investigated. Of particular interest was the net energy balance, the ratio of the energy value of the fuel produced (such as ethanol) to the energy consumed in producing, harvesting, and processing the biomass. Not only should this ratio exceed unity, so that there is a net output of energy from the biomass conversion process, but also it must be sufficiently greater that nonenergy costs of production can be covered by the sales revenues generated.[8] Such requirements could only be met by utilizing a high-energy crop like sugar cane and incorporating the energy supply of the crop residue into the fuel synthesis process. Currently, these energy and economic constraints make fuels derived from biomass food crops, such as grains or sugar cane, noncompetitive with fossil fuels at current world market prices.

---

[8]For fossil fuels, the energy consumed in production, processing, and distribution is less than 10% of the energy of the end product.

### 8.3.2.1   Bioethanol

As an example, in the United States corn is currently employed as a feedstock to manufacture ethanol for use as an additive in motor vehicle fuel (gasoline). Fossil fuel is consumed in the production and harvesting of corn and the production of ethanol in the fermentation and distillation process; the fossil fuel energy so consumed in both process about equals the ethanol energy produced.[9] However, a by-product, distillers' dry grain, suitable for animal feed and having an energy content of 20% of that of the ethanol produced, is also produced. Subtracting the energy input allocated to this by-product, the ethanol energy output is 1.25 times the energy input allocated to the ethanol production, for a net energy benefit of 20%.

As a consequence, the substitution of ethanol for gasoline as a vehicle fuel results in only a 20% reduction in fossil fuel energy use and a commensurate reduction in greenhouse gas emissions, compared with the use of only fossil fuel. On the other hand, the market price of corn energy versus that of oil and the net production cost of ethanol determine whether the ethanol fuel cost is less or more than that of the gasoline it displaces.

Corn is an unusual food grain in that large inputs of fossil energy per unit grain mass are required to maximize the grain mass per unit area of arable land. Much of the plant energy is stored in the nongrain plant mass, a cellulosic material called stover. Conversion of nonfood cellulosic plant mass to ethanol is seen as a future alternative to corn grain ethanol.

Cellulosic plant mass, such as the leaves and stems of grasses and the branches of woody plants and trees, can also be used to produce liquid fuels. Fermentation of such biomass to produce ethanol is much more difficult and expensive than that of grains such as corn or wheat or sugar from cane. But cellulosic plant mass can be grown under much more varied conditions and on soils of lesser productivity than that used for food grain production, lowering its production cost and consumption of fossil fuel with attendant carbon emissions. Gasification followed by fermentation or catalytic synthesis is an alternate path for the production of alcohols or other liquid fuels from cellulosic biomass, but economical large-scale production has yet to be demonstrated.

### 8.3.2.2   Biodiesel

Another biofuel that can be produced from food plants is biodiesel. It consists primarily of oil from soy beans, other vegetable oils, or palm oil; it is readily produced from the parent plant with only small inputs of fossil fuel energy. Biodiesel may be mixed with fossil diesel fuel, having equivalent combustion characteristics. Soybean diesel has a relatively greater by-product energy than does corn ethanol; the soybean meal energy is two thirds that of the soybean diesel produced in the process. The fossil fuel energy share for producing the soybean diesel is just about 25% of the soybean diesel energy. Substitution of biodiesel for fossil diesel results in a 75% reduction of fossil fuel energy use.

### 8.3.3   Wood as Biofuel

The use of wood harvesting residues to generate process heat and electric power in paper mills, eliminating their use of fossil fuels, is now common in the paper industry. The residues are collected along with the wood used to feed the pulping operation. While the residues supply some or all of the

---

[9]J. Hill, et al. *Proceedings of the National Academy of Sciences* 103 (2006): 1206–7.

**FIGURE 8.3**   A gasifier at a power plant in Burlington, Vermont, processes about 200 tons per day of wood chips and waste, generating a gaseous fuel that provides 8 MW of electric power from a steam plant. (By permission of DOE/NREL-PIX.)

mill's energy requirements, there is generally no significant surplus that might otherwise generate electric power for other purposes. (The same is true for wood harvested for lumber.) In contrast to this use of an otherwise valueless by-product, electric power plants fueled by wood harvested for fuel-only purposes have been constructed. The harvesting and transportation costs of the wood fuel for these plants, added to the market value of the wood component as pulp feed or lumber (in contrast to the residue component), generates a costly fuel that can be more expensive than fossil fuel used to generate electric power. But wood-based fuel generates no net carbon emissions, an advantage that may offset a higher production cost when fossil carbon emissions are regulated to prevent global warming.

Figure 8.3 shows a gasifier that converts wood waste to gaseous fuel for use in an electric power plant. By converting from a solid to a gaseous form, the fuel can be used in a combined cycle power plant, significantly increasing the thermal efficiency of about 20% obtained in conventional wood-burning power plants.

## 8.3.4   Environmental Effects

The use of crops or their residues to supply fuel to replace fossil fuel creates environmental impacts similar to those associated with agriculture and silviculture: consumption of manufactured fertilizers, spreading of pesticides and herbicides, soil erosion, consumption of irrigation water, and interference with natural ecosystems. Ordinary air emissions from the combustion of biomass fuels are not always less than those from the fossil fuels they replace, but fossil carbon emissions can be much less.

There is controversy over possible indirect environmental damage from biofuel production from food crops. Diversion of food production to fuel production can induce conversion of forest land to replace the lost food production. The global scarcity of productive arable land limits the world's food supply; it may be both more expensive and environmentally damaging to reduce fossil carbon emissions by producing biofuels than to limit such emissions by other technologies, such as carbon capture and storage.

Agricultural plants store carbon from the atmosphere during the growing season. Most of this is returned to the atmosphere during the rest of the year, but an essential residue needed to replenish soil viability for future crops remains. When one crop is replaced by another, soil carbon is lost to the atmosphere until the new crop can replace it, which may take many years. Studies show that the carbon reduction expected from introducing a biofuel crop may not be realized fully until years later when the biofuel crop has rebuilt the soil carbon to its former level.

## 8.4   GEOTHERMAL ENERGY

Fossil fuel deposits, which form the bulk of the world's current energy supply, are the residue of biomass formed millions of years ago, in which solar energy was stored in living matter. Made accessible by industrial technology, fossil fuel is the most easily and cheaply exploitable form of energy. But the energy stored in the hot interior of the earth is vastly greater in magnitude, and potentially exploitable, yet is almost entirely inaccessible. This energy is what remains from the gravitational collapse of the interplanetary material from which the earth was formed.

The earth's interior consists of a core of mostly molten material at a temperature of about 4000°C, extending to a little more than half the earth's radius and surrounded by a mantle of less deformable and cooler material. The outer edge of the mantle is covered by a crust of solid material of thickness between 5 and 35 km. Within the crust there is an outward flow of heat from the earth's interior of approximately 50 mW/m$^2$, which is accompanied by a temperature gradient of about 30 K/km.[10] No practical use can be made of such a feeble flow of heat unless it can be substantially amplified. By drilling wells to depths of 5–10 km and pumping from them water or steam heated to 200–300°C, enough heat may be extracted to generate hundreds of megawatts of electrical power in a single geothermal plant.

The most economical sites at which to develop geothermal energy are those where the subsurface temperatures are highest and underground water and steam deposits are closest to the surface. Such sites are found mostly at the borders of the earth's tectonic plates, near active or recently inactive volcanos, hot springs, or geysers. Favorable sites of this type occupy only on a small fraction of the earth's land area.

The principal countries that have installed geothermal electrical power generation systems are listed in Table 8.3, together with the amounts of installed electrical power. Of the world total of nearly 6,000 MW, the United States accounts for 44%, most of which is located in northern California. In 1977, the average capacity factor for U.S. geothermally generated electricity was 15.6%.

Geothermal heat is recovered in many locations where the reservoir temperature is insufficient for economical power generation but where the low-temperature heat may be used for space heating

---

[10]This heat flux is tiny compared with the average solar energy flux to the earth's surface of about 500 W/m$^2$. It is the latter that determines the earth's surface temperature.

**TABLE 8.3**   Installed Electrical and Thermal
Power of Geothermal Systems in 1993[a]

| Country | Electrical (MW) | Thermal (MW) |
|---|---|---|
| United States | 2,594 | 463 |
| Philippines | 888 | — |
| Mexico | 752 | 8 |
| Italy | 637 | 360 |
| New Zealand | 285 | 258 |
| Japan | 270 | 3,321 |
| Indonesia | 144 | — |
| El Salvador | 105 | — |
| Other | 240 | 6,802 |
| Total | 5,915 | 11,204 |

[a] Data from Dickson, Mary H., and Mario Fanelli, eds. *Geothermal Energy.* Chichester: John Wiley & Sons, 1995.

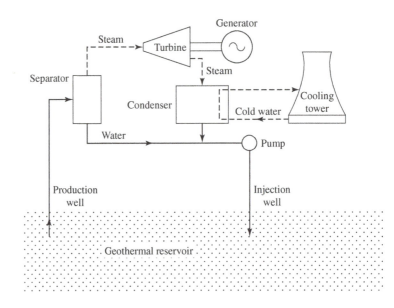

**FIGURE 8.4**   A diagram of the flow of fluids in a geothermal electric power plant.

or industrial processing. The heat recovery capacity of this type of geothermal energy supply is listed in the last column of Table 8.3. Assuming an average thermal efficiency of 20% for geothermal electric power generation, the figures in Table 8.3 suggest that nearly three quarters of the worlds geothermal heat recovery is used to generate electricity.

A typical geothermal electric power plant is sketched in Figure 8.4. A production well supplies pressurized warm water, a water and steam mixture, or only steam, depending upon the temperature and pressure conditions in the underground deposit. This fluid is fed to a separator, where steam is separated from liquid. The liquid water is pumped back into the ground via an injection well, while the steam is fed to a low-pressure turbine that generates electrical power. The turbine exhaust steam is condensed in a condenser, where cool water from a cooling tower removes heat from the exhaust stream. The condensed steam is reinjected into the ground. These systems typically supply steam at 160–180°C and a pressure of 0.6–0.9 MPa and have thermal efficiencies of 20–25%. More elaborate systems than that shown in Figure 8.4, which improve the thermal efficiency, have been constructed, albeit at higher capital cost.[11]

A geothermal plant transfers heat from a hot underground reservoir to the earth's surface, where it can be profitably utilized to produce electrical power or for other heating purposes. If it is to be a source of renewable energy, this heat must not be removed at a faster rate than it can be replenished by conduction from deeper in the earth. In the regions where geothermal resources have been developed, this heat flux is of the order of 0.3 W/m$^2$ = 3E(5) W/km$^2$. A 100-MW power plant would require a geothermal field of 1,700 km$^2$ to ensure a steady source of hot fluid. Most current geothermal plants remove heat at more than 10 times the replenishable rate so as to reduce the cost of collecting and reinjecting the hot fluid. Consequently, the underground reservoirs for these plants are slowly being cooled down, but this cooling is not significant over the useful economic life of the plants.

The problem of developing an economical geothermal resource is similar to that for developing a petroleum or natural gas resource. A reservoir of underground water of sufficiently high temperature must be located at depths that are accessible by drilling wells. It must be sufficiently large to supply the required heat for a period of 40 years or more. The resistance to underground fluid flow to and from the supply and injection wells must be low enough to produce the volume flow of hot fluid needed for the plant without requiring excessive pumping power. For these reasons it is never certain that a promising resource can be developed until sufficient exploratory drilling and testing has been conducted, which adds to the costs of geothermal plants.

An alternative geothermal system, known as "hot dry rock," would utilize impermeable rock formations lacking underground water, a more common geological condition that would greatly increase the potential availability of geothermal energy in proximity to prospective users. The plan is to fracture the formation surrounding the producing and injection wells by hydraulic pressurization or explosives. This would provide sufficient permeability to enable the pumping of water between the wells, heating it by contact with the hot rock. So far, exploratory attempts to develop these systems have failed, for various reasons related to the establishment of a satisfactory production potential. It appears that the cost of creating a permeable, water-filled underground resource equivalent to the natural ones that have already been or can be exploited is too great, using current technology and geological development practices. The economics of these systems is currently unfavorable to their development.

One recent development that utilizes the heat storage capacity of an underground aquifer is called the geothermal heat pump. In this system, water from a well supplies heat to a heat pump that then releases a greater amount of heat to a building space during cold weather. Since the well water is warmer than the ambient winter air, the heat pump produces much more heat per unit of electrical work required to run it than would be the case if the heat pump used cold winter ambient

---

[11] See Dickson, Mary H., and Mario Fanelli, eds. *Geothermal Energy.* Chichester: John Wiley & Sons, 1995.

air as its heat source.[12] In climates where air conditioning is needed in summertime, this same system can be reversed to deposit the heat rejected by the air conditioner to the well water, which is more saving of power than rejecting it to the hot summer ambient air. This system is practical only for central air conditioning and heat pump systems and then involves the additional expense of well drilling. Depending upon the price of electricity, the savings in power may repay the additional investment.[13]

### 8.4.1   Environmental Effects

Geothermal fluids contain dissolved solids and gases that must be disposed of safely. The solids are returned to the underground reservoir in the reinjected fluid. (Without reinjection, the discharge of geothermal water to surface waters would adversely affect water quality and possibly produce subsidence in the geothermal field.) Of the dissolved gases, some of which must be removed from the condenser by steam ejectors (an energy loss), hydrogen sulfide, sulfur dioxide, and radon, all of which are toxic to humans, must be safely vented.

### 8.5   SOLAR ENERGY

A prodigious source of radiant energy, the sun emits electromagnetic radiation whose energy flux per unit area, called *irradiance*, decreases as the square of the distance from the sun center. At the mean sun–earth distance of 1.495E(8) km, the solar irradiance is 1,367 W/m$^2$.[14] The sun's light is nearly, but not exactly, parallel; the sun's disc, viewed from the earth's mean orbit, subtends a plane angle of 9.3E($-$3) radian = 31.0 minutes and a solid angle of 5.40E($-$6) stearadian. It is this energy stream from the sun that maintains the earth at a livable temperature, far above the cosmic background temperature of 2.7 K that would exist if the sun's irradiance were zero.

Not all of this solar radiation reaches the earth's surface. Some is reflected from the earth's atmospheric gases and clouds; some is absorbed by air molecules (principally oxygen, carbon dioxide, water vapor, and ozone), clouds, and atmospheric dust; some is scattered by air molecules and dust. As a consequence, only a fraction of the solar irradiance impinging on the earth's atmosphere (called the *extra-terrestrial irradiance*) actually reaches ground level. Of that fraction, some retains its solar direction, called *beam irradiance*, while the remainder, called *diffuse irradiance*, has been scattered through large angles, approaching the ground from directions quite different than that of the sun.[15] Of the sunlight reaching the ground, some is absorbed by the ground and the rest

---

[12]Near-surface underground temperature in temperate climates is about 10°C.

[13]Strictly speaking, this is not a geothermal system in that net geothermal energy is not being extracted on an annual basis. To produce space heating and cooling, electric power from some other source is used to provide space conditioning as in conventional systems.

[14]Because the earth's orbit around the sun is elliptical rather than circular, the solar irradiance varies by $\pm3.3\% = \pm45.1$ W/m$^2$ during the year, being at the maximum on December 26 and minimum on July 1, when the earth is closest and farthest from the sun, respectively.

[15]Atmospheric molecules scatter blue light more than red light, making a clear sky uniformly blue and sunsets red.

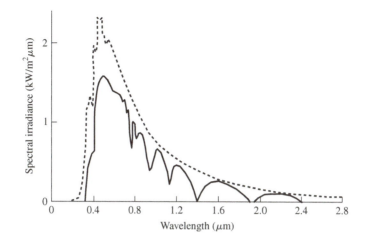

**FIGURE 8.5**    The spectral distribution of the solar irradiance as a function of wavelength. The dotted line is the extraterrestrial irradiance and the solid line is the clear-sky ground-level irradiance for an overhead sun.

reflected upward, with the latter undergoing scattering and absorbtion in the atmosphere on its way out to extraterrestrial space.[16]

The solar radiation approaching the earth comprises a wide band of wavelengths, but most (94%) of the radiant energy lies between 0.3 and 2 micrometers in wavelength. The distribution of the extraterrestrial solar irradiance among its various wavelengths is shown as a dotted line in Figure 8.5. The main components of the solar spectrum are the ultraviolet ($<0.38\,\mu$m), the visible ($0.38\,\mu$m [violet] to $0.78\,\mu$m [red]), and the infrared ($>0.78\,\mu$m). The energy content of these portions are 6, 48, and 46% of the total irradiance, respectively. The energy of an individual photon, the quantum packet of electromagnetic radiation, is proportional to its frequency and hence inversely proportional to its wavelength. Ultraviolet photons, having shorter wavelengths and hence higher individual energies than visible or infrared photons, are potentially more damaging to living organisms, even though their aggregate share of solar energy is much smaller.

The portion of the incoming beam irradiance that is absorbed or scattered by the atmosphere depends upon the wavelength of the incoming light. Scattering predominates for short wavelengths (blue) and molecular absorption for long wavelengths (red). The net beam irradiance at ground level (i.e., what remains after scattering and absorption) is shown as a solid line in Figure 8.5. The sharp dips in spectral irradiance in the red and infrared regions are caused by molecular absorption, principally by carbon dioxide, water vapor, and ozone (which are greenhouse gases), while the ultraviolet absorption is caused by diatomic oxygen and ozone. Under the best of conditions, for

---

[16]A small portion of absorbed incoming sunlight causes photochemical reactions in the atmosphere, mainly ozone formation in the stratosphere and smog formation in the lower troposphere. Also, some of the ground-level incident radiation absorbed by plants results in photosynthesis. Nevertheless, the solar energy invested in chemical change in the atmosphere and plants is tiny compared with the heating of the atmosphere, hydrosphere, and geosphere by absorbed sunlight.

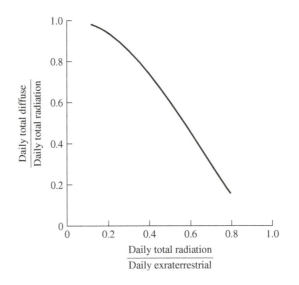

**FIGURE 8.6** The ratio of daily diffuse to total radiation as a function of the ratio of daily total to extraterrestrial radiation, showing that cloudiness increases the diffuse portion of total radiation.

a clear sky and the sun overhead, about 80% of the extraterrestrial beam irradiance reaches the earth's surface.

The distinction between beam and diffuse radiation is important to the design and functioning of solar energy collection systems. If the collector is a flat surface exposed to the sky, it will collect both beam and diffuse radiation. But if an optical focusing system is used to intensify the solar radiation impinging on the collector, only beam radiation is collected, which is a lesser amount than the total of beam and diffuse radiation. The proportions of beam and diffuse radiation depend upon the amount of cloud cover, with a clear sky providing mostly beam radiation and a heavy cloud cover resulting in entirely diffuse radiation.[17] Summed over daytime periods, this relationship is exhibited in Figure 8.6. The abscissa is the ratio of daily total solar radiation on a horizontal flat plate to the extraterrestrial value and is called the *clearness index*, while the ordinate is the ratio of the daily total diffuse radiation to daily total radiation. As the sky becomes less clear, Figure 8.6 shows that both the beam and the diffuse radiation decrease, but the former decreases proportionately faster. Focusing systems collect little energy on cloudy days.

The amount of solar energy falling upon a solar collector depends upon its orientation with respect to the sun's direction, as viewed from the earth's surface. The solar direction varies with the local hour of the day and the day of the year. Fixed collectors remain in the same position year round. Common types of fixed collectors are horizontal (e.g., solar pond), tilted at an angle above the horizontal, and vertical (windows). On the other hand, focusing collectors may move in the vertical and azimuthal directions to direct the collector aperture to be normal to the solar

---

[17] If one can't see the sun through the clouds and no shadows form, the light is all diffuse.

**TABLE 8.4**   Clear-Sky Irradiance at 40° N Latitude

| Position | Horizontal | 40° tilt | Vertical | Direct normal |
|---|---|---|---|---|
| **Daily total (MJ/m$^2$)** | | | | |
| 21 Mar. | 21.02 | 26.44 | 16.84 | 33.09 |
| 21 June | 30.05 | 25.24 | 6.92 | 36.08 |
| 21 Sept. | 20.29 | 25.28 | 16.08 | 30.73 |
| 21 Dec. | 8.87 | 18.54 | 18.68 | 22.44 |
| Annual average | 20.06 | 23.88 | 14.63 | 30.58 |
| **Maximum hourly (MJ/m$^2$)** | | | | |
| 21 Mar. | 810 | 1027 | 656 | 968 |
| 21 June | 958 | 911 | 309 | 879 |
| 21 Sept. | 785 | 987 | 630 | 914 |
| 21 Dec. | 451 | 867 | 829 | 898 |
| Annual average | 751 | 948 | 606 | 915 |

direction. Each type of collector receives a different amount of solar energy per unit of aperture area.

Table 8.4 lists the clear day solar irradiance at 40°N latitude[18] incident upon four common types of collectors: a horizontal surface (first column); a south-facing flat plate tilted 40° above the horizontal (second column); a vertical south-facing surface (third column); and a focusing collector that is oriented both vertically and horizontally to collect direct (or beam) irradiance (fourth column). The daily total of energy incident upon these collectors is listed in the upper part of the table for 4 days of the year: the spring and fall equinoxes (21 Mar. and 21 Sept.) and the summer and winter solstices (21 June and 21 Dec.). On the equinoxial days the sun's position in the sky is essentially identical, while the solstices mark the annual extremes in solar irradiance. The lower half of the table lists the maximum hourly solar irradiance, which for the collectors selected occurs at local noon. The annual average value of these items is also listed in the table.

Several important conclusions about clear sky solar irradiance may be drawn from the entries in Table 8.4. For the horizontal collector, the summer daily total is more than three times the winter value.[19] By tilting the collector to 40°, the difference between summer and winter is much reduced, to a ratio of 1.5:1, so that the daily total is more nearly uniform year round. For a vertical collector, such as a window, the winter daily total irradiance is much greater than that in summer, making a south-facing window a good source of winter space heat and a small source of summer air-conditioning load. The focusing collector's daily irradiance is about 30% higher than the tilted flat plate, which may be hardly enough in itself to justify the complication of moving the collector to track the sun. On an annual average basis, there is only a factor of 2 difference between the greatest

---

[18]This is the median latitude for the lower 48 U.S. states, which generally lie between 30° and 50° N latitude. A lower latitude will experience a greater solar irradiance and a lesser variability between summer and winter and vice versa for a higher latitude.

[19]The daily irradiance on a horizontal surface maintains the surface atmospheric temperature level. The difference between summer and winter irradiance accounts in part for the atmospheric temperature difference between these seasons.

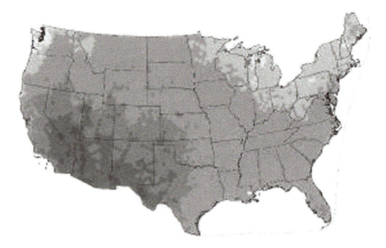

**FIGURE 8.7**  Annual average solar irradiance on a fixed flat plate collector (south facing, tilt angle = latitude) as a function of location in the contiguous United States. The gray scale shading covers the range of 170–270 W/m$^2$, northeast to southwest.

and least values of the daily totals for the various collectors of Table 8.4. Similar distinctions apply to the maximum noontime hourly irradiance, listed in the lower half of Table 8.4. Under most circumstances, these values lie within the range 800–1000 W/m$^2$, or 60–75% of the extraterrestrial irradiance of 1,367 W/m$^2$.

The values of Table 8.4 apply only to clear sky irradiance. The effect of clouds and atmospheric dust will reduce the available irradiance to lesser values. Figure 8.6 illustrates how cloudiness, in reducing the irradiance on a horizontal surface, has a greater effect on beam irradiance than diffuse, or scattered light. The average value of the clearness index, the ratio of daily total irradiance to the extraterrestrial value, varies with geographic location and season of the year. As a consequence, seasonal or annual average irradiances can be as low as half of the values shown in Table 8.4.

The effect of U.S. geographic location on the annual average insolation on a fixed flat plate collector (south facing, tilt angle = latitude) is shown in Figure 8.7. Areas of greater darkness correspond to greater irradiance. The darkest region, in the southwest where clear skies predominate, corresponds to an annual average of 250 W/m$^2$, with peaks at 270 W/m$^2$. In the northeast, where the irradiance is a minimum, of 170 W/m$^2$, skies are very often cloudy. A large region of higher irradiance of 190 W/m$^2$ extends from the upper west to the southeast. The high irradiance area in the arid southwest is the preferred site for both solar thermal and PV farms.

As will be seen below, only part of the incident solar energy flux can be collected by a well-designed solar thermal collector system. While a collector can absorb a high fraction of the solar irradiance, it also loses heat to the ambient atmosphere, the more so as it attempts to store the collected energy at a temperature higher than atmospheric. Practical solar collectors for domestic hot water heating may average 7 GJ/m$^2$ y at typical U.S. locations. If this energy were used to replace electric heating, whose current U.S. price to home owners is 20–50 $/GJ, then the economic value of the solar heat would be 140–350 $/m$^2$ y. This energy cost saving would pay for the capital cost of the solar thermal heating system in only a few years. On the other hand, if the solar heating

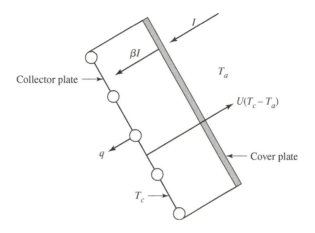

**FIGURE 8.8**  A flat plate collector absorbs incident sunlight on a thermally conducting collector plate that is cooled by a flow of liquid delivering the absorbed heat to a storage tank. A glass cover plate reduces the heat loss to the ambient environment.

system replaced fossil fuel, whose cost would be smaller than that of electricity by a factor of 3 or 4, the solar heating system would be less economical.

### 8.5.1   Flat Plate Collector

The most common and economical form of solar thermal energy collector is the flat plate collector, illustrated in Figure 8.8. It consists of a thermally conducting collector plate equipped with passages through which a heat transfer fluid passes, transferring heat from the collector plate to a fluid storage tank. The collector plate surface facing the incoming sunlight is treated to absorb as much of that sunlight as possible. A transparent cover (usually glass) is placed parallel to the collector plate, forming an enclosed space that reduces the heat loss to the surrounding atmosphere, whose temperature $T_a$ is less than that of the collector plate, $T_c$.

Of the incoming solar irradiance $I$ that falls on the cover plate, some is reflected and absorbed by the cover plate, and only a portion of what is transmitted through the cover plate is absorbed by the collector plate. Overall, if the fraction of the incoming solar irradiance $I$ that is absorbed by the collector plate is $\beta$, then the solar heat flux to the plate is $\beta I$.[20] The warm collector plate will lose heat to the surrounding atmosphere by heat conduction, convection, and radiation at a rate that is proportional to the temperature difference $T_c - T_a$ between the plate and the ambient atmosphere. The proportionality constant $\mathcal{U}$, called the overall heat transfer coefficient, depends in part upon the heat transfer properties of air and the radiant heat transfer properties of the collector and cover plates. As a consequence, the net unit heat flux $q$ that is collected in the storage system is the

---

[20]The value of $\beta$ depends upon the angle of incidence of the incoming solar radiation, the refractive and transmissive properties of the glass cover plate (see Figure 8.8), and the absorptive properties of the collector surface for solar wavelengths. For the best designs, about 80% of the incident solar radiation is absorbed by the collector plate.

difference between the absorbed irradiance and the loss to the atmosphere,

$$q = \beta I - \mathcal{U}(T_c - T_a). \tag{8.3}$$

Note that the higher the temperature of collection, the less heat is collected, there being a maximum collector temperature $(T_c)_{max}$ at which no heat is collected:

$$(T_c)_{max} = T_a + \frac{\beta I}{\mathcal{U}}. \tag{8.4}$$

The fraction of the incident irradiance that is collected, $q/I$, is called the collector efficiency $\eta$:

$$\eta \equiv \frac{q}{I} = \beta - \frac{\mathcal{U}(T_c - T_a)}{I}. \tag{8.5}$$

The collection efficiency $\eta$ is a linearly decreasing function of the ratio $\mathcal{U}(T_c - T_a)/I$ and the heat collection rate $q$ decreases with increasing collector temperature. The maximum collector efficiency is $\beta$ when heat is collected at ambient temperature. The efficiency falls to zero when the collector temperature reaches $(T_c)_{max}$. Neither of these limits of operation has practical value because either no heat is collected or it is collected at ambient temperature, which has no useful function. Practical collector systems will function at intermediate conditions where $\eta < \beta$ and $T_c < (T_c)_{max}$.

Flat plate collectors operate at lower efficiency in winter than in summer. Using typical collector values of $\beta = 0.8$ and $\mathcal{U} = 5$ W/m$^2$K, winter and summer clear day noontime values of $I = 870$ W/m$^2$, $T_a = 21°C$ (70°F) and $I = 910$ W/m$^2$, $T_a = 0°C$ (32°F), and a collector temperature of $T_c = 60°C$ (140°F) suitable for domestic water heating, the corresponding winter and summer collector efficiencies are $\eta_w = 45.5\%$ and $\eta_s = 58.6\%$. At other hours, where $I$ is less than these noontime values, the collector efficiencies will be less. This will also be the case when the sky is cloudy. Year-round collector efficiencies are likely to be in the range of 30–50%.

Because of the vagaries of solar irradiance from day to day, a solar collector, no matter how big, can never completely satisfy the demand for year-round heat for domestic hot water or space heating, and a backup supply must be available for satisfactory operation. The relationship between annual heat collection and collector area is sketched in Figure 8.9a. A very small collector accumulates

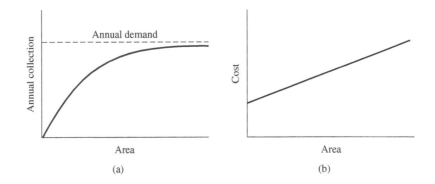

(a)                                      (b)

**FIGURE 8.9**    The characteristics of solar heat collection systems as a function of collector area: (a) annual heat collection and (b) capital cost.

but little heat, whereas an oversized one collects enough heat to meet the daily demand on all but a few very cold, cloudy days, when auxiliary heat must be supplied. The capital cost of a collector system is plotted in Figure 8.9b, showing how the cost increases with collector area, but is finite for small areas because the storage, piping, and control system needed constitute an irreducible cost. An optimum design would minimize the ratio of cost to heat collection and would have an area intermediate between a small and large value because these extremes have either small heat collection or large cost, making the ratio larger than the optimum.

## 8.5.2  Focusing Collectors

The purpose of employing focusing solar collector systems is to increase the intensity of the solar radiation falling on the collector, thereby making it possible to collect solar energy at a higher temperature and with a smaller collector area than for a simple flat plate system. The factor by which the solar irradiance is increased is called the concentration ratio, CR.

The principle of light concentration is sketched in Figure 8.10. A curved parabolic mirror whose axis points in the sun's direction will form an image of the sun at its focal point, a distance $F$ from the mirror called the focal length. The image dimension $D_i$ is equal to the product of the focal length $F$ and the small plane angle $\alpha = 9.3E(-3)$ radian that the sun subtends when viewed from the earth. The ratio of mirror dimension $D_m$ to image dimension $D_i$ is thus found to be

$$\frac{D_m}{D_i} = \frac{D_m}{\alpha F} = 107.5 \left(\frac{D_m}{F}\right). \tag{8.6}$$

For a mirror of circular shape that forms an image as does a camera (called spherical), the concentration ratio is equal to the ratio of the area of the mirror to that of the image, or $(D_m/D_i)^2$. For a cylindrical mirror, which focuses light only in one dimension, the concentration ratio is $(D_m/D_i)$. Thus,

$$CR = 1.156E(4) \left(\frac{D_m}{F}\right)^2 \quad \text{(spherical)}$$

$$= 1.075E(2) \left(\frac{D_m}{F}\right) \quad \text{(cylindrical)}. \tag{8.7}$$

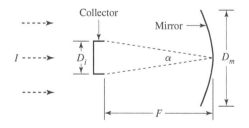

**FIGURE 8.10**  A focusing mirror ($D_m$) concentrates the solar irradiance $I$ (dashed lines) on a smaller collector ($D_i$) located at the focal point of the mirror.

In camera lenses, the ratio of focal length to lens diameter is called the *f number*. The smaller its value, the greater is the lens' light-gathering power, or concentration ratio, permitting film exposure in dim lighting conditions. For focusing solar collectors, it is not practical to construct mirrors with $F/D_m$ less than about 2.

The collectors for focusing systems, which operate at higher temperatures, are of different design than those for flat plate systems. Nevertheless, their collection efficiency and maximum temperature obey the same relationships of that of Equations (8.5) and (8.4), but with the collector irradiance increased by the factor (CR)

$$\eta \equiv \frac{q}{I} = \beta - \frac{\mathcal{U}(T_c - T_a)}{I(\text{CR})} \tag{8.8}$$

$$(T_c)_{\max} = T_a + \frac{\beta I(\text{CR})}{\mathcal{U}}, \tag{8.9}$$

in which CR is to be taken from Equation (8.7), and the absorbed light fraction $\beta$ and heat transfer coefficient $\mathcal{U}$ are those for the focusing system collector. It should also be noted that $I$ is the solar beam irradiance and does not include the diffuse light, which cannot be focused by the mirror. But given the large values of the CR that are available to focusing systems, they can operate at high efficiency even with high collector temperatures.

Focusing systems collect solar energy at a sufficiently high temperature to use that energy in a heat engine cycle to generate electric power efficiently. Figure 8.11 shows a spherical focusing system consisting of individual mirrors attached to a movable frame whose optical axis tracks the sun's position in the sky. The mirrors focus on a Sterling cycle heat engine that generates 25 kilowatts of electrical power, about 24% of the solar beam irradiance falling on the mirrors.

**FIGURE 8.11**   A parabolic mirror system focuses sunlight on a Sterling cycle heat engine that produces electrical power. (By permission of DOE/NREL-PIX.)

**FIGURE 8.12**  A spherical mirror system consisting of individual focusing mirrors arrayed around a central tower that collects the focused light. (By permission of DOE/NREL-PIX.)

An alternative spherical system is shown in Figure 8.12 where individual mirrors arrayed about a central focusing axis each track the sun's motion so as to focus the sun's image on a collector at the top of a central tower. In this system, 2 megawatts of electrical power are generated in a Rankine cycle engine that is heated by hot fluid circulating through the collector at the tower's top. A cylindrical focusing system can be seen in Figure 8.13 where the collector is a pipe through which fluid circulates.

### 8.5.2.1  Solar Thermal Farms

To develop electric power plants utilizing solar thermal energy collection, in the range of 10–100 MW of electric power output, focusing collectors must be arrayed over a large area of 2–10 hectares per MW. The characteristics of several examples of such systems are listed in Table 8.5. The largest, in the United States (Nevada), utilizes cylindrical trough reflectors such as those shown in Figure 8.13. Two smaller plants having two-axis tracking, of the type shown in Figure 8.12, are also listed. All four are located at favorable sites where clear skies predominate, so as to maximize the collection of beam irradiance. The operating farms have a capacity factor of 0.24 and capital cost of ~4 (2007$)/W.

### 8.5.3  Photovoltaic Cells

Photovoltaic (PV) cells are solid-state devices that generate electric power when irradiated by solar light. They provide a direct method of electric power production from sunlight without the thermomechanical complications of solar thermal systems. Like solar thermal systems, only a fraction

**FIGURE 8.13** A cylindrical collector mirror focuses sunlight on a glass pipe containing fluid. (By permission of DOE/NREL-PIX.)

**TABLE 8.5** Solar Thermal Farms

| Location | Type | Rated power (MW) | Area (Hectare) | Power/area (W/m$^2$) | Capacity factor | Cost ((2007$)/W) |
|---|---|---|---|---|---|---|
| United States (Nevada) | Trough | 64 | 130 | 49 | 0.24 | 3.9 |
| United States (Florida) | Trough | 75 | 200 | 37 | — | 6.4 |
| Australia $^a$ | Tower | 10 | — | — | 0.34 | 1.9 |
| Spain | Tower | 11 | 95.2 | 11.6 | 0.24 | — |

$^a$ Proposed.

of the solar irradiance incident upon a solar cell can be converted to electric power. Nevertheless, their mechanical and electrical simplicity, small operating cost, and ability to produce power at any scale make the use of PV power systems very attractive. A major obstacle to widespread use of PV cells is their current high initial cost per unit of electrical power output and the intermittency of the output power.

The process whereby the energy flux in solar radiation is converted to electrical power is quantum mechanical in origin. The energy of sunlight is incorporated in packets of electromagnetic radiation called photons, each of which possesses an energy of amount $hc/\lambda$, where $h$ is Planck's constant, $c$ is the speed of light, and $\lambda$ is the wavelength of light.[21] Sunlight contains photons of

---

[21] See Table A.3.

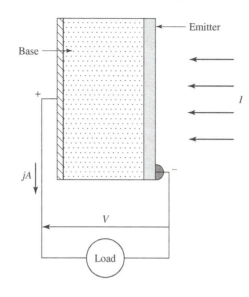

**FIGURE 8.14**   The elements of a photovoltaic cell of area $A$ exposed to a solar irradiance $I_s$ and supplying a total current $jA$ to an external load at an electric potential difference $V$. The current density, electric current per unit area, in the cell is $j$.

all wavelengths and energies, but most of the solar energy content lies in the wavelength range of 0.3–2.5 $\mu$m (see Figure 8.5). If a photon can deliver all its energy to an electron in a semiconductor, for example, it can move the electron to an energy state of electric potential higher by the amount $hc/e\lambda$, where $e$ is the electron charge.[22] For solar photons, the corresponding electric potential range is 4.1–0.5 volts. If the energetic electron can flow through an electric circuit while experiencing this potential change, it can deliver electrical energy to an external load.

The components of a photovoltaic cell and its accompanying electric circuit are sketched in Figure 8.14. The cell consists of a *base* layer of a p-type semiconductor, about 250 $\mu$m in thickness, joined to an extremely thin *emitter* layer of an n-type semiconductor about 0.5 $\mu$m thick that is exposed to the solar irradiance. Solar radiation is absorbed mostly in the thin region close to the junction of the two materials. Current generated by the cell is collected at positive and negative electrodes attached to the exterior of the p-type and n-type semiconductor layers, respectively. The external circuit that receives the electrical power generated by the cell is connected to these electrodes.

The n-type and p-type semiconductor materials consist of a pure semiconductor, such as silicon, doped with a small amount of another element. In the n-type layer, silicon is doped with P or As, elements that readily give up an electron from the outer shell and are thereby called *donors*. In the p-type layer, silicon is doped with B or Ga, which accept an electron into their outer shell and are called *receptors*. The mobile electrons in the n-type layer and the mobile "holes" in the p-type layer provide charge carriers that permit an electric current to flow through the cell. At the junction

---

[22]The factor $hc/e$ has the value of 1.239 volt $\mu$m.

of the two layers, there exists an electric potential difference between them required to maintain thermodynamic equilibrium between the charge carriers in either layer.

When sunlight falls upon the cell, some photons penetrate to the region of the interface and can create there an electron–hole pair, provided that the photon energy equals or exceeds the gap energy $E_g$ (measured in electron volts) needed to move an electron from the valence band to the conduction band (i.e., provided the wavelength $\lambda$ is less than $hc/E_g$). The electron and hole then move to the negative and positive electrodes, respectively, and provide a current that moves through the external circuit from the positive to the negative electrodes with an accompanying electric potential drop, both sustained by the flow of photons into the cell.[23]

In these processes only a fraction of the solar energy flux is utilized in creating electrical power to feed into the external circuit. Long wavelength photons ($\lambda > hc/E_g$) have insufficient energy to create an electron–hole pair, so their absorption merely heats the cell. Short wavelength photons ($\lambda < hc/E_g$) have more energy than needed, with the excess ($hc/\lambda - E_g$) appearing as heating, not electrical power. Typically, only about half of the solar irradiance is available to produce electrical power. Of this amount, only a fraction eventually results in electric power flowing to the external circuit because of various additional internal losses in the cell.

This upper limit to the fraction $f$ of the transmitted solar irradiation is determined by the black body spectrum of the sunlight falling on the PV cell, according to the relation

$$f = \frac{15x_g}{\pi^4} \int_{x_g}^{\infty} \frac{x^2}{e^x - 1}\, \mathrm{d}x\,;$$

$$x \equiv \frac{hc}{\lambda kT}\,;\quad x_g \equiv E_g/kT, \tag{8.10}$$

where the integral evaluates the energy $E_g$ delivered to the external circuit by all photons having an energy exceeding that value. Figure 8.15 depicts $f$ as a function of $E_g/kT_s$, where $T_s = 5900$ K is the temperature of the solar spectrum. It can be seen that the maximum value of $f$ is about 0.54, and this occurs when $x_g = 2.3$. The corresponding values of $E_g$ and $\lambda = hc/x_g kT_s$ are 1.17 eV and 1.06 $\mu$m, respectively.

When exposed to sunlight, a PV cell generates electric current and a cell potential difference, depending upon the solar irradiance level and the electrical characteristics of the load in the external circuit. In the limit where the external circuit is not closed, no current can flow but an open circuit voltage $V_{oc}$ is generated that increases with solar irradiance, but not proportionately so, as sketched in Figure 8.16a. At the opposite limit of an external short circuit, where the cell voltage is zero, a short circuit current, of current density $j_{ss}$, flows through the cell in an amount proportional to the solar irradiance (see Figure 8.16a). In both these extremes, no electrical power is generated since either the current or the voltage is zero. For intermediate cases where the cell delivers electrical power to a load, the cell voltage and current density, $V$ and $j$, are each less than the limiting values of $V_{oc}$ and $j_{ss}$. For a given value of the irradiance, the relationship between $V$ and $j$, sketched in Figure 8.16b, depends upon the electrical characteristics of the load. The power output is the product $Vj$, which reaches a maximum $(Vj)_{max}$ at a point intermediate between the open and short

---

[23]A related effect is the emission of light when electric power flows into the cell from an external source, as in the light-emitting diode.

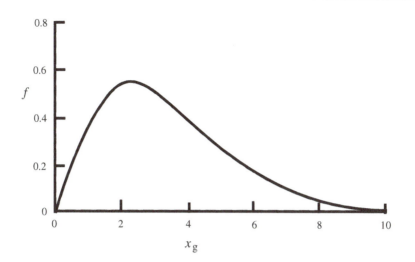

**FIGURE 8.15** The maximum fraction $f$ of the transmitted solar irradiance that can be converted to electric power in the external circuit as a function of the dimensionless photovoltaic cell energy gap $x_g$ of Equation (8.10).

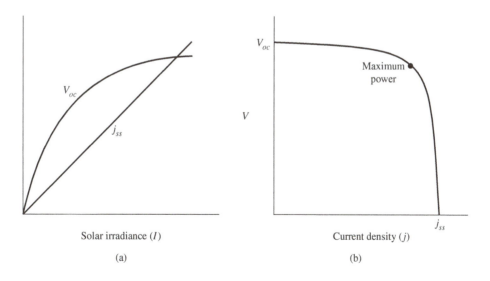

**FIGURE 8.16** (a) The open circuit voltage $V_{oc}$ and short circuit current density $j_{ss}$ of a photovoltaic cell as functions of solar irradiance. (b) The voltage versus current density characteristics of a photovoltaic cell for a fixed value of solar irradiance, showing the point of maximum power.

circuit conditions. The maximum power per unit area that the cell can deliver, $(Vj)_{max}$, is generally about 60–80% of the product $V_{oc}j_{ss}$.

The efficiency $\eta$ of a PV cell is the ratio of the electrical power generated per unit area, $Vj$, to the solar irradiance $I_s$ impinging on the cell surface:

$$\eta \equiv \frac{Vj}{I_s}. \tag{8.11}$$

Given the properties sketched in Figure 8.16, the efficiency increases with increasing irradiance and, for any level of irradiance, is a maximum at the maximum of output power. PV cell efficiencies are rated for a solar irradiance $I_s = 1$ kW/m$^2$ having a spectral irradiance of that of sunlight, under the conditions for maximum power. Typical efficiencies are in the range of 15–25%, with $V_{oc} \sim 0.5$ V and $j_{ss} \sim 300$–400 A/m$^2$.

The PV cell is but one component of a practical system for converting sunlight directly to useful electric power. Many individual cells must be wired together to produce output power at desirable voltage and current levels, and the array of cells must be mounted in a manner that optimizes the solar irradiance input. PV power is DC power and must be converted to AC form for most uses. These other constituents of a practical PV power supply add to its capital cost, although at the present time the basic cell cost is so high that it constitutes the major component of system cost.

### 8.5.3.1  Photovoltaic Farms

As a source of electric power feeding into an electric transmission and distribution system supplying a regional network, PV receivers need to be arrayed in a field of units comprising hundreds of megawatts of generating capacity. Compared with fossil- or nuclear-fueled power plants of equal generating capacity, much greater land areas are needed, and these must be economical to rent or buy. PV farms are built in rural regions having low land costs.

Table 8.6 lists some recently built or proposed PV power plant farms, in the size range of 1–200 MW rated power. The columns list the farm area in hectares, the ratio of rated power to farm area (watts per square meter), the annually averaged capacity factor, and the capital cost (2007$ per rated watt). Note that the values in the last three columns are almost independent of power plant size. Rated power per unit area is usually about 20–40 watts per square meter (0.2–0.4 megawatts per hectare), reflecting the rated solar insolation (one kilowatt per square meter of cell surface), cell efficiency of collection and conversion to electricity, and the area needed to deploy the cells without shadowing each other. The capacity factors are surprisingly nearly identical. The range of capital costs is small: 7–8 2007$ per rated watt.[24] The last row of the table, listing the properties of a small photovoltaic plant retrofitted to the flat roof of an existing building, illustrates the higher unit cost and lower capacity factor of systems that do not maximize the benefit/cost ratio or include the advantages of large scale.

The low capacity factor of $\sim 0.2$ is a compound of several effects: solar insolation falls on the cell for only half the time on average, its average over that time is only $2/\pi$ of its noon time maximum (for a fixed flat plate receiver), and the annual average of the peak insolation is always less than the rated power. An advantage of PV power is that it produces electricity during daylight hours when system demand is higher than its daily average; but this production is weather dependent, falling when clouds cover an otherwise clear sky. Since cloud cover can be predicted quite reliably

---

[24]The proposed Australian plant utilizes concentrating collectors that reduce cell area per watt and cost.

**TABLE 8.6**  Photovoltaic Farms

| Location | Rated power (MW) | Area (Hectare) | Power/area (W/m$^2$) | Capacity factor | Cost ((2007$)/W(rated)) |
|---|---|---|---|---|---|
| United States (California)[a] | 250 | 600 | 42 | — | 3.5 |
| Australia[a,b] | 154 | — | — | 0.2 | 2.7 |
| South Korea[a] | 20 | — | — | — | 7.8 |
| United States (Nevada) | 14 | 57 | 25 | — | — |
| Portugal | 11 | 37 | 30 | 0.22 | 6.8 |
| Portugal | 11 | 61 | 18 | — | 7.1 |
| Bavaria | 10 | 25 | 40 | — | — |
| Spain | 1.2 | 12 | 10 | 0.2 | — |
| United States (MIT) | 0.012 | 0.023 | 51 | 0.14 | 30 |

[a] Proposed.
[b] Concentrating collector.

a day in advance, the scheduling of substitute reliable conventional power can provide substitute system power for the proverbial rainy day.

To lower the economic cost of PV power, several improvements have been proposed. One is to increase the conversion efficiency $\eta$, thereby reducing the cell area and cost per rated watt. A second is to use focusing mirrors to increase the intensity of radiation falling on the cell, increasing the rated output per unit cell area (but not per unit collector area) and thereby reducing the cell cost per unit rated watt. An alternative approach is to reduce the manufacturing cost per unit area by using multicrystalline or noncrystalline semiconductor materials or processes that do not degrade the efficiency more than they decrease the cost per unit area.

One way to increase cell efficiency is to use three layers of semiconductors having two interfaces each tuned to different wavelengths, thereby increasing the yield of a greater range of the solar photon spectrum, but this is inherently more expensive to manufacture. For space applications, this type of cell has been used with focusing collectors to reduce the mass per unit rated watt, much more important than manufacturing cost for space systems. These systems can have concentration ratios of 300–1,000 that require high-grade optical focusing; they also have somewhat higher efficiency. They have been proposed for use in hot dry regions with two-axis motion that tracks the solar image across the sky from dawn to dark. At locations with considerable cloudiness where most solar insolation is diffuse rather than direct, they do not perform as well as flat plate collectors. It remains to be seen whether their capital cost is significantly below that of flat plate collectors.

Much progress is being made in utilizing different materials and manufacturing processes to reduce the cost per rated watt. Manufacturing cost generally decreases as manufacturing volume grows. Flat plate cells are the only type that can be easily deployed for individual homes or business consumers. When the owner/investor is located where annual solar insolation and end-consumer prices are high and where excess electric power can be sent into the regional grid for credit against consumption at that same price (called net metering), the economics become more favorable.[25]

---

[25]Retail electricity prices include charges for generating, transmitting, and distributing electricity, where the latter two are usually comparable to the generating cost. End consumer generation need only be cheaper than the total retail cost to be economically justified.

### 8.5.4  Environmental Effects

The principal environmental effect of solar thermal or photovoltaic farms is the total encumbering of the land area occupied by the receivers, which cannot be otherwise used, except in the case of rooftop receivers on commercial or residential buildings. The farm area needed is $\sim 20–100 \, m^2/kW$. The ecological and amenity values of the land would most likely be lost, as they would for most other industrial uses.

## 8.6  WIND POWER

The use of wind to provide mechanical power for grinding grain and propelling ships and boats dates from ancient times, 3 or 4 millennia ago. Substantial improvement in wind technology within the past thousand years, especially for transoceanic vessels, made possible the migration of populations to western and southern hemisphere continents and the initiation of intercontinental trade. The technology of sailing vessels had reached a high level in the mid to late 19th century when it was suddenly displaced by steel-hulled, fossil fuel powered steamships that greatly transformed intercontinental travel and trade.

At one time wind-powered mills for grinding grain and pumping water were common in western Europe, numbering some 10,000 by the 12th century. By present standards their design was rudimentary and their use, like that of sailing vessels, was displaced by the advent of industrial power in the 19th century.

The modern wind turbine is a development of the past few decades that utilizes the latest technology in design and manufacture. Currently used exclusively to produce electrical power, wind turbines are usually tied into electrical transmission and distribution systems, although some turbines are used to power remote installations. Despite the many ingenious forms of wind turbines that have been developed, the predominant type used today is the horizontal axis machine, mounted on a support tower, that is free to rotate so as to align its axis with the wind direction. The turbine rotor, invariably a three-bladed propeller-like structure, drives an electric power generator through a speed increasing gear. Because the turbine-generator must be able to rotate with respect to its supporting tower, special provisions are needed to transmit the generated power to the stationary supporting tower. Since the turbomachinery is mounted at the top of a tower 50 or more meters in height, the wind turbine rotating and the whole power unit swivelling into the wind, a reliable and strong supporting structure that will withstand the highest expected storm winds is required.

Below in Section 8.6.1 we begin by explaining the aerodynamic principles governing the output of a wind turbine and how that is related to the turbine diameter and wind speed. In Section 8.6.2 on electromechanical aspects of wind turbines, we show how the turbine mechanical power output is matched to the electrical generator that sends that power to the transmission line of the regional power network. Since a wind turbine converts wind power to useful electrical power, we next examine the nature of the wind resource in Section 8.6.3, showing how the great variability in wind speed produces variable electrical power output from the wind turbine. This has important consequences to the usefulness of wind turbines to satisfy electric power demand, to the effectiveness of utilization of the wind resource, and to the capital costs of wind turbine plants (Section 8.6.4). Finally, in Section 8.6.5, we explain the constraints on the design of wind farms and the array of wind turbines needed to provide commercially viable sources of electrical power from wind turbines.

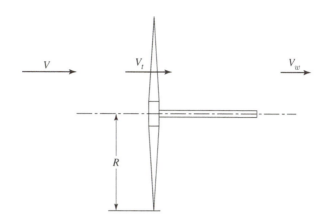

**FIGURE 8.17** The flow of wind having a speed $V$ and passing through a wind turbine whose blade radius is $R$ is slowed to a speed $V_t$ at the turbine disk and even further to a speed $V_w$ in the wake region downstream of the turbine.

## 8.6.1 Aerodynamics of Wind Turbine Operation

A wind turbine more closely resembles an airplane propeller than it does a steam or gas turbine rotor. The wind turbine blades are long and slender; the tip of the blade moves at a speed much greater than the wind speed. An airplane propeller is designed to produce a large thrust while the wind turbine must produce power, which is the product of the torque that the wind applies to the turbine rotor times its rotational speed. Nevertheless, the wind turbine blade shape is quite similar to that of an airplane propeller. The slender blades, twisted to conform to a helical surface, occupy a small fraction of the turbine disk area (the circular plane area swept by the rotating turbine blades), yet they absorb kinetic energy from the entire flow passing through the turbine disk area.

The source of power from the wind is the flow of air through the wind turbine. If $V$ is the wind speed, each unit mass of air possesses a kinetic energy of amount $V^2/2$.[26] If $A = \pi R^2$ is the area subtended by the rotating blades of length $R$, then the mass flow rate of air through an area $A$ of the undisturbed wind stream is $\rho VA$, where $\rho$ is the air density.[27] The maximum rate at which the wind kinetic energy could be supplied by the wind flow passing through the turbine is the product of the mass flow rate, $\rho VA$, and the kinetic energy per unit mass, $V^2/2$, for a value of $\rho V^3 A/2$. In practice, the wind turbine power available from aerodynamically perfect design is less than this value because the action of the turbine modifies the surrounding wind flow, reducing the mass flow rate below the value of $\rho VA$.

To illustrate this effect, Figure 8.17 shows the ideal wind flow past a turbine. Because the turbine extracts some of the kinetic energy of the wind flow, the wind speed is reduced in the vicinity of the turbine to a value $V_t$ at the turbine and to an even lower value of $V_w$ downstream of the turbine, called the wake region. The turbine power $\mathcal{P}$ is then the product of the mass flow rate

---

[26]In SI units, the unit of kinetic energy per unit mass is $m^2/s^2 = J/kg$.

[27]During a small time interval $dt$, a fluid particle moves a distance $V dt$ normal to the wind turbine disk, so that a volume $AV dt$ and mass $\rho AV dt$ of air pass through the turbine disk, where $\rho$ is the air mass density (SI units = $kg/m^3$).

through the turbine, $\rho V_t A$, and the reduction of kinetic energy of the wind, $V^2/2 - V_w^2/2$, or

$$P = \frac{1}{2}\rho V_t(V^2 - V_w^2)A. \tag{8.12}$$

As the air slows down both upstream and downstream of the wind turbine, it undergoes a pressure rise that is very small compared with atmospheric pressure. This provides a pressure drop across the wind turbine of amount $\rho(V^2 - V_w^2)/2$ and a corresponding axial thrust force $T$ equal to $\rho(V^2 - V_w^2)A/2$. But this force must also equal the reduction in momentum of the wind flow, $\rho V_t A(V - V_w)$,

$$T = \rho V_t A(V - V_w) = \frac{P}{V_t}, \tag{8.13}$$

where $P = V_t T$ is the relation between power and thrust, as for an airplane propellor. It then follows from (8.12) and (8.13) that $V_t$ is the average of $V$ and $V_w$:

$$V_t = \frac{1}{2}(V + V_w). \tag{8.14}$$

The power $P$ and thrust $T$ can now be expressed in terms of $V$ and $V_w$ as

$$P = \frac{1}{2}\rho V^3 A \left[ \frac{(1 + V_w/V)^2(1 - V_w/V)}{2} \right] \tag{8.15}$$

$$T = \frac{1}{2}\rho V^2 A \left[ 1 - \left(\frac{V_w}{V}\right)^2 \right] = \frac{P}{V}\left[ \frac{2}{(1 + V_w/V)} \right]. \tag{8.16}$$

The factor in brackets in (8.15) has a maximum value of 16/27 when $V_w = V/3$, in which case 8/9 of the wind's kinetic energy has been removed by the wind turbine. As a consequence of (8.15), the wind turbine power cannot exceed the limit[28] of

$$P \le \frac{16}{27}\left[ \frac{1}{2}\rho V^3 A \right]. \tag{8.17}$$

The wind turbine support structure must be able to withstand the thrust $T$, which is comparable in magnitude to the drag force on a sphere of cross-sectional area A.

The foregoing considerations do not reveal the detailed mechanism whereby the wind flow exerts a torque on the wind turbine rotor in the direction of its rotation, thereby generating mechanical power. In Figure 8.18 is sketched the amount and direction of the wind flow relative to a section of the turbine blade at a radius $r$ from the turbine axis. In the tangential direction, the velocity component is $2\pi r f$, where $f$ is the rotational frequency of the turbine shaft.[29] In the direction of

---

[28] If the flow in Figure 8.17 is reversed in direction, with an incoming speed of $V_w$ and a leaving speed of $V$, the flow is that through a propellor at a flight speed $V_w$. The relations (8.15) and (8.16) then define the propellor power and thrust.

[29] Strictly speaking, the relative speed in the tangential direction is somewhat less than $2\pi r f$ because the blade tangential force induces a small amount of tangential flow. This is generally negligible for efficient wind turbine designs.

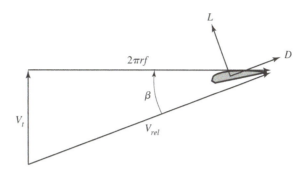

**FIGURE 8.18**  The motion of the wind relative to a turbine blade consists of an axial speed $V_t$ and a tangential speed $2\pi rf$ that generates a lift force $L$ and drag force $D$.

the turbine axis the wind speed is $V_t$. The net relative speed $V_{rel}$ is thereby $\sqrt{V_t^2 + (2\pi rf)^2}$, and this velocity lies at an angle $\beta = \arctan(V_t/2\pi rf)$ to the tangential direction, as shown in Figure 8.18.[30] The airfoil-shape section of the turbine blade, pitched at a small angle of attack to this direction, develops a lift force $L$ and drag force $D$ that are, respectively, perpendicular and parallel to the direction of the relative velocity. The net force exerted on the blade section in the direction of its motion is $L \sin \beta - D \cos \beta$, and the rate at which this force does work on the turbine, which is the power developed by the wind, is thereby $2\pi fr(L \sin \beta - D \cos \beta)$.

There are two important deductions that can be made from this analysis. The first is that it is necessary that the airfoil lift $L$ should greatly exceed its drag $D$, which can be secured by utilizing efficient airfoil shapes in forming the turbine blades. The second is that the turbine power is enhanced by turbine tangential speeds that are large compared with the wind speed, so as to come as close as possible to the power limit of Equation (8.17). Optimum designs incorporate a speed ratio (turbine tip speed/wind speed) of about 5 to 7.

The performance of a wind turbine may be described in terms of a dimensionless power coefficient $C_p$ defined as

$$C_p \equiv \frac{\mathcal{P}}{\frac{1}{2}\rho V^3 A}.$$  (8.18)

The power coefficient cannot exceed the ideal value of 16/27, but is otherwise a function of the ratio of tip speed to wind speed, as sketched in Figure 8.19, reaching a maximum at the design speed ratio. At lower speed ratios the airfoil angle of attack is too great and the aerodynamic drag $D$ increases greatly, reducing turbine power. At higher speed ratios, the airfoil angle of attack decreases, reducing the lift $L$ and also the turbine power.

---

[30]The angle between the blade chord and tangential direction is called the pitch angle. It varies between 90 degrees at the hub and a very small value at the tip. The blade surface is a helix, as in a machine screw; hence, the name "screw propellor" is used for both airplanes and ships.

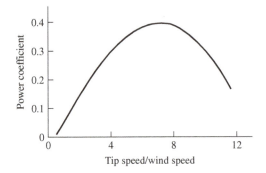

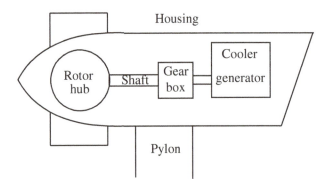

**FIGURE 8.19** The power coefficient of a wind turbine of fixed pitch is a function of the ratio of the tip speed to wind speed. Its maximum value is less than the theoretical value of $16/27 = 0.593$.

**FIGURE 8.20** A sketch of the electromechanical system for converting wind turbine power to electric power, housed in the turbine hub pod.

### 8.6.2  Mechanical and Electrical Components

Compared with a steam or gas turbine, a wind turbine of equal power revolves much more slowly, usually only a few revolutions per minute (rpm), but has much higher torque.[31] Sixty-cycle AC electric generators rotate at speeds up to 3,600 rpm. To match the high speed and low torque characteristics of the electric generator that the wind turbine drives, a speed increase gear is required. The diameters of the gear and the electric generator are much less than that of the turbine and they are housed in a relatively small pod that supports the turbine hub atop the support tower. This housing and its turbine must pivot to face the wind, yet still transmit the very large aerodynamic thrust to the support tower.

Figure 8.20 shows a sketch of the electromechanical system that converts the mechanical power of the wind turbine into AC electrical power that can be fed to transmission or distribution

---

[31]This difference is a consequence of the low value of the energy flux per unit area, $\rho V^3/2$, in a wind turbine compared with the very high value in steam or gas turbines, where $\rho$ and $V$ are so much greater.

lines serving users. The principal components are the speed-increasing gear box and the electric power generator with its cooler and power controls. The asynchronous generator operates over a range of frequencies, while the power control fashions the output to match the exact line frequency and voltage required by the transmission system. In addition, mechanical systems for adjusting the wind turbine blade pitch and rotor yaw angle are provided. Electric controls provide for generator operation, which accepts the wind turbine speed and torque conditions arising from the large range of wind conditions within which the wind turbine operates while providing 60-cycle voltage and current that transports the power to an electric transmission line.

Wind turbines are designed to produce power over a range of wind speeds, from a low *cut-in* speed where very little net power is generated to a high *cut-out* speed above which aerodynamic forces would overstress the turbine blades. The ratio of maximum to minimum speed is about 4; the ratio of maximum to minimum available wind power is thus about 64. It is not economical to build a wind turbine to utilize fully this large dynamical range of wind speed and power; instead, the design is based upon matching the power of an aerodynamically efficient turbine to the maximum electric power output of the generator at an intermediate wind speed, called the *rated* speed $V_r$, which is approximately half the cut-out speed. When the wind speed $V$ is higher than $V_r$, the turbine can deliver much more power than the rated power $\mathcal{P}_r$ (since $\mathcal{P}$ varies approximately as $V^3$) but the electric generator cannot absorb this additional power without overheating, so the rotor pitch angle is adjusted to deliver only the rated power output to the generator. On the other hand, at wind speeds less than $V_r$, less power than $\mathcal{P}_r$ is generated, by a factor of about $(V/V_r)^3$, and the electrical output is reduced below the rated value. The economical optimum design is one that maximizes the ratio of the average electrical power output to the capital cost of the wind turbine installation, given the wind speeds available at the site. For efficient designs used at desirable sites, the ratio of time-averaged power output to the rated output, called the *capacity factor*, is about 30–40%.

Given these constraints on wind turbine operation, the expected electrical output power $\mathcal{P}$ as a function of $V$ is sketched in Figure 8.21, where the ratio $\mathcal{P}/\mathcal{P}_r$ is plotted as a function of $V/V_r$. For $V \geq V_r$, $\mathcal{P} = \mathcal{P}_r$ because the power output is limited by the generator. For $V \leq V_r$, where

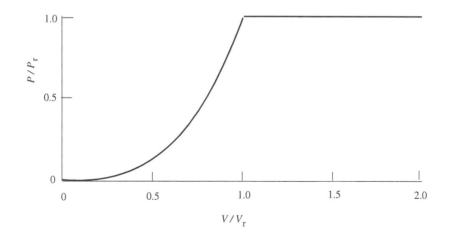

**FIGURE 8.21**   A plot of the ratio of wind turbine power $\mathcal{P}$ to rated power $\mathcal{P}_r$ as a function of the ratio of wind speed $V$ to rated wind speed $V_r$, as in (8.19).

**TABLE 8.7**    Properties of Wind Energy Turbine Systems[a]

| | | | | |
|---|---|---|---|---|
| Rated electrical power (kW) | 3,600 | 2,500 | 2,000 | 1,500 |
| Rotor diameter (m) | 104 | 100 | 80 | 70.5 |
| Rated wind speed (m/s) | 14 | 12.5 | 15 | 13 |
| Cut-in wind speed (m/s) | 3.5 | 3.5 | 4.0 | 4.0 |
| Cut-out wind speed (m/s) | 27 | 25 | 25 | 25 |
| Rotor speed (rpm) | 8.5–15.3 | — | 9–19 | 12–22 |
| Rated power/area (kW/m$^2$) | 0.424 | 0.318 | 3.98 | 0.384 |
| Rated power coefficient | 0.257 | 0.270 | 0.196 | 0.290 |
| Tip speed ratio | 3.3–6.0 | — | 2.5–5.3 | 3.4–6.2 |

[a] Data from http://www.gewindenergy.com and http://www.vestas.com.

the power output is limited by the turbine, $\mathcal{P} = (V/V_r)^3 \mathcal{P}_r$, assuming a constant value of power coefficient in this range,

$$\frac{\mathcal{P}}{\mathcal{P}_r} = \left(\frac{V}{V_r}\right)^3 ; \qquad \text{if } V \leq V_r$$
$$= 1; \qquad \text{if } V \geq V_r. \tag{8.19}$$

Table 8.7 displays the design characteristics for four wind turbines of 3.6, 2.5, 2 and 1.5 MW of rated electrical power that span a narrow range of rated wind speeds. Despite the range of rotor diameters and power outputs, the dimensionless rated power coefficients and tip speed ratios are nearly the same. This implies that the designs are aerodynamically and electromechanically similar.

What is striking about the entries in Table 8.7 is the large size of wind turbines needed to generate modest amounts of power: a 100-m-diameter turbine is rated at about 3 MW. Typical steam or gas turbines used in power plants generate hundreds of megawatts and have diameters of several meters. The high density and velocity of powerplant steam or gas makes it possible to generate a thousand times more power per unit flow area of turbine at correspondingly much lower cost per rated MW. There are inherent structural problems in building wind turbines of much greater power than those illustrated in Table 8.7. Instead, large amounts of power are generated by deploying an array of very many wind turbines in a "wind farm."

### 8.6.3  Wind Resources

A desirable site for a wind power installation is one having a high time-averaged wind speed $\overline{V}$. Sites are commonly rated by the average value of the energy flux per unit area, $\rho \overline{V^3}/2$.[32] High values of wind energy flux require smaller, less costly turbine rotors for a given electrical power output. Figure 8.22 identifies U.S. regions with high wind energy levels. These are located mostly in open unforested plains, mountainous regions, and coastal areas. Similar ranges of wind energy are encountered on other continents.

---

[32]Site energy fluxes are classified in one of five levels: class 1 (0–100 W/m$^2$), class 2 (100–200 W/m$^2$)... class 5 (more than 400 W/m$^2$).

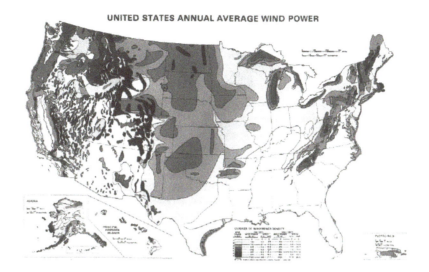

UNITED STATES ANNUAL AVERAGE WIND POWER

**FIGURE 8.22**  A U.S. map showing areas graded by the annual average wind energy flux $\rho V^3/2$, with darker areas indicating higher wind energy flux. (Elliot, D. L., *et al. Wind Energy Resource Atlas of the United States*. DOE/CH 10093-4. Golden, Colo.: Solar Energy Research Institute, 1987.)

At any one site, wind speeds are quite variable in magnitude, duration, and direction. Daytime wind speeds are higher than nighttime levels because daytime solar heating of the atmosphere promotes mixing with the higher speed flow aloft. Average winter season wind speeds exceed those of the summer. While the time-averaged annual wind speed $\overline{V}$ approximately measures the availability of wind energy, a more detailed knowledge of wind speed statistics is needed for an accurate assessment of a site's wind energy potential and the selection of a suitable wind turbine design.

The variability of the wind speed with time is an important factor in determining how much of the wind's available power can be delivered to an electric transmission line and how well that power can be utilized by customers. This variability can be expressed in the form of the probability distribution function $p\{V\}$ of wind speed, such as

$$p\{V\} = \frac{\pi}{2}\frac{V}{\overline{V}^2}\exp\left[-\frac{\pi}{4}\left(\frac{V}{\overline{V}}\right)^2\right], \tag{8.20}$$

where $p\{V\}\,dV$ is the probability that the wind speed $V$ lies in the range between $V$ and $V+dV$ and where $\overline{V}$ is the time-averaged wind speed.[33,34] Plotted in Figure 8.23, it can be seen that a very low or high wind speed, compared with the average, is very unlikely, and that the most probable wind speed is $(2/\pi)\,\overline{V} = 0.637\,\overline{V}$. In addition, the average value of the cube of the wind speed, $\overline{V^3}$, to

[33]Equation (8.20) is called a Rayleigh distribution. It satisfies the integral conditions that $\int_0^\infty p\{V\}\,dV = 1$ and $\int_0^\infty Vp\{V\}\,dV = \overline{V}$.

[34]An alternate interpretation of $p\{V\}\,dV$ is that it is the fraction of the time that the wind speed lies in the range $V$ to $V+dV$.

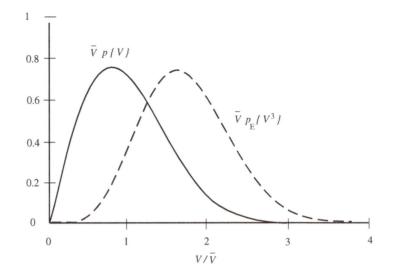

**FIGURE 8.23**   A probability distribution $p\{V\}$ of wind speed (solid line) at a site for which the average speed is $\overline{V}$ shows low values at low and high speeds and a maximum near the average speed. The corresponding probability distribution of $V^3$ is shown as a dashed line.

which the average wind power is proportional, can be calculated to be $\overline{V^3} = (6/\pi)\overline{V}^3 = 1.91\,\overline{V}^3$. Thus the average value of the wind energy flux $\rho\overline{V^3}/2$ is nearly double its value at the average wind speed, $\rho(\overline{V})^3/2$.

It is also of interest to determine the probability distribution of the energy flux, which is proportional to $V^3$ and is denoted by $p_E\{V^3\}$,

$$p_E\{V^3\} = \frac{\pi^2}{12}\frac{V^4}{\overline{V}^5}\exp\left[-\frac{\pi}{4}\left(\frac{V}{\overline{V}}\right)^2\right]. \tag{8.21}$$

This is plotted in Figure 8.23 as a dashed line, showing a distribution that is spread over a wider range of $V/\overline{V}$ than is $p\{V\}$ and having a maximum at $V/\overline{V} = \sqrt{8/\pi} = 1.596$.

The energy flux probability distribution $p_E\{V^3\}$ shows how most of the available wind power is found when the wind speed is between $\overline{V}$ and $2.5\overline{V}$, and this occurs less than half the time. As we shall see below, this makes the average power generated much less than the rated power.

### 8.6.3.1   Capacity Factor

Over a long period of time such as a year or season, as the wind waxes and wanes, a wind turbine produces power that varies between 0 and $\mathcal{P}_r$. Although the economic value of the power depends upon the time of the day it is available, being higher at times of peak demand than low demand, the time-averaged power $\overline{\mathcal{P}_r}$ is a useful measure of the plant output. The ratio of time-average to

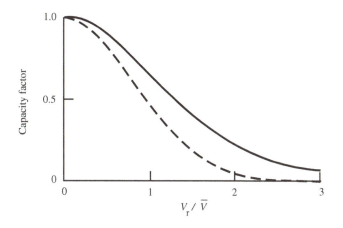

**FIGURE 8.24**  The capacity factor (solid line) of a wind turbine as a function of the ratio of rated to average wind speed, $V_r/\overline{V}$. The fraction of the capacity factor attributed to rated power output is shown by the dashed line.

rated power is called the *capacity factor* CF,

$$\text{CF} \equiv \frac{\overline{\mathcal{P}}}{\mathcal{P}_r}. \tag{8.22}$$

To determine the capacity factor for a wind turbine with the characteristics of (8.19), shown in Figure 8.21, we multiply (8.19) by the energy probability distribution (8.20) and integrate over $V$ to find

$$\text{CF} = \exp(-\pi\alpha^2/4) + (\pi/2\alpha^3)\int_0^\alpha v^4 \exp(\pi\alpha^2/4)\,dv; \qquad \alpha \equiv V/V_r, \ \ v \equiv V_r/\overline{V}. \tag{8.23}$$

The first term is the contribution to CF when $V \geq V_r$ and the power is limited by the generator ($P = P_r$); the second is the remaining contribution when the turbine limits the power ($P \leq P_r$). The capacity factor (8.23) is plotted in Figure 8.24. Also shown in Figure 8.24 is the first term of (8.23) (dashed line), which is the power supplied during the time when the output power is the rated power. This not only decreases with increasing $V_r/\overline{V}$, but is a decreasing fraction of CF.

To deliver electric power reliably to meet consumer demand, it is desirable to have a capacity factor close to unity, so that the wind turbine can supply its rated power nearly all the time. But for this to be so, $\overline{V} \ll V_r$, which means that the turbine area, and thereby cost, is much more than is needed to run the generator over much of the wind speed range. At the other extreme, $\overline{V} \gg V_r$, the generator output is much less than its rated value, so that the generator is oversized. A more balanced design would result if the rated wind speed were to be chosen about equal to the wind speed for the most probable energy flux, or $V_r/\overline{V} \simeq \sqrt{8/\pi} = 1.6$. If this choice is made, the capacity factor of (8.23) is 0.35, of which 0.13 is operation at the rated power (see Figure 8.24). In other words, the turbine produces rated power 13% of the time; the remaining 87% of the time its average output is 25% of the rated value.

At any site, wind speed and energy flux increase gradually with distance $z$ above ground level. While it is theoretically advantageous to place a wind turbine on a high tower so as to take advantage of the higher energy flux available there, the greater cost of a higher tower may not be offset by a less costly turbine. Generally, tower height is proportional to turbine diameter, with the turbine axis being about one to two diameters above ground.

Wind turbines need to be protected from damage by storm or hurricane level winds and, in northern climates, from icing conditions in winter months.

### 8.6.3.2  Effectiveness

On average, a wind turbine does not capture all the power that flows through it. The ratio of the average power collected to the maximum possible power that can be collected is called the *effectiveness*, EFF. Wind power, like oil, is a scarce resource; we would like to recover as much of it as possible. The maximum power per unit turbine flow area is $(16/27)(1/2)\rho V^3 = (8/27)\rho V^3$ and its time-averaged value is $(8/27)\rho \overline{V^3} = (16/9\pi)\rho \overline{V}^3$. The time-averaged output of the turbine is $(CF)\, \mathcal{P}_r = (CF)(C_p \rho V_r^3)/2$, so that the effectiveness becomes

$$\text{EFF} = (CF)\left(\frac{9\pi C_p}{32}\right)\left(\frac{V_r}{\overline{V}}\right)^3. \tag{8.24}$$

Figure 8.25 is a plot of effectiveness versus capacity factor for the particular case of $C_p = 16/27$, the highest possible value. It can be seen that it is not possible to have both a high effectiveness and a high capacity factor. For the particular case of $V_r/\overline{V} = 1.6$, a typical value, the effectiveness is at most 0.74.

### 8.6.3.3  Wind Variability and Predictability

The probability distribution of wind speed, Equation (8.20), determines the long-term average performance of a wind turbine, as embodied in the capacity factor (see Figure 8.24). But for shorter time periods, such as an hour or day, the average behavior is very important to the delivery of power to an electric system grid.

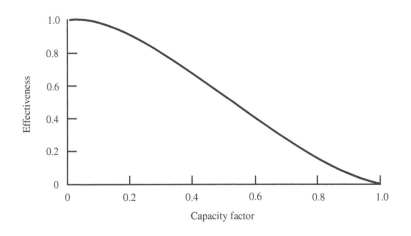

**FIGURE 8.25**   A plot of effectiveness as a function of capacity factor for a wind turbine having $C_p = 16/27$.

Because there is essentially no storage of electric energy in the transmission and distribution system, electric power generation has to match consumption demand exactly, minute by minute. In a large network supplied by many power plants, most operate at rated power continuously (base load plants), some are not producing power but will be turned on when demand increases, and a few produce variable amounts of power less than their rated output so as to match the changing demand. The outputs of all plants are ordered by dispatchers who balance the supply to the demand.

A wind turbine cannot dispatch more power than is blowing past its blades at any moment and cannot reliably provide power to the system when it is needed. Instead, wind turbines provide whatever power they can produce from the wind available at any time, reducing the amount needed from the variable load plants, but not reducing the number of such plants needed to be available for times when the wind speed drops to zero. To the system, wind turbines present a negative demand, increasing the variability of demand with which their system operators must contend.

To help in the regulation of power production, it is helpful to be able to predict the wind speed at future times. Weather forecasters can predict approximate wind speeds related to developing weather patterns for several days in advance. There is a standard statistical methodology for analyzing the ability to forecast the future wind speed at any location, for any time in the future.

We begin by defining the random velocity difference $u\{t\}$ between the instantaneous wind speed $V\{t\}$ and its average value $\overline{V}$ as

$$u\{t\} \equiv V\{t\} - \overline{V}, \tag{8.25}$$

so that the time average of $u\{t\}$ is identically zero. Of more significance is the time average of $(u\{t\})^2$,

$$\overline{(u\{t\})^2} = \overline{(V\{t\} - \overline{V})^2} = \overline{(V\{t\})^2} - 2V\{t\}\overline{V} + (\overline{V})^2 = \overline{(V\{t\})^2} - (\overline{V})^2$$

$$= (4/\pi - 1)(\overline{V})^2, \tag{8.26}$$

where $\overline{(V\{t\})^2} = \int V^2 p\{V\}dV = (4/\pi)(\overline{V})^2$ and thus $\sqrt{\overline{(u\{t\})^2}} = 0.273\overline{V}$.[35]

In turbulent fluid flow, such as in a pipe or along a flat surface, the turbulence intensity $\sqrt{\overline{(u\{t\})^2}}$ is usually only about 5–10% of the mean flow speed $\overline{V}$. The much higher value of about 30% for the earth's wind field encountered by wind turbines is a consequence of the largely unsteady geostrophic flow in the atmosphere generated by differential heating by the sun between high and low latitudes. Weather events of extremes in temperature or wind speed extend over regional areas of several thousand kilometers in lateral extent and require days to drift across a particular turbine site. The uneven diurnal cycle of solar heating also produces significant unsteadiness in the local wind speed.

Statistically, the changes in wind speed with time are measured by the autocorrelation coefficient ACC,

$$\text{ACC}\{\tau\} \equiv \frac{\overline{(V\{t\})(V\{t + \tau\})}}{\overline{(V\{t\})^2}}, \tag{8.27}$$

------

[35]The time average of $(V\{t\})^2$ is called the *variance* of $V\{t\}$ and $\sqrt{\overline{(u\{t\})^2}} = \sqrt{\overline{(V\{t\})^2} - (\overline{V})^2}$ is called the *standard deviation* of $V\{t\}$.

where the time-averaged product $\overline{(V\{t\})(V\{t+\tau\})}$ represents the average relationship between $V\{t\}$ and its value at a time $\tau$ later, $V\{t+\tau\}$. For example, when $\tau = 0$, ACC $= 1$; the wind speed deviation $(V\{t\})$ is unchanged over the next instant. At the other extreme, for large enough $\tau$, ACC $= 0$; there is no consistent relationship between $V\{t\}$ and its value a very long time $\tau$ later.

The wind unsteadiness associated with the diurnal and regional atmospheric motion described above are made evident in measurements of ACC$\{\tau\}$ over the range of $D/\overline{V} < \tau <$ year. There is a contribution in the region of $\tau = D/\overline{V}$ from the local turbulence at the site. A higher contribution occurs at $\tau = 24$ hours from the diurnal wind cycle. But the highest contribution is a consequence of weather patterns, for $\tau = 3$–5 day. Seasonal effects are also noticeable for the annual weather cycle, $\tau = 1$ year.

For the purpose of dispatching wind electric power to an electric utility system, weather forecasting models can provide wind speed, and thereby power, projections that are as reliable as short term weather forecasts. Nevertheless, standby conventional power must be available to replace some or all of the wind power not being produced 87% of the time.

### 8.6.4   Economical Turbine Designs

A wind turbine is composed of two subsystems, an aeromechanical system and an electric generating system. The former converts the kinetic energy of the wind blowing through the rotating turbine blades into mechanical power turning a drive shaft that energizes the electric generator, just as a steam or gas turbine drives the electric generator in a steam or gas turbine power plant. But the wind turbine rotor/blade system is markedly different from that of a steam or gas turbine; the former has a few very long slender blades, while the latter has many short stubby blades. Nevertheless, both types of turbines must satisfy two aerodynamic and mechanical constraints: the turbine must convert the power in the fluid stream to mechanical power in a rotating shaft and withstand the stresses in the blade materials imposed by the aerodynamic flow over the blades. Thus, there is a power requirement and a strength requirement, both of which limit the turbine performance and cost.

In Section 8.6.1 we found that the there is an upper limit to the turbine power $\mathcal{P}$, given by Equation (8.17), that is proportional to the product of the turbine disk area $A \equiv \pi D^2/4$ and the cube of the wind speed $V$. The wind speed $V_r$ at which the wind turbine can generate the maximum electrical (rated) power $\mathcal{P}_r$ is called the rated wind speed. While the potential power available at higher wind speeds than the rated winds speed is greater than $\mathcal{P}_r$, it cannot all be converted to electrical power because the electric generator cannot produce more power than $\mathcal{P}_r$. Instead, when $V \geq V_r$, the wind turbine pitch is reduced until the mechanical power output just supplies that needed by the electric generator at rated capacity. The extra power in the wind when $V \geq V_r$ is not collectable because it exceeds the ability of the electric generator to utilize it.

The wind turbine structure must withstand the thrust force $T$ exerted by the wind, as given in Equation (8.16), which is proportional to $\mathcal{P}/V$ and whose upper limit is $\mathcal{P}_r/V_r$. The thrust force exerts a bending moment on the turbine blades and on the support tower of the wind turbine and a corresponding bending stress $\sigma$ in these components. If we characterize the geometric scale of the wind turbine by its diameter $D$, then the order of magnitude of the stress $\sigma$ is $T/tD$, where $t$ is the thickness of the wall of the turbine blade or tower. The volume $\mathcal{V}$ of stressed material in the aeromechanical structure is proportional to $tD^2 \sim \mathcal{P}_r D/\sigma V_r$. This may be expressed in terms of the rated power $\mathcal{P}_r$, air density $\rho$, rated wind speed $V_r$, and the working stress $\sigma$ by

$$\frac{\mathcal{V}}{\mathcal{P}_r} \sim \left(\frac{\mathcal{P}_r}{\rho\sigma^2 V_r^5}\right)^{1/2}. \tag{8.28}$$

Equation (8.28) enables us to estimate how the cost of the aerodynamic component of the wind turbine system depends upon turbine rated power ($\mathcal{P}_r$) and rated wind speed ($V_r$) of the turbine design.

The capital cost of a wind turbine system is a major factor in the cost of producing electricity from wind energy. There are two main components of the capital cost: (a) the turbine component and its support tower that converts wind power into mechanical power of the rotating turbine shaft and (b) the electromechanical system that converts the turbine power into electrical power sent out to the high-voltage transmission line, with the latter consisting of the gear box, electric generator, and power conditioning equipment. Each of these these components has a capital cost that is related to its individual technology and size. To a first approximation, the turbine component cost is proportional to the material volume $\mathcal{V}$, as in Equation (8.28), while the electrical equipment cost is directly proportional to the rated power, independent of the turbine design parameters. The total plant capital cost CC may be written as

$$CC = [\$/W_r]_{el}\, \mathcal{P}_r + \left( k\frac{\mathcal{P}_r}{\rho\sigma^2 V_r^5} \right)^{1/2} \mathcal{P}_r, \tag{8.29}$$

where k is a dimensional constant independent of turbine design and $[\$/kW]_{el}$ is the specific dollar cost per kilowatt of generator power $\mathcal{P}_r$.

What we want to minimize is the ratio of the capital cost CC to the average power output $\overline{\mathcal{P}} = (CF)\mathcal{P}_r$. We rewrite (8.29):

$$\frac{CC}{\overline{\mathcal{P}}} = \frac{1}{CF}\left( [\$/W_r]_{el} + \left( k\frac{\mathcal{P}_r}{\rho\sigma^2 (V_r/\overline{V})^5} \right)^{1/2} \overline{V}^{-5/2} \right). \tag{8.30}$$

In Section 8.6.3.1 it was noted that the average power $\overline{\mathcal{P}}$ is maximized when $V_r/\overline{V} = 1.6$ and for which CF = 1/3. It therefore follows from (8.30) that the aeromechanical system capital cost per unit of average power grows as $\overline{V}^{-5/2} = (\overline{V}^3)^{-5/6}$, or the $-5/6$ power of the wind energy flux. It is inherently more expensive to collect a unit of average power as the average wind energy at the site decreases. This is an illustration of the general rule of resource recovery: the least expensive resource to recover is that site with the highest concentration of the resource.

In some cases, wind turbines optimized for one site are deployed at another site having a lower average wind speed. Although the capacity factor is increased, the $CC/\overline{\mathcal{P}}$ ratio increases to a value even greater than that of (8.30) because a smaller fraction of the resource is recovered.

The wind turbine examples of Table 8.7 cover a modest range of power and rated wind speed, but have nearly the same rated power coefficient. It is likely that their design provides for minimizing the capital cost per average power output for the same value of $V_r/\overline{V}$.

## 8.6.5  Wind Farms

Wind turbines are customarily installed adjacent to each other at sites called "wind farms," an example of which is shown in Figure 8.26. Currently, wind turbines are manufactured in the power range of a few hundred kilowatts to several megawatts, so hundreds of them must be deployed to equal the output of a typical steam electric power plant. Maintenance of this many units at a remote or not easily accessible site, in an unsheltered and sometimes hostile environment, presents practical difficulties that may lead to high operating costs. Nevertheless, grouping many units

**FIGURE 8.26**  A row of wind turbines at a wind farm in Palm Springs, California. (By permission of DOE/NREL-PIX.)

together is advantageous for reducing unit operating costs and achieving economies of scale in connecting the array to transmission lines.

The wind turbines are usually located at the intersections of a rectangular grid, so that they are aligned in columns and rows, each column and row being perpendicular to prevailing wind directions.[36] The spacing between adjacent turbines is in the range of 5 to 10 rotor diameters. Thus the land area occupied by each turbine, called its footprint, FP, is proportional to the square of its diameter and thereby proportional to its rated turbine power. It follows that the power developed per unit of land area is independent of the turbine size. For example, a turbine of diameter $D$ with a rated power coefficient of $C_p$ and interunit grid spacing $\beta D$ would have a rated power $\mathcal{P}_r$ per unit land area FP of

$$\frac{\mathcal{P}_r}{\text{FP}} = \frac{\pi \rho V_r^3 C_p}{8\beta^2}. \tag{8.31}$$

For the 3.6-MW turbine of Table 8.7 and assuming $\beta = 8$, the calculated $\mathcal{P}_r/\text{FP} = 5.2$ W/m$^2$ = 52 kW/ha.[37]

---

[36]In the mainland United States, the prevailing winds in summer and winter are southwest and northwest, respectively.

[37]Assuming a capacity factor of 1/3, the time averaged value of $\mathcal{P}/\text{FP}$ would be 1.7 W/m$^2$. Compared with an annual average solar input at the earth's surface of 150 to 230 W/m$^2$ for the continental United States, this extraction of electromechanical energy per unit area of the wind farm is very small compared with the solar input.

It would seem that very close spacing of wind turbines in a farm would be economically advantageous in reducing operating costs and land costs. But close spacing reduces the power output of downwind turbines compared with that of an upwind turbine because of turbine wake effects.

Consider the flow of wind through a turbine as sketched in Figure 8.17. Immediately downwind of the turbine, the flow speed has been reduced to $V_w$ from $V$ upstream. The difference, $V - V_w$, is called the *velocity deficit*. The energy flux per unit area in this nearby wake region is smaller than that upwind by the factor $V_w^3/V^3$. At maximum power, $V_w = V/3$ so that the downwind flux is only 1/27 as much as the upwind flux. A second turbine this close behind the upwind turbine would produce negligible power.

However, further downwind of the turbine of Figure 8.17, the fluid speeds up because it mixes with the higher speed fluid that flowed outside the turbine at speed $V$. This mixing causes the wake region to grow in diameter (and area $A$) and the velocity deficit to decline in inverse proportion to the area, so that $A(V - V_w)$ is a constant. In the turbulent atmospheric flow the wake width $\sqrt{A}$ grows linearly with the distance $x$ behind the turbine, so that

$$\frac{V - V_w}{V} \propto \left(\frac{D}{x}\right)^2; \qquad x \gg D. \qquad (8.32)$$

Now consider a row of turbines spaced a distance $\beta D$ apart. The velocity defect at the second turbine will be proportional to $1/\beta^2$. The velocity defect at the third turbine will be the sum of that from the second turbine, $1/\beta^2$, and that from the first turbine, a distance of $2\beta D$ or $(1/\beta^2)(1 + 1/2^2)$. For the fourth turbine, the velocity defect is proportional to $(1/\beta^2)(1 + 1/2^2 + 1/3^2)$. This series converges to $\pi^2/6\beta^2 = 1.645/\beta^2$ for an infinite line of turbines. By selecting a large enough spacing $\beta D$, the average power loss per turbine in a large array, compared with an isolated turbine, can be held to a tolerably small percentage.

It has been suggested that an array of wind farms, each spaced a sufficiently large distance apart from its nearest neighbor, would supply in total a nearly steady output because low farm output at low wind speed locations would be counterbalanced by high farm output at high wind speed locations. But as the discussion in Section 8.6.3.3 notes, the size of such a wind farm array would have to be of the order of thousands of kilometers and would require a transmission line network moving power across such distances to average out the varying farm outputs. Until wind farms become widespread sources of electrical power, utility grid management is likely to depend upon fossil-fueled plants to supplement wind power when wind speed is too low or too high.

Wind farms have been constructed in coastal continental waters where wind speeds tend to be higher than on nearby land.[38] Offshore wind farms can interfere with other water uses, such as fishing and vessel navigation, as well as being a hazard for water fowl migration. Operating and construction costs are generally higher than on land.

An example of an offshore wind farm is one proposed by Cape Wind for the southeast coast of Massachusetts.[39] The farm consists of 130 turbines of 3.6 MW of rated power each with the following characteristics: rotor diameter $D = 104$ m; rated wind speed $V_r = 14$ m/s; rated power

---

[38] On average, water is smoother than land and retards the wind flow less, leading to higher wind speeds and more power per unit capital cost.

[39] http://www.capewind.org.

coefficient $C_p = 0.257$; site average wind speed $\overline{V} = 8.89$ m/s ($V_r/\overline{V} = 1.57$); and annual average power per turbine $\overline{\mathcal{P}} = 1.31$ MW (capacity factor = 0.364). The farm area footprint per turbine is 6.3 E(5) m$^2$; the ratio of turbine spacing to diameter is $\beta = 7.63$; and the ratios of rated and average power per unit footprint area are 5.7 and 2.1 W/m$^2$, respectively. These values are close to those discussed in the sections above.

Other offshore wind farms have been built or proposed in the Irish Sea (rated power per unit area = 3.3 W/m$^2$), the North Sea (rated power per unit area = 8.6 W/m$^2$, capacity factor = 0.20, capital cost = 1,630 $/kW), and the Gulf of Mexico (rated power per unit area = 3.3 W/m$^2$, capital cost = 2,000 $/kW).

On land, wind farms in the United States (Texas) have a rated power density of 2.4 W/m$^2$ and a capital cost of 1,500–2,000 $/kW; in Colorado, the capital cost is 1,600 $/kW. For 163 farms in 10 countries in North America, Europe, and North Africa the average capacity factor is 0.15.

## 8.6.6  Integrating Wind Farms into the Electric Power Network

High-quality wind farms require large amounts of open land area, of the order of 1/3 km$^2$/MW$_a$. This is much larger than the area needed for a fossil or nuclear power plant. The latter are sited in or near heavily populated cities with concomitant short transmission distances between supply and demand locations. Any substantial sources of wind power would have to be sited much further from the users, on land of much lower economic value and economic productivity. The wind power would have to be gathered into major transmission lines that deliver the power to distant population centers for distribution to consumers. This would be analogous to the collection of oxygenated blood in the human lung by the venous system that delivers it to the heart for distribution in the arterial system to all the cells of the body that consume the oxygen.

The operators of central electricity power plant systems have very high standards for the performance of these systems. They require that contributing power plants closely control the transmission voltage and frequency to assure that varying demand for power is met by supplying the required current at the system voltage and frequency. The transmission and distribution system that links the small number of central power suppliers to the myriad number of consumers, large and small, must maintain a very close balance between the supply and demand at all times because the system has very little energy storage capacity to smooth out the peaks and troughs of the differences between demand and supply. Given the characteristics of power plant operation, in which steam or gas turbines power electric alternators, this can only be accomplished by strict adherence to line voltage and frequency.

We have seen that an individual wind turbine produces a time-variable power output that lies between an upper limit of its rated power and a lower limit of zero power. While the turbine power conditioning system can maintain the voltage and frequency standards of a central power generation plant, the variable and intermittent wind turbine output cannot match the instantaneous demand of the numerous consumers drawing from the distribution system.

How variable is the output of a wind turbine? One way of quantifying this variability is to measure the time history ($P\{t\}$) of the output from a wind turbine at a particular site or to calculate it from a measurement of the time history of the wind speed ($V\{t\}$) measured at a site.[40] Mathematically, the variability of the power history can be defined by its standard deviation SDEV,

---

[40]The latter is called a simulated or modeled power history.

as shown in Section 8.6.3.3,

$$\mathrm{SDEV}\{P\{t\}\} \equiv \sqrt{\overline{P\{t\}^2} - (\overline{P})^2}, \tag{8.33}$$

where $\overline{P}$ and $\overline{P\{t\}^2}$ are the time averages of $P\{t\}$ and $P\{t\}^2$. The standard deviation is a measure of how much $P\{t\}$ differs from its average value $\overline{P}$. This is usually expressed as the *coefficient of variance* (COV), which is the ratio SDEV/$\overline{P}$:

$$\mathrm{COV} = \sqrt{\frac{\overline{P\{t\}^2} - (\overline{P})^2}{(\overline{P})^2}} = \sqrt{\frac{\overline{P\{t\}^2}}{(\overline{P})^2} - 1}. \tag{8.34}$$

In electrical circuits, the numerator and the denominator of the right side of (8.34) are called the noise power and signal power, respectively. We may regard (8.34) as defining the noise-to-signal power of a wind turbine. If wind turbines are to supply electric power to a distribution system, it is desirable for them to have a very small noise-to-signal ratio, or COV $\ll 1$.

It is possible to estimate COV for wind turbines by employing a simplified model of how wind turbines operate. Figure 8.21 portrays how wind turbine power is related to wind speed. If we neglect the power produced when the wind speed is less than the rated wind speed, then $P\{t\} = P_r$ when the wind speed exceeds the rated wind speed and zero at all other times. For this type of operation, the capacity factor $\mathrm{CF} \equiv \overline{P}/P_r$ is equal to the fraction of the time that power is being produced. We can now evaluate (8.34) entirely in terms of CF:

$$\mathrm{COV} = \sqrt{\frac{1 - \mathrm{CF}}{\mathrm{CF}}}. \tag{8.35}$$

This is bad news! To have smooth, low-noise power requires that CF be only slightly less than unity. As explained in Section 8.6.3.1, this would significantly increase the cost of the electricity produced and fail to utilize most of the wind resource available at a site. Most wind turbines at windy sites have a capacity factor of about 1/3, for which (8.35) predicts COV $= \sqrt{2}$, a very noisy output indeed. While (8.35) is only approximate, a more exact calculation will not change this dismal problem very much.

### 8.6.6.1 Averaging an Array of Wind Farms

Averaging the output of a large number of wind farms dispersed geographically over a wide area has been investigated as a way to smooth out the collective stream of power they produce. The idea is that the sum of the power outputs of a large number of wind turbines whose individual outputs are $P_i\{t\}$ would be less noisy than that of any individual turbine or that the average COV of the ensemble would be much less than the average of the individual COVs.

For simplicity, consider two wind farms, $i$ and $j$, separated from each other by a given distance geographically. The total instantaneous power $\sum$ generated by both turbines would be $P_i\{t\} + P_j\{t\}$, or $P_i + P_j$ for short. The terms on the right side of (8.35) involve time averages of $\sum$ and $\sum^2$:

$$\overline{\sum} = \overline{P_i} + \overline{P_j}$$

$$\overline{\sum^2} = \overline{(P_i + P_j)^2} = \overline{P_i^2} + \overline{P_j^2} + 2\overline{P_iP_j}. \tag{8.36}$$

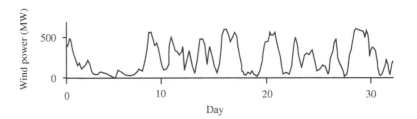

**FIGURE 8.27**  The total power output from all wind farms in the Republic of Ireland for the month of February 2007. (Data from MacKay, David. *Sustainable Energy—Without the Hot Air.* Cambridge, UK: LIT Cambridge Ltd., 2009.)

The term $\overline{P_i P_j}$ in (8.36) is called the covariance of $P_i$ and $P_j$. If the distance between two wind farms is 100 km or more, the covariance is negligible and the value of COV for the ensemble is the same as it would be for a turbine having the ensemble averages value of $\overline{P}$ and $\overline{P^2}$. On the other hand, if the turbines (or farms) are closely spaced, then the value of COV is $\sqrt{2}$ times larger.[41]

An example of the variability of the output from a wide array of wind farms is shown in Figure 8.27, which shows the total electric power output of wind farms in the Republic of Ireland for the month of February 2007. Their total rated power is 745 MW in about 60 sites. Note that there are about 10 periods of peak power of about 500 MW at an average period of 3 days, typical of the period passing of weather patterns drifting through the region. None of these events provided sufficient wind speed in excess of the rated speeds at all sites for this month. The peak output events are separated by troughs of nearly zero output everywhere in the region. The collective variability of the entire array would appear to be not much different from that of the component farms, as expected from the discussion above.

## 8.6.7  Environmental Effects

There are several environmental drawbacks of wind energy systems. Wind turbines generate audible noise, somewhat akin to that of helicopters, but much less intense because of their much lower power levels. Nevertheless, wind turbines are unwelcome noisy neighbors in populated areas but have limited adverse effects on agricultural operations on land or fishing offshore. They can, however, kill migrating birds that attempt to fly through the turbine. To some observers they provide visual blight, especially if located in otherwise undeveloped natural areas.

## 8.7  TIDAL POWER

The regular rise and fall of the ocean level at continental margins has been used in past centuries to produce mechanical power for grinding grain or sawing wood by damming water in coves or river mouths at high tidal levels and then releasing it to the sea through turbines or water wheels when

---

[41] In midlatitudes the weather patterns that determine the wind variations at any particular site drift from west to east. The wind flow within these drifting weather patterns is somewhat correlated over a distance that is small compared with the size of the weather pattern.

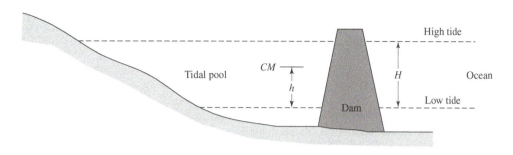

**FIGURE 8.28**   A sketch of the cross-section of a tidal power pool with dam.

the tide is at its lowest level. Intermittent tidal power proved practical only where the tidal range (the difference in elevation between high and low tides) exceeded a few meters. In recent years, tidal plants have been constructed at the La Rance estuary on the Brittany coast of France and at Annapolis Royal, Nova Scotia, in the Bay of Fundy, both of which generate electrical power.

The tidal rise and fall in the ocean is a consequence of the differential gravitational force exerted on opposite sides of the earth primarily by the moon and secondarily by the sun, which gives rise to simultaneous higher ocean surface elevations on the earth's sides facing toward and away from the moon (or sun) and lower elevations halfway in between.[42] At a given point in the ocean, the period $T$ of the lunar tide is one half of the lunar day ($T = 12$ h, 24 min $= 4.464\text{E}(4)$ s); that of the sun is just 12 hours. The solar and lunar effects reinforce each other twice a month, at the time of a full moon and a new moon, giving rise to maximum, or *spring*, tides. Halfway between the spring tides, at the time of the first and last quarters of the moon's monthly cycle, the solar effect is least, resulting in the minimum, or *neap*, tides.[43]

The tidal range of the ocean at the continental margins is far from uniform worldwide. At a few locations, such as the Bay of Fundy and Cook Inlet in North America and the Bristol Channel and Brittany coast in western Europe, resonance effects in the tidal motion lead to ranges as high as 10 m, which contrasts with an average oceanic range of 0.5 m. This large amplification of the oceanic tidal range is associated with a resonant motion of water movement across the shallow waters of the continental shelf into restricted bays or estuaries. About two dozen sites in North and South America, Europe, East and South Asia, and Australia have been identified as possessing this tidal amplification.

The principle of tidal power is illustrated in Figure 8.28, showing in elevation a cross-section of a *tidal pool* separated from the ocean by a *dam* (or *barrage*). A *sluiceway* in the dam allows water to flow quickly between the sea and the tidal pool at appropriate times in the tidal cycle, as does a turbine in a *power house*. If the tidal pool is filled to the high tide level and then isolated from the sea until the latter reaches its low tide level, a volume of water is impounded within the tidal pound of amount $\bar{A}H$, where $\bar{A}$ is the time-averaged surface area of the tidal pond and $H$ is the tidal range. If the center of mass CM of this tidal pool water is a distance $h$ above the low tide

---

[42]The solar effect is about 40% of that caused by the moon.

[43]There are additional perturbations of the tidal range associated with the position of the moon relative to the solar ecliptic plane and the variation of both the moon–earth and the sun–earth distances.

**TABLE 8.8**  Tidal Power Plant Characteristics[a]

| Site location: | Annapolis Royal | La Rance | Sihwa Lake |
|---|---|---|---|
| Pool area, $\overline{A}$ (km$^2$) | 4.8 | 12.9 | 14.2 |
| Tidal range, $H$ (m) | 6.3 | 8.5 | 8 |
| Rated turbine flow rate (m$^3$/s) | 378 | 3,240 | 4,820 |
| Rated turbine head (m) | 5.5 | 8 | 5.8 |
| Rated electric power (MW) | 17.8 | 240 | 254 |
| Average electric power (MW) | 5.71 | 54 | 64 |
| Average electric power/pool area (MW/km$^2$) | 1.19 | 4.19 | 4.5 |
| Sluice gate flow area (m$^2$) | 230 | 1,530 | 1,440 |
| Number of turbines | 1 | — | 10 |
| Capacity factor | 0.321 | 0.225 | 0.25 |
| Effectiveness | 0.171 | 0.331 | 0.34 |

[a] Data from Fay, J., and M. Smachlo. *J. Energy*, 7 (1983): 529; 19th World Energy Congress, 2004.

level, work can be extracted from this volume at the time of low tide by allowing it to flow through a turbine until the pool level reaches that of low tide. The maximum amount of this work is equal to the product of the mass of fluid $\rho\overline{A}H$, the acceleration of gravity g, and the average distance $h$ by which it falls during outflow to the sea, for a total maximum energy of $\rho g\overline{A}Hh$. If the emptied pool is then again closed off from the sea until high tide occurs, an additional amount of energy, $\rho g H\overline{A}(H - h)$, may be reaped for a total energy of

$$\text{Ideal Tidal Energy} = \rho g\overline{A}H^2 \tag{8.37}$$

and total average power of

$$\text{Ideal Tidal Power} = \frac{\rho g\overline{A}H^2}{T}. \tag{8.38}$$

The ideal tidal power per unit of tidal pool surface area, $\rho g H^2/T$, which equals 0.22 W/m$^2$ for $H = 1$ m, increases as the square of the tidal range, showing the importance of tidal range in an economical power plant design.

If power is generated during both inflow to and outflow from the tidal pool, the design is called a *double effect* plant. Because it is difficult and expensive to design a turbine and power house that operates in both flow directions, most tidal power plants operate in the outflow direction only, called a *single effect* plant.[44] Since $h$ is usually more than half of $H$, more energy can be recovered on the outflow than the inflow to the pool.

For a typical single effect plant, Figure 8.29 shows the pool and sea surface levels and power output during one tidal cycle. Beginning at midtide, the sluice gates are opened and sea water flows into the pool, filling it to the high tide level in the first quarter period, at which point the sluice is

---

[44]The La Rance plant (see Table 8.8) was built to operate as a double effect plant, including the possibility of incorporating pumped storage, but now operates exclusively as a single effect outflow tidal plant for various practical reasons.

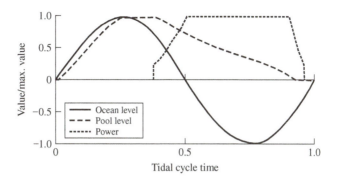

**FIGURE 8.29**   The pool and ocean levels, as well as power output, of a single effect tidal power plant during one tidal cycle. (Data from Fay, J., and M. Smachlo. *J. Energy*, 7 (1983): 529.)

closed. Shortly thereafter, the turbine inlet is opened and the turbine power output rises quickly to its rated power for nearly half of the tidal cycle. When the difference in water level between pool and ocean becomes small, near the end of the cycle, the turbine is closed down.

The hydroturbine of a tidal power plant operates at a low head that is somewhat less than the tidal range $H$ (see Figure 8.29). This requires an axial flow machine with a low flow velocity of about $\sqrt{gH}$. Because the turbine axis must be located below the lowest water level in the cycle so as to avoid cavitation, the turbine diameter is generally less than $H$. The maximum power from a single turbine is of the order of $\rho(gH)^{3/2}H^2$, which works out to about 10 to 30 MW for practical plants. For large installations, multiple turbines are required.

There are two measures of performance of a tidal power plant. One is the *capacity factor*, the ratio of the average turbine power output to the rated turbine output. A high capacity factor provides more annual revenue per unit of capital investment in the turbine/generator and a better ability to match consumer electric power demand. A second measure is the plant *effectiveness*, the ratio of the average power output to the ideal power of Equation (8.38). A high effectiveness allows the plant to utilize most of the tidal power available at the site and thereby justify the capital investment in site development. Unfortunately, it is not possible to obtain high values for both measures simultaneously because increasing one causes the other to decrease. Figure 8.30 shows the relationship of these parameters for single effect plants.[45]

Despite the variable output of a tidal power plant, the predictability of the tidal cycle enables power pool dispatchers to incorporate this variable but predictable contribution to a regional power supply much more easily than that from less predictable wind turbines.

The characteristics of two plants that are currently in operation are listed in Table 8.8. The smaller plant, at Annapolis Royal (Figure 8.31), has a higher capacity factor but lower effectiveness than the larger one at La Rance, in agreement with the trends of Figure 8.30. The average electrical power outputs per square kilometer of tidal pool area are 1.19 and 4.19 MW/km² (W/m²), respectively, for the Annapolis Royal and La Rance plants. Also listed in Table 8.8 are the characteristics of a tidal power plant under construction in Sihwa Lake, South Korea. This plant is quite similar to that at La Rance.

---

[45]For the cycle shown in Figure 8.29, the capacity factor is 0.4 and the effectiveness is 0.25.

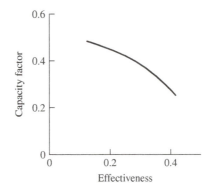

**FIGURE 8.30**    The turbine capacity factor as a function of power plant effectiveness for single effect tidal power plants. (Data from Fay, J., and M. Smachlo. *J. Energy*, 7 (1983): 529.)

**FIGURE 8.31**    The tidal power plant at Annapolis Royal, Nova Scotia, Canada. In the foreground is the entrance to the power house from the tidal pool; in the background is the exit into the tidal waters of the Annapolis Basin.

## 8.7.1   Tidal Current Power

The tidal flow from the deep ocean onto a continental shelf not only gives rise to an enhanced tidal range but also is accompanied by strong tidal currents. These currents are especially swift where the tidal flow into a bay or estuary is constrained to pass through a narrow channel at the entrance. The maximum current speed in such situations would be of the order of $\sqrt{gH}$, where $H$ is the tidal range, but is generally much less than this value. Nevertheless, there have been proposals to deploy the marine underwater equivalent of wind turbines to capture the kinetic energy of this water flow for generating electric power.

**FIGURE 8.32**  Twin tidal current power turbines and support structure. (Courtesy of Marine Current Turbines, Ltd.)

An example of a proposed tidal current power installation is shown in Figure 8.32. Twin power turbines are mounted on a pylon anchored to the sea bed. The pylon extends above the sea surface, enabling above-water access to the turbines for servicing and repair.

The tidal current velocity $V\{t\}$ is approximately sinusoidal,

$$V\{t\} = V_m \sin\left(2\pi \frac{t}{T}\right), \tag{8.39}$$

where $V_m$ is the maximum current speed and $T$ is the tidal period, usually about 12.4 hours. As in the case of wind turbines, the tidal current power that is available for a turbine to capture is proportional to the cube of the current speed $V\{t\}$, which is a fraction of the maximum,

$$\text{Fraction of maximum power} = \frac{V\{t\}^3}{V_m^3} = \sin^3\left(2\pi \frac{t}{T}\right). \tag{8.40}$$

Equation (8.40), plotted in Figure 8.33, shows how the available tidal current power is concentrated near the time of peak current. The time-averaged value of current speed cubed, $\overline{V\{t\}^3}$, can be calculated to be $4/3\pi = 0.424$ times the maximum value, $V_m^3$. This ratio is the capacity factor of a tidal current turbine designed for the maximum current speed; its value is considerably higher than that for wind turbines. Because the maximum current speed during one tidal cycle varies during the lunar month, a month-long average capacity factor will be less than 0.424.

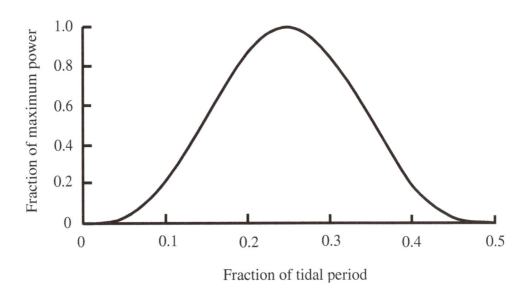

**FIGURE 8.33**   The ratio of tidal current power to its maximum value as a function of the fraction of the tidal period, as given in Equation (8.40).

A tidal current turbine will have two power-producing cycles per tidal period, each producing significant power for about half of each cycle half-period, as illustrated in Figure 8.33. As in the case of tidal power plants, this variable output is easily incorporated into regional power pool management.

### 8.7.2   Environmental Effects

The principal environmental effects of tidal power plants are a consequence of the changes to the tidal flow in the pool and, to a lesser extent, the ocean exterior to the pool. For a single effect plant, the flow into the pool is reduced by about half, decreasing the intertidal zone by about the same amount and reducing the average salinity of the pool waters when the pool is an estuary, both of which alter the nature of the original marine ecosystems. The plant also imposes an impediment to the movement of marine mammals and fish. The patterns of siltation, often a significant natural process at sites having large tidal ranges, is changed by the presence of a tidal plant, as is the tidal motion on the ocean side of the tidal dam. Ship navigation into the pool is prevented unless a lock is built into the dam. The unpredictable and possibly adverse changes to existing ecological systems, including fisheries and other wildlife, have weighed heavily in the assessment of proposals for new tidal power plants.

Tidal current power plants, deployed in the marine equivalent of wind farms, would have similar environmental effects as tidal power plants, but dispersed over a wider area.

### 8.8   OCEAN WAVE POWER

The damage to ocean shorelines caused by storm-driven waves is ample evidence of the dynamic power released when waves impinge on a coastal shore. It would seem that harnessing the power

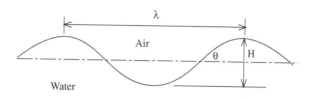

**FIGURE 8.34**   The profile of an ocean wave of wavelength $\lambda$ and wave height $H$.

of waves could provide a source of electrical energy analogous to that of the wind. The technology for doing so, however, presents both economic and engineering challenges.

## 8.8.1   Ocean Wave Energy and Power

Ocean waves are formed on the ocean's surface by action of the wind. By friction and pressure forces exerted at the air–water interface, wind energy is transferred to the energy of gravity waves moving across the ocean surface. The wave energy moves with the wave, but the ocean water has no net motion in the direction of the wave propagation. By suitable mechanical devices, the energy in a wave may be converted to mechanical power.

The energy of an ocean wave consists of two parts: the kinetic energy of the moving water particles in the wave and the gravitational potential energy of the fluid displaced from its equilibrium position of a flat horizontal surface. It is this motion and displacement imparted to the ocean surface by the wind that constitutes the source of power that can be abstracted by a wave power system.

The ocean wave pattern generated by the wind consists of a broad spectrum of surface waves of different wave lengths, frequencies, and heights. Each component of this pattern involves a periodic motion of the water at or close to the ocean surface that is sustained by gravity and that may be represented as a sinusoidal deformation of the air–sea interface, as shown in Figure 8.34. The cyclic frequency $f$ of such a component is related to its wavelength $\lambda$ by[46]

$$f = \sqrt{\frac{g}{2\pi\lambda}}, \tag{8.41}$$

where g is the acceleration of gravity.[47] The velocity $c$ at which the wave pattern moves across the surface of the water, called the phase velocity, is the product of the wavelength times the frequency:

$$c = f\lambda = \sqrt{\frac{g\lambda}{2\pi}} = \frac{g}{2\pi f}. \tag{8.42}$$

Long-wavelength waves move faster than those of shorter wavelength, leading to dispersion, or spreading of the energy in an ocean wave system. Another wave property of interest is the maximum wave slope $\theta$, shown in Figure 8.34, which is a measure of the steepness of the wave and its

[46]For the properties of gravity waves on water see Fay, James A. *Introduction to Fluid Mechanics*. Cambridge, MA: MIT Press, 1994.

[47]This relationship is identical to that for the frequency of a pendulum of length $\lambda$.

propensity to break:

$$\tan \theta = \frac{\pi H}{\lambda}. \tag{8.43}$$

The motion of the water in an ocean wave is confined mostly to a layer of depth $\lambda/2\pi$ below the surface. If $H$ is the wave height, measured from crest to trough as in Figure 8.34, then the velocity of a water particle in the wave is about $Hf$ and the kinetic energy of the water in this layer, per unit of surface area of the wave, is the product of the water mass density $\rho$ times the depth $\lambda/2\pi$ times the kinetic energy per unit mass, $(Hf)^2/2$, for a total of $\rho\lambda H^2 f^2/4\pi = \rho g H^2/8\pi^2$. Thus the average kinetic energy per unit surface area of a wave system is of the order of $\rho g H^2$ and is independent of the wavelength or frequency. The potential energy of the wave, per unit surface area, is approximately the product of the displaced mass per unit area $\rho H$, its vertical displacement $H$, and the gravitational acceleration g, for a total of $\rho g H^2$. An exact calculation of the kinetic and potential energies per unit of wave surface area reveals that they are equal and their sum can be expressed as[48]

$$\text{Wave energy/area} = \frac{\rho g H^2}{8}. \tag{8.44}$$

The wave energy of Equation (8.44) is propagated in the direction of motion of the wave, which is generally the direction of the wind, at the group velocity of an ocean surface wave. The latter equals half the phase velocity $c$, or $g/4\pi f$. The flux of wave energy, or power, per unit of distance normal to the direction of wave propagation is the product of the energy per unit area times the group velocity:

$$\text{Wave power/length} = \frac{\rho g^2 H^2}{16(2\pi f)} = \frac{\rho g^{3/2} H^2 \lambda^{1/2}}{16(2\pi)^{1/2}}. \tag{8.45}$$

Unlike the wave energy, the wave power depends upon the wavelength, albeit as the half power of the wavelength.

Depending upon the wind speed, the surface of the ocean is covered with a random collection of waves of different height and wavelength. As the speed $V$ of the wind increases, both the mean square wave height $\overline{H^2}$ and the mean wavelength increase, while the mean frequency $\overline{f}$ decreases. Oceanographers have measured these quantities as a function of wind speed and found them to be empirically related by the following formulas[49]:

$$\overline{H^2} = 2.19\text{E}(-2)\frac{V^4}{g^2} \tag{8.46}$$

$$2\pi\overline{f} = \frac{0.70\,g}{V}. \tag{8.47}$$

---

[48]This relationship between the energy density and the square of the wave amplitude is found to hold in many wave phenomena, such as sound waves in air and electromagnetic waves in a vacuum.

[49]These relations hold under conditions where the wind duration is long enough over a sufficiently great unimpeded fetch of ocean surface.

In addition, we may define the mean wave slope $\bar{\theta}$ using (8.43) as

$$\tan \bar{\theta} = \frac{\pi \sqrt{\overline{H^2}}}{\bar{\lambda}} = \frac{2\pi^2 \bar{f}^2 \sqrt{\overline{H^2}}}{g}$$

$$= 0.073; \quad \bar{\theta} = 4°, \tag{8.48}$$

where the last line of (8.48) follows from (8.46–8.47). Thus the average wave slope is independent of the sea state as characterized by the wind velocity.

Substituting these relations into Equations (8.44) and (8.45) for wave energy and power, we find that

$$\text{Wave energy/area} = 2.74\text{E}(-3)\frac{\rho V^4}{g} \tag{8.49}$$

$$\text{Wave power/length} = 3.89\text{E}(-3)\frac{\rho \bar{V}^5}{g}. \tag{8.50}$$

For example, a wind speed of 20 knots (10.29 m/s) will generate an energy density of 3.13 kJ/m$^2$, a wave power of 45.8 kW/m, an average wave height of 1.6 m, and an average wave frequency of 0.106 Hz, which is a wave period of 9.4 s.

The dependence of wave power on wind speed, $V^5$, is steeper than that for wind power, $V^3$ [see (8.17)]. As a consequence, most of the wave power available at a site is collectible only when the wind exceeds its average value, as can be seen in Figure 8.35, showing the probability distribution of wave power with wind speed, assuming the wind speed probability distribution of Figure 8.23. Figure 8.35 contrasts sharply with Figure 8.23 for the distribution of wind speed. The average wave power for the distribution of Figure 8.35 is $60/\pi^2 = 6.08$ times the wave power available at the mean wind speed $\bar{V}$.

Economical wave power devices do not generate the full amount of wave power available at high wind speeds. No matter how much is available, the output of electric power is limited by the

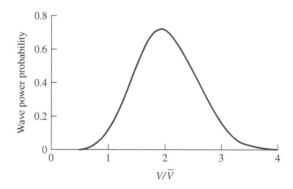

**FIGURE 8.35**   The probability of wave power per unit of relative wind speed $V/\bar{V}$ as a function of relative windspeed.

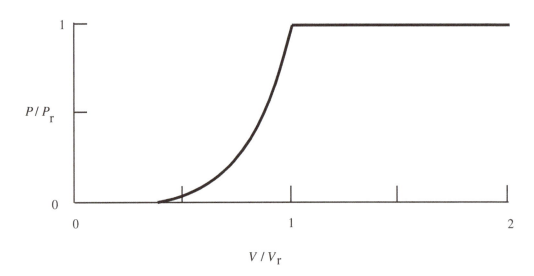

**FIGURE 8.36**  A plot of the ratio of wave power to rated power versus the ratio of wind speed to rated wind speed.

capacity of the installed electromechanical equipment, as defined by the rated power $P_r$ produced at a rated wind speed $V_r$. As shown in Figure 8.36, the power output equals $P_r$ whenever $V \geq V_r$ and is approximately $P_r(V/V_r)^5$ whenever $V \leq V_r$.

Like wind energy, most of the wave energy at a site is available only a fraction of the time. Capacity factors for practical wave energy systems are small, less than 50%. This is unfavorable to the economic viability of wave power and its suitability as a reliable source of electrical power.

The most favorable sites for ocean wave power are the continental or island margins exposed to prevailing winds. For mid- to high latitudes these winds are westerly, while for tropical latitudes they blow from the east. At favorable locations, the annual average wave power available is in the range of 40–60 kW/m.

Figure 8.37 shows annual average wave power, in units of kW/m, along coastal regions of western Europe. These reflect strong prevailing westerly winds of the North Atlantic Ocean at high latitudes.

### 8.8.2   Ocean Wave Power Systems

Many ingenious mechanisms have been proposed, and some of them tested, to produce mechanical or electrical power from wave systems. They are of two types, floating bodies that are anchored in place and structures fixed to the sea bed. Mechanical power is produced by the relative displacement of parts of the structure or by the flow of fluid within it, caused by the impingement of waves on the device. The power that is produced is a fraction of the power available in the waves that are intercepted by the device.

One of the devices, which has been tested to confirm a high efficiency of wave power conversion, is illustrated in Figure 8.38. Called the Salter cam, the device consists of a horizontal cylindrical float, of asymmetrical cross-section shown in Figure 8.38, that in still water floats with

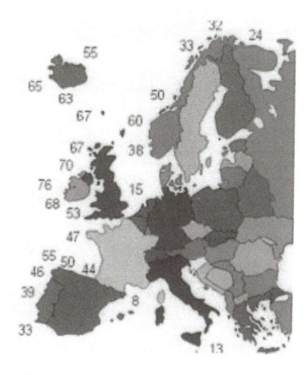

**FIGURE 8.37**  The annual average wave power along coasts of western Europe (numbers are values in kW/m).

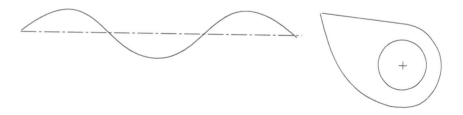

**FIGURE 8.38**  A sketch illustrating the operation of the Salter wave power system.

most of its volume below the sea surface. When waves impinge on the float, it rotates about a stationary circular spline in an oscillatory manner. An electrical power system applies a restraining torque to the float that absorbs most of the power in the incident wave system. The cam structure must have dimensions greater than the wave height and be tuned to the average wave frequency of the waves corresponding to its power capacity. In effect, the volume of the device will be proportional to its rated power.

Other devices utilize buoy-like floats that ride up and down on the wave surface. To produce power, air is confined in a chamber open to the sea at its bottom. As a wave moves by, air is pumped

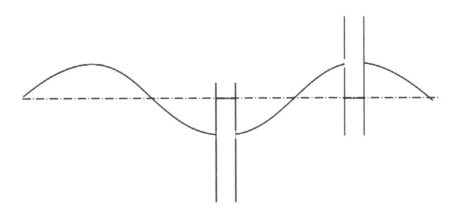

**FIGURE 8.39**  A sketch illustrating the principle of a pneumatic wave power generator.

in and out of the chamber by the movement of water into the chamber bottom. Air flow from the top of the chamber passes through an air turbine, generating power.

Figure 8.39 illustrates the physical principle of pneumatic generation of wave power. A schematic float in the form of a vertical pipe, open at the top to the air and at the bottom to the sea, rides up and down on the wave surface. In either position, the air–sea interface is at the same elevation as the undisturbed sea surface. The volume of air sucked into the pipe from trough to crest is $HA$, where $A$ is the cross-sectional area of the pipe. The maximum possible pressure drop across an air turbine at the top of the pipe is $\rho gH$. Thus the power is the product of the volume times the pressure drop times the wave frequency, or

$$\text{Power} \sim \rho gH^2 Af$$

$$\frac{\text{Power}}{A} \sim \rho gH^2 f; \quad A \ll \lambda^2, \tag{8.51}$$

where the restriction on the base area $A$ ensures that the device rises and falls the distance $H$.[50]

Another system, nick-named "Pelamis,"[51] consists of four long cylindrical floats joined at three locations, much like a train, that rides the surface of the ocean while held in place by anchors. Power is produced by four hydraulic pumps at each joint that oppose the deflection of the cylindrical floats as they ride over the oncoming wave train, in both horizontal and vertical planes. The high-pressure hydraulic fluid thus generated passes through a hydraulic motor that powers an electric generator.

The motion of the floats in a seaway is illustrated in Figure 8.40 for a float length equal to half the wavelength. In the position shown, there is a vertical balance of the wave buoyancy force and the gravity force on each float. But the floats are not in rotational equilibrium unless a couple is exerted on each float by the power-producing hydraulic system at the common joint that opposes

---

[50]Note that expression (8.51) has the same form as that for a tidal power plant, (8.38). For the latter, the very low frequency of the tidal motion is not counterbalanced by a greater value of $H^2$, so tidal power plants have less power per unit area than optimal wave power generators.

[51]http://www.oceanpd.com.

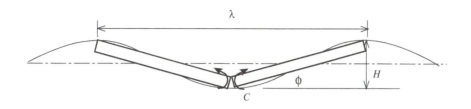

**FIGURE 8.40** The relative deflection of two floats of the Pelamis system is in vertical equilibrium if a couple (shown by arrow) is exerted at the common joint by hydraulic pumps.

**FIGURE 8.41** The four floats of the Pelamis wave power train oscillate as it rides over an oncoming wave pattern, generating electric power. (Copyright Pelamis Wave Power, Ltd.)

the relative rotational motion there. The power produced by the hydraulic system is a maximum when the float length is about one half the wavelength; it is much smaller when the wavelength is much greater or much less than the float length. Figure 8.41 shows the Pelamis train responding to an oncoming waves.

The size of the Pelamis system is impressive. The overall length is 100 m (4 × 25 m) and the overall mass is 700 tons (4 × 175 tons); the float diameter is 3.5 m. The rated power is 750 kW (3 × 250 kW) at the rated wave power of 55 kW/m. At a site with this annual average wave power, the plant will produce 2.7 GWh of electric energy; the corresponding capacity factor (average power/rated power) is 0.41.

It is evident that the mass of the Pelamis system per unit rated power output, about 1 ton per kilowatt, will be an important factor in the specific capital cost of such a system.

An approximate determination of the wave power from a Pelamis system, analogous to that of (8.51) for a pneumatic system, may be derived using the principles of hydrostatic equilibrium of floating bodies.[52] The couple $C$ needed to tip a cylinder of length $L$ and diameter $D$, floating on

---

[52]Fay, James A. *Introduction to Fluid Mechanics.* Cambridge, MA: MIT Press, 1994.

the surface of the ocean, through a small angle $\phi$, is approximately

$$C \sim g\Delta\frac{L^2}{D}\phi, \tag{8.52}$$

where $\Delta$ is the displacement of the cylinder (i.e., the mass of water displaced by the cylindrical float when in static equilibrium). The power produced by this motion equals the product of $C$ and the angular velocity of the tipping motion, which can be taken as $\sim \phi\sqrt{g/L}$,[53] so that the power becomes

$$\text{Power} \sim \frac{\Delta(gL)^{3/2}}{D}\phi^2. \tag{8.53}$$

But the amplitude of the pitching motion must be approximately equal to the mean wave slope $\bar{\theta}$, so the wave power may be expressed as

$$\text{Power} \sim \frac{\Delta(gL)^{3/2}}{D}\bar{\theta}^2. \tag{8.54}$$

The significance of (8.54) is that the wave power generated is mainly proportional to the displacement $\Delta$. Large wave power collection requires large floating segments. If one evaluates the right side of (8.54) using (8.48) and the dimensions of the Pelamis system given above, the power is 1 MW, which compares favorably with the rated power of 750 kW for each 175-ton segment.

### 8.8.3   Wave Power Farms

A proposed 30-MW (rated) wave energy farm would consist of 40 Pelamis units arranged in a grid of three rows, encompassing an area of 1 km$^2$, for a rated power density of 30 W/m$^2$ of sea surface and an annual average power density of 12.3 W/m$^2$. The alongshore spacing between trains is 160 m; the ratio of annual average electrical power output of the wave energy farm to the annual average wave power incident upon it (effectiveness) is 0.10.

An ocean wave system carries momentum as well as energy. If the wave energy is absorbed by a wave power system, a horizontal force per unit length will be exerted on the power system of amount equal to the energy per unit surface area, Equations (8.44) and (8.49). A corresponding restraining force must be supplied by the anchoring gear that holds the power system in place.

### 8.8.4   Environmental Impacts

The construction and operation of wave power systems immediately seaward of the ocean shorefront would have some adverse environmental impacts. They must be anchored to the sea bottom, disturbing some part of it. They may impede navigation and recreational use of the ocean front and no doubt would have aesthetic drawbacks.

---

[53]The term $\sqrt{g/L}$ is proportional to the wave frequency for a wavelength $\lambda \sim L$.

## 8.9  OCEAN THERMAL POWER

The concept of generating mechanical power using warm surface water from a tropical ocean site was first advanced by Jacques d'Arsonval in 1881. The warm water would provide heat to a Rankine cycle heat engine while cool, deep water would supply the cooling needed to condense the working fluid. In low latitude regions of the ocean, within 20 degrees of the equator, water near the surface, where sunlight is absorbed, is warmer by about 20 K than water at depths greater than a kilometer. (The cool, deep ocean water originates in the polar regions.) Commonly called *ocean thermal energy conversion* (OTEC), such a plant would necessarily have a small thermodynamic efficiency since it would operate with a temperature difference that is small compared with the absolute temperature of the heat source.[54] Although the heat that can be extracted from the ocean surface is great, mechanical power is required to circulate the surface and deep water through the power plant, subtracting from the power generated by the Rankine cycle turbine. This and other practical difficulties have prevented OTEC plants from developing beyond the demonstration stage.

Two types of OTEC plants have been tried. The first, an open cycle, utilizes water evaporated from the warm stream under low pressure to supply a turbine discharging to an even lower pressure condenser cooled by a spray of cool water (see Figure 8.42a). The second, a closed or hybrid cycle, employs heat exchangers that supply or remove heat from a closed Rankine cycle working fluid, usually ammonia, permitting a more economical and efficient turbine than that for the open cycle, but requiring a more expensive evaporator and condenser (see Figure 8.42b). In both cycles, large volumes of warm and cool water must be pumped from the ocean through long, large-diameter ducts and then returned to the sea via a third duct. These systems must be constructed to minimize the amount of power required to circulate the fluid streams from and to the ocean via the power plant.

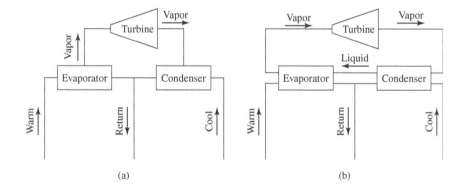

**FIGURE 8.42**  Diagrams of the fluid flow in an ocean thermal power plant. (a) An open cycle employs the water evaporated from warm sea water as the working fluid in a low-pressure steam cycle. (b) A closed cycle uses ammonia as the working fluid, with the latter being vaporized and condensed in heat exchangers supplied with warm and cool ocean streams.

---

[54]A Carnot cycle operating with a temperature difference of 20 K and supplied with heat from a warm source at 25°C would have a thermodynamic efficiency of $20/(25 + 273) = 6.7\%$. A practical Rankine cycle would have even lower efficiency than the ideal Carnot cycle.

In 1979, a small, closed-cycle demonstration plant mounted on a barge anchored off Keahole Point in Hawaii generated 52 kW of gross electrical power but only 15 kW of net power, the difference being consumed in pumping losses. The heat exchangers were made from titanium plate. The cool-water polyethylene pipe was 0.6 m in diameter and drew water from a depth of 823 m. Currently, experiments are continuing with a 50-kW closed-cycle plant located on shore at the Natural Energy Laboratory of Hawaii, with testing of heat exchangers that will be more economical.

In 1981, Japan demonstrated a shore-based closed-cycle plant utilizing a fluorocarbon working fluid. Cool water was drawn from a depth of 580 m. The gross and net powers from this plant were 100 and 31 kW, respectively.

In 1992, an open-cycle plant was tested at the Natural Energy Laboratory of Hawaii that generated 210 kW gross electrical power and 40 kW net. The turbine rotor for this plant weighed 7.5 tons.

There are several problems to be overcome if ocean thermal power systems are to become practical. Fouling of the warm water supply pipe and heat exchanger by marine organisms and the corrosion of heat exchangers need to be eliminated. The mechanical power needed to pump fluids through the system must be kept well below the gross output of the turbine. All components of the system need to be constructed economically so that the capital cost per unit of net power output becomes competitive with other types of renewable energy systems.

## 8.10  CAPITAL COST OF RENEWABLE ELECTRIC POWER

It was mentioned in Section 8.1 that the capital cost of renewable energy systems was a dominant factor in the cost of production of renewable energy. In this section we examine the capital costs of all renewable energy systems for which reliable cost estimates are currently available. In a future in which fossil carbon emissions to the atmosphere will need to be reduced drastically and in which fossil-fueled electric power plants will capture and sequester the fuel carbon, the preponderance of electricity will be generated in capital intensive systems, of which nuclear plants are currently the prime example. These zero- or low-carbon plants are base load plants; they cannot readily supply peak power, which requires a plant that can be started and stopped as peak demand waxes and wanes, such as conventional gas-fired combined-cycle plants. Renewable energy, which is only available when the natural supply exists and not necessarily when the demand needs it, must be supplemented by standby, reliable fossil fuel plants. Thus the cost of renewable energy supplied to an electric power system will be more than the cost of production at specific renewable plants, to the extent that the supplemental power needed may be more expensive than that required for peak load purposes.

Under the new regulatory regime being instituted in some states of the United States and which perhaps will be extended subsequently to all states, the price of electricity paid by consumers is the sum of several parts. One part is the price charged by the electricity producer for electricity leaving the power plant. Other parts include the price for transmission by high-voltage power lines and distribution networks to the consumer's location. One might say that the electricity producer sells a product, electric energy (measured in kilowatt hours), while the transmission and distribution utility sells a service, transmitting the electric energy to the customer. Individual producers will compete to sell electric energy to consumers for the lowest price, but the transmission and distribution system will remain a public utility monopoly whose price of service will be regulated by public authorities. In a competitive market, electricity producers will price their product to cover the costs of production, plus a profit for privately owned producers.

Most of the currently installed renewable energy systems in the United States generate electric power (see Table 8.1). These systems collect naturally available forms of renewable energy and convert them to electrical power. The economic cost of doing so stems principally from the capital investment required to produce and install the equipment that accomplishes this energy conversion. In contrast to fossil fuel power plants, where the cost of fuel, operations, and maintenance may equal or exceed the annualized capital cost, renewable energy plants have small operating costs and, of course, no fuel cost. Because renewable energy plants have higher capital costs and lower utilization than fossil fuel plants, their cost of producing electricity is usually higher than that of fossil fuel plants.

To compare the renewable energy technologies with each other and with fossil fuel plants, we computed the component of the cost of producing electricity that is allocated to financing the capital investment needed to construct the facility for each type of renewable energy plant listed in Table 8.9. The capital cost per unit of rated power ($/kW_r$), the capacity factor (ratio of average power/rated power), the capital cost per unit of average power ($/kW_r$), and an annualized cost of capital (cents/kWh, assumed to be 15% of the capital cost) are shown in the respective columns.

For the renewable energy systems listed in Table 8.9 the capital cost component of electricity covers a large range, 3,000–20,000 $/kW_a$. Part of this variability reflects the difference in unit capital cost (1,200–14,000 $/kW_r$); the rest is a consequence of different capacity factors (0.2–0.6). Except for photovoltaic systems, these systems require turbine-generator machinery that constitutes a large share of the capital cost, which is in the range of 1,000–2,000 $/kW. The capacity factors are mostly low, for reasons inherent in the variability of the renewable energy resource.

In contrast, the capital cost for modern efficient fossil fuel plants lies in the range of 500–1,200 $/kW_r$ and the corresponding capital cost of electricity is 1.0–2.3 cents/kWh. Taking into account the costs of fuel and operations, these fossil fuel plants produce electricity at a cost of 4–6 cents/kWh. If carbon capture and storage are added, the range is 1,100–1,400 $/kW_r$, or 2–2.7 cents/kWh electricity capital cost and 6–8 cents/kWh total cost. Of the renewable systems, only hydropower, biomass, geothermal, and wind are close in cost to carbon-free fossil systems, with tidal and wave within striking distance if carbon capture costs prove greater than current estimates. But solar thermal and photovoltaic systems are not at all competitive.

**TABLE 8.9**  Capital Cost of Renewable Electric Power

| Type | Capital cost / Rated power ($/kW_r$) | Capacity factor | Capital cost / Average power ($/kW_a$) | Capital cost of electricity (cents/kWh) |
|---|---|---|---|---|
| Hydropower | 2,000 | 0.50 | 4,000 | 6.8 |
| Biomass[a] | 1,800 | 0.60 | 3,000 | 5.1 |
| Solar thermal | 4,000 | 0.25 | 16,000 | 27 |
| Geothermal | 1,800 | 0.6 | 3,000 | 5.1 |
| Wind | 1,200 | 0.35 | 3,400 | 5.8 |
| Photovoltaic | 4,000 | 0.20 | 20,000 | 34 |
| Tidal | 2,000 | 0.30 | 6,700 | 11 |
| Wave | 3,000 | 0.40 | 7,500 | 13 |

[a] Wood-fueled steam plant.

It has been argued that research and production subsidies may so enlarge the market for renewable energy technologies that their capital costs will be reduced to the point that they become economically competitive with fossil fuel plants. In addition, if fossil fuel plants must reduce their $CO_2$ emissions by capturing and sequestering the $CO_2$, the extra cost of doing so may make renewable sources more economically attractive. In either case government intervention, either economic or regulatory, will be necessary to increase the market share for renewable electric power.

## 8.11  CONCLUSION

The renewable energy technologies that have been demonstrated to be technologically practical sources of power are hydro, biomass, geothermal, wind, solar thermal, photovoltaic, ocean tidal, ocean wave, and ocean current. Of these, hydro and biomass comprised almost the entire renewable installed capacity in the United States in 2007, but the proportion of wind is increasing, especially worldwide. Neither ocean wave nor ocean thermal electric power has yet emerged from the development stage.

Hydro, biomass, and geothermal plants are able to supply electric power dependably on a daily and annual basis. Wind, solar thermal, photovoltaic, and ocean tidal power have diurnal rhythms and seasonal changes that do not necessarily match the demand for electric power. When linked to an integrated power transmission system, they displace fossil fuel consumption, reducing air pollutant and carbon emissions.

Renewable energy technologies are capital intensive, having capital costs per installed kilowatt that are two to eight times those of fossil fuel plants. Furthermore, these plants may have low capacity factors (the ratio of average to installed power), which increases the capital cost portion of the price of electricity they generate compared with conventional plants. The economic competitiveness of renewable energy plants, compared with fossil fuel plants, is strongly dependent upon the costs of capital and fuel.

The adverse environmental effects of renewable energy plants are generally less than those of conventional power plants, but are by no means negligible. Hydro and tidal plants require extensive changes to the natural hydrologic system, and biomass plants emit air pollutants. Because of the low intensities of renewable energy, the size of a renewable power plant is larger than that of a conventional plant of equal power, requiring significantly more land area that may adversely affect other needed land uses.

# PROBLEMS

### Problem 8.1

A proposed hydroelectric power station would produce maximum power when the volume flow rate reached 100 m$^3$/s at a head of 10 m. (a) If the turboelectric machinery operates with an efficiency of 85%, calculate the maximum electric power output of the facility. (b) If the annual average capacity factor of the plant is 65%, calculate the annual income from the sale of electric power if the selling price is $0.03/kWh. (d) If the capital cost of the power plant is $100/kW, calculate the ratio of annual income to capital cost, expressing the result as %/y.

## Problem 8.2

The annual average daily solar irradiance falling on agricultural land in the United States is about 10 $MJ/m^2$ per day. (a) If 0.1% of this is converted to biomass heating value, calculate the annual rate of biomass crop heating value that may be harvested per hectare of crop land. (b) If the biomass heating value is converted to electric power at an efficiency of 25%, calculate the annual average electric power generated per hectare of cropland. (c) If electric power is sold at 0.03 $/kWh, calculate the annual income, per hectare of land, from electricity sales of this biomass energy.

## Problem 8.3

A large oak tree produces 2.2 tons of wood in 50 years of growth. The tree has a canopy of 10 m in diameter and collects solar energy for 6 months each year at an average rate of 177 $Watt/m^2$. What is the tree's efficiency for converting solar energy to wood heating value? (Assume a wood heating value of 20 MJ/kg.)

## Problem 8.4

Draw a schematic diagram for a geothermal heat pump that works to supply space heat in winter and air conditioning in summer and explain how it works.

## Problem 8.5

The rate of heat $q$ collected by a flat plate solar collector is given by Equation (8.3). According to the limits of the second law of thermodynamics, the maximum electric power $p_m$ that could be generated from this heat flux $q$ would be lower by the factor $(1 - T_a/T_c)$, where $T_c$ and $T_a$ are the temperatures of the collector fluid and the atmosphere, respectively. (a) Derive an expression for the temperature ratio $T_c/T_a$ that will maximize $p_m$, in terms of the collector parameters $\beta I$ and $\mathcal{U}$, and the temperature $T_a$. (b) Calculate the numerical values of $T_c/T_a$, $T_c$, and $p_m$ when $\beta = 0.8$, $I = 900\ W/m^2$, $\mathcal{U} = 5W/m^2K$, and $T_a = 300$ K.

## Problem 8.6

Calculate the collector surface area $A$ required to heat 500 liters of water a day from 15 to 80°C under conditions where the daily insolation on a slanted collector is 1.13 E(7) $J/m^2$, assuming 33% collector efficiency.

## Problem 8.7

Explain why a solar flat plate collector used for domestic water heating can work even in subfreezing ambient temperatures.

## Problem 8.8

A spherically focusing solar collector is being designed to generate electric power from a heat engine, similar to that shown in Figure 8.10. For this design, $\beta = 0.9$, the design solar irradiance is $I = 700\ W/m^2$, and the concentration ratio CR = 2,000. Assuming that the heat loss rate from the collector, $\mathcal{U}(T_c - T_a)$, is equal to the black body radiation from the collector, or $\sigma T_c^4$ (where $\sigma$ is the Stefan–Boltzmann constant [see Table A.3]), calculate the maximum collector temperature, $(T_c)_{max}$.

## Problem 8.9

If incident solar radiation averages 700 W/m$^2$ for 8 hours, estimate the area $A$ of 80% efficient heliostats needed to provide 10 MW of electrical power (as in Solar One). Assume that the efficiency of converting solar heat to electric energy is 35%.

## Problem 8.10

Calculate the land area in km$^2$ that would be needed for a solar thermal power plant delivering 1,000 MW of electrical power under the following conditions: the concentrating mirrors receive 700 W/m$^2$; each concentrating mirror and its platform requires a land area twice the area of the mirror itself; the efficiency of converting solar heat to electric energy is 35%.

## Problem 8.11

An earth satellite photovoltaic power system is oriented toward the sun to intercept the solar irradiance of 1367 W/m$^2$. (a) If the photovoltaic cell efficiency is 17%, calculate the electrical power output per unit of surface area. (b) If 90% of the solar irradiance is absorbed by the cell, calculate the heat flow rate, per unit area, radiated to space that is needed to maintain the cell at a fixed temperature. (c) If this radiant heat flow is that of a black body ($\sigma T^4$, where sigma is the Stefan–Boltzmann constant [see Table A.3] and $T$ is the cell absolute temperature), calculate the cell temperature.

## Problem 8.12

If solar irradiation is 700 W/m$^2$ and the efficiency of a photovoltaic cell is 10%, calculate the area (in m$^2$ and ft$^2$) of PV cell needed to run simultaneously, and without a grid or battery backup, a typical refrigerator, toaster, TV, and stereo set. Use electric power consumption data from your own appliances or from the literature.

## Problem 8.13

A single-story retail store wishes to supply all its lighting requirement with batteries charged by PV cells. The PV cells will be mounted on the horizontal rooftop. The time-averaged lighting requirement is 20 W/m$^2$; the annual average solar irradiance is 150 W/m$^2$; the PV efficiency is 10%; the battery charging–discharging efficiency is 80%. What percentage of the roof area will the photocells occupy?

## Problem 8.14

Calculate the power produced per m$^2$ of wind turbine disk area for a wind velocity of 20 miles per hour, assuming a power coefficient $C_p$ of 0.5 and an air density of 1.2 kg/m$^3$. Calculate the diameter of a wind turbine that would supply 1 MW of electrical power under these conditions.

## Problem 8.15

Calculate the mechanical power produced by a wind turbine with rotor diameter 40 m when the wind speed is 8 m/s, assuming a power coefficient $C_p$ of 0.3 and an air density of 1.2 kg/m$^3$. How many households can this wind turbine supply, assuming an average power requirement of 3 kW/household? How many wind turbines would be needed for a city of 100,000 population with 4 people per household? Calculate the land area for this population, in m$^2$ and acres, needed for the wind turbine farm, assuming that the wind turbines should be placed 2.5 rotor diameters apart perpendicular to the wind and 8 rotor diameters apart parallel to the wind. Place 10 windmills perpendicular to the wind and the rest parallel. Compare the wind farm area with that of a typical city of 100,000 (single family homes, no high rises).

### Problem 8.16

An inventor proposes to utilize a surplus ship's propeller, of diameter equal to 2 m, as an underwater "wind turbine" at a location where a tidal current is 2 m/s. Calculate the maximum power that this turbine could produce from the underwater flow.

### Problem 8.17

For the Annapolis Royal tidal power plant, Table 8.8 lists the pool area $\bar{A} = 4.8$ km$^2$, tidal range $H = 6.3$ m, rated electric power $P_{el} = 17.8$ MW, and capacity factor $\eta = 0.321$. (a) Using Equations (8.37)–(8.38) and these values, calculate the ideal tidal energy and ideal tidal power. (You may assume the tidal period equals 4.46E(4) s.) (b) Calculate the average power and the effectiveness $\epsilon =$ (average power)/(ideal power), and compare it with the value of Table 8.8.

### Problem 8.18

A wave energy system is being evaluated for a site where the design wind speed $V$ is 20 knots. Using Equations (8.46), (8.47), and (8.50), calculate the expected mean wave height $\sqrt{\overline{H^2}}$, mean wave frequency $\bar{f}$, and the wave power per unit length.

# BIBLIOGRAPHY

Archer, M., and R. Hill, eds. *Clean Electricity from Photovoltaics.* London: Imperial College Press, 2001.

Armstead, H., and H. Cristopher. *Geothermal Energy.* London: E. F. N. Spon Ltd., 1978.

Avery, William H., and Chih Wu. *Renewable Energy from the Ocean.* Oxford: Oxford University Press, 1994.

Baker, A. C. *Tidal Power.* London: Peter Peregrinus Ltd., 1991.

Boyle, Godfrey, ed. *Renewable Energy. Power for a Sustainable Future.* Oxford: Oxford University Press, 2004.

Boyle, Godfrey, ed. *Renewable Electricity and the Grid. The Challenge of Variability.* London: Earthscan, 2007.

Bryden, I. G., T. Grinsted, and G. T. Melville. "Assessing the potential of a simple tidal channel to deliver useful energy." *Applied Ocean Research* 26 (2004): 198–204.

Dickson, Mary H., and Mario Fanelli (eds.). *Geothermal Energy.* Chichester: John Wiley & Sons, 1995.

Duffie, John A., and William A. Beckman. *Solar Engineering of Thermal Processes*, 2d ed., New York: John Wiley & Sons, 1991.

Fay, James A. *Introduction to Fluid Mechanics.* Cambridge, MA: MIT Press, 1994.

Kreith, Frank, and Jan F. Kreider. *Principles of Solar Engineering.* Washington: Hemisphere Publishing Co., 1978.

Lewis, Tony. *Wave Energy Evaluation for C.E.C.* London: Graham & Trotman Ltd., 1985.

Lunde, Peter J. *Solar Thermal Engineering.* New York: John Wiley & Sons, 1980.

MacKay, David. *Sustainable Energy—Without the Hot Air.* Cambridge: LIT Cambridge Ltd., 2009.

Manwell, J. F., J. G. McGowan, and A. L. Rogers. *Wind Energy Explained. Theory, Design and Application.* Chichester: John Wiley & Sons, 2002.

Markvart, Thomas, ed. *Solar Electricity.* New York: John Wiley & Sons, 1994.

McCormick, Michael. *Ocean Wave Energy Conversion.* New York: John Wiley and Sons, 1981.

Merrigan, Joseph A. *Sunlight to Electricity*, 2d ed. Cambridge, MA: MIT Press, 1982.

National Research Council. *Electricity from Renewable Resources: Status, Prospects, and Impediments.* Washington DC: National Academies Press, 2010.

Probstein, Ronald F., and R. Edwin Hicks. *Synthetic Fuels.* Cambridge: pH Press, 1990.

Selzer, H. *Wind Energy. Potential of Wind Energy in the European Community. An Assessment Study. Solar Energy R&D in the European Community*, Series G, Volume 2. Dordrecht: Reidel, 1986.

Seymour, Richard J. *Ocean Energy Recovery: The State of the Art.* New York: American Society of Civil Engineers, 1992.

Simeons, Charles. *Hydropower. The Use of Water as an Alternative Source of Energy.* Oxford: Pergamon Press, 1980.

Smil, Vaclav. *Energy in Nature and Society.* Cambridge, MA: MIT Press, 2008.

Warnick, C. C. *Hydropower Engineering.* Englewood Cliffs: Prentice Hall, 1984.

Wortman, Andrze J. *Introduction to Wind Turbine Engineering.* Boston: Butterworth, 1983.

# Automotive Transportation

## 9.1 INTRODUCTION

Near the end of the 19th century in the United States, a predominantly steam-powered, coal-fueled transportation system had aided the transformation of what had been a mostly agricultural economy at the century's beginning into one dominated by industrial activity. Railroads, coastal ships, and river barges moved farm produce and forest products to urban consumers, mineral ores to refineries and steel mills, and coal to industrial, commercial, and residential consumers. Except for the affluent who could afford horse-drawn carriages, most urban and rural workers walked from home to their place of employment. In the major cities, however, horse-drawn or electric-powered street cars provided passenger transport for daily travel where residential areas had spread beyond the urban core in response to population growth. Long-distance passenger travel was exclusively provided by trains and ships. Nevertheless, horse-drawn wagons were necessary for the distribution of food and goods within urban and rural regions alike, and animal power provided the source of much agricultural energy.

The development of the internal combustion engine (ICE) and its supply of liquid fossil fuel transformed both urban and rural communities in the 20th century, most markedly by greatly expanding the kind and function of transportation and work vehicles. By the century's end practically all adults had the use of an automobile for commuting to work or other personal daily travel. Passenger travel to and in central urban areas by bus and electric powered rail (above, at or underground) complemented the need for personal travel where the density of travel could not be accommodated by automobiles alone. Most freight moved by truck, but bulk commodities were shipped by rail and river barge. The rapid development of commercial air travel, beginning at midcentury, expanded long-distance travel availability, superseding railroads, which had already lost market share to intercity buses and private automobiles. ICE-powered tractors revolutionized agriculture, and rural farm population declined drastically. Work machines (bulldozer, chain saw, construction crane, etc.) greatly increased human productivity in tasks that had previously consumed large amounts of hard human labor.

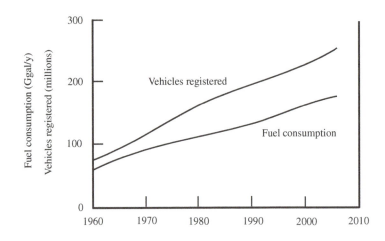

**FIGURE 9.1** The number of registered highway vehicles and their annual fuel consumption in the United States for the period 1960–2006. (Data from the National Highway Traffic Administration, 2009.)

The growth of transportation in the 20th century paralleled that of electric power, with both becoming major factors in the economies of developed nations at the century's end. In the United States in 2006, transportation and electricity generation accounted for about 29 and 45% of the total primary energy consumption, respectively. The availability of both electric power and transportation in modern industrial societies is an important factor in human productivity, although the efficiency of their use may well be improvable.

The rapid growth in the number of U.S. highway transportation vehicles in the period 1960–2006 is graphically depicted in Figure 9.1. Over this period, the population of road vehicles in use grew at a rate of about 3.8 million vehicles per year, reaching 251 million in 2006, of which 64% are automobiles, 36% are trucks (light, medium, and heavy duty), and 0.3% are buses.[1] The number of these vehicles is closely approaching the number of eligible drivers, but the annual rate of increase in vehicles exceeds that of the U.S. population increase by 1 million per year.[2]

The growth in the number of highway vehicles, mostly privately owned, has been accompanied by a growth in highways and roads, publicly financed and maintained. In the United States there are currently 3.93 million miles of highways (79% rural, 21% urban), about 40 vehicles per highway mile. In the period 1990–2000, highway miles increased at an annual rate of about 0.1% per year, much less than the 1.5% per year growth in the vehicle population. This slow growth in highway miles reflects both the difficulty of siting new roadways, especially in urban regions, and the scarcity of public funds for infrastructure improvement. The concentration of national vehicle ownership in urban areas, which account for only 21% of highway miles but 75% of the population, and the faster growth of vehicle population than highway miles has aggravated the problem of increasing highway congestion in U.S. cities.

---

[1]About 16 million new vehicles are sold each year, 12 million of which are needed to replace the aging members of the 251 million vehicles in use. The approximate vehicle life expectancy, equal to the ratio of vehicle population to new vehicle sales, is 21 years.

[2]The United States has the highest ratio of vehicle to human population of any nation. Most developing countries currently have vehicle/human ratios about the same as that of the United States in the 1920–1940 period.

**TABLE 9.1**   U.S. Transportation Vehicle Use, 1995[a]

| Type | Number (million) | Annual miles (Gmile/y) | Miles per vehicle (mile/y) | Miles per gallon (mile/gal) |
|---|---|---|---|---|
| Passenger cars | 132.2 | 1,448 | 10,958 | 21.3 |
| Light trucks, sports utility vehicles | 65.7 | 790 | 12,018 | 17.4 |
| Trucks | 6.72 | 178 | 26,500 | 6.2 |
| Buses | 0.69 | 6.4 | 9,312 | 6.6 |
| Transit | 0.12 | 3.6 | 30,900 | 4.7 |
| Rail | 1.22 | 30.4 | 24,900 | 8.7 |
| Commercial air | 0.0056 | 4.4 | 790,000 | 0.35 |

[a]Data from Bureau of Transportation Statistics, U.S. Department of Transportation, 1999.

Highway vehicle fuel consumption shows a growth proportionate to the vehicle population (see Figure 9.1). In 2006, U.S. highway vehicle fuel consumption was 175 Ggal/y (664 GL/y), averaging out to 700 gallons per year (2532 L/y) per vehicle. Weighted by the fuel heating value, this annual consumption is 0.70 TW or 2.8 kW per vehicle.

A breakdown of the use of all transportation vehicles in the United States in 1995 is given in Table 9.1. Small passenger vehicles (cars, light duty trucks, sports utility vehicles, vans) account for 96% of all highway vehicles, 92% of highway vehicle miles, and 76% of highway fuel. These privately owned vehicles clearly dominate highway travel and energy use. Public passenger travel, by bus and transit on the ground and airplane above, accounts for only 0.6% of the total transportation vehicle miles annually.

Small passenger vehicles are used but a small fraction of the time, mostly for short trips. Their average yearly mileage, about 12,000 miles per year, amounts to 33.7 miles per day, about an hour per day of operating time. The average one-way trip length is less than 10 miles, so vehicles do not move far from their home locations on average.

In the United States, the manufacture, operation, and maintenance of highway vehicles and the refining and distribution of their fuels is pervasively regulated by both federal and state governments. Two of the regulatory objectives are of interest here: control of exhaust and evaporative emissions and vehicle fuel economy.[3] Other objectives include vehicle and passenger safety, operator competence, and owner fiscal liability. The regulation of emissions and fuel economy falls principally upon the vehicle manufacturer, to a lesser extent on the fuel supplier, and hardly at all upon the vehicle owner, whose principal responsibility is to maintain control equipment during the vehicle lifetime.

The problem of air pollutant emissions from transportation vehicles is primarily that associated with the private passenger vehicle. Their emissions, principally carbon monoxide (CO), oxides of nitrogen ($NO_x$), hydrocarbons (HC), sulfur dioxide ($SO_2$), and particulate matter (PM), are distributed geographically in proportion to vehicle usage, which is concentrated in urban regions. But the secondary pollutants, especially ozone, photochemically formed from the direct emissions

---

[3]U.S. federal regulation of vehicle fuel economy was instituted after the oil crisis of the 1970s so as to reduce the dependence of the national economy on the supply of imported oil. In 2008, there was a pending legislative requirement that vehicle fuel economy should be regulated to reduce vehicular $CO_2$ emissions.

of $NO_x$ and HC can reach elevated levels downwind from the vehicular sources and outside the urban region.[4] As a consequence, all primary emissions are regulated to ensure that primary and secondary air pollutant levels do not exceed harmful levels, either locally or regionally.

In this chapter we discuss the technology of the automobile as it affects the efficiency of fuel use and the emission of atmospheric pollutants. The chapter emphasizes the cause and amelioration of ordinary air pollutant emissions as well as the reduction of $CO_2$ emissions by improvements in vehicle fuel efficiency. We summarize the current state of development of alternative vehicle systems that show promise of significant emission reductions, including electric-drive vehicles (battery or fuel-cell powered) and hybrid electric/ICE-drive vehicles.

## 9.2   INTERNAL COMBUSTION ENGINES FOR HIGHWAY VEHICLES

The most common engine in road vehicles is the gasoline-fueled spark ignition (SI) reciprocating engine. It is economical to manufacture and maintain, provides ample power per unit weight, and has a useful life that equals that of the typical passenger vehicle. The less common alternative, the diesel-fueled compression ignition (CI) reciprocating engine, is more durable and more fuel efficient, but is more expensive to manufacture and provides less power per unit weight (unless turbocharged).[5] Each engine type has different air pollutant emissions and requires different technology to reduce them.

The SI and CI engines, developed at a time when the dominant steam engine was a reciprocating device, utilize the same mechanism of a movable piston within a closed-end cylinder linked by a connecting rod to a rotating crankshaft that converts the reciprocating motion of the piston to the rotary motion of the crankshaft (see Figure 9.2a). High-pressure gas in the cylinder exerts an outward force on the piston, doing mechanical work on the crankshaft as the piston recedes on the power stroke. On the return inward stroke, where the cylinder is filled initially with low-pressure gas, less work is done by the crankshaft on the piston than was done on the crankshaft during the outward stroke, so the crankshaft delivers net positive work to the mechanism the engine is turning. A flywheel is required to smooth out the rotational speed of the crankshaft while the piston exerts variable amounts of torque during its reciprocating motion.

As previously explained in Section 3.9.3 on the Otto cycle, a high pressure is generated in the cylinder just at the beginning of the outward expansion stroke by burning a fuel with air in the cylinder. In the SI engine, a fuel–air mixture, prepared outside the cylinder, flows into it prior to the inward compression stroke, eventually being ignited by a spark when the piston reaches its innermost position (called top center, TC). On the other hand, for the CI engine, pure air is ingested prior to the compression stroke and the fuel is sprayed into the air at the end of compression, whereupon it ignites and burns quickly, without the necessity of a spark, to produce a pressure rise.

To make a repetitive cycle out of this process it is necessary to replace the burned gases in the cylinder with a fresh charge (fuel–air mixture for the SI engine, pure air for the CI engine). This is accomplished by opening ports in the cylinder connected to ducts (called manifolds) that either

---

[4] See Section 10.2.6.

[5] The CI engine, in contrast to the SI engine, does not lend itself to light, low-power uses such as motorcycles, chainsaws, lawn mowers, and so on or for use in small aircraft, where engine weight is important. On the other hand, large reciprocating engines for construction equipment, ships, and railroad locomotives are invariably CI engines.

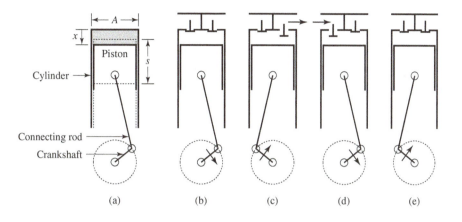

**FIGURE 9.2**    A sketch (a) of the mechanical action of a reciprocating internal combustion engine and the four piston strokes of the four-stroke cycle engine: (b) outward power stroke, (c) inward exhaust stroke, (d) outward intake stroke, and (e) inward compression stroke.

conduct the fresh charge into the cylinder or collect the combustion gases leaving the cylinder at the end of the expansion stroke. There are two mechanical schemes for effecting this replacement of burned gas by a fresh charge. In one, called the *two-stroke cycle*, the fresh charge displaces the combustion products during a short time interval when the piston is near its outermost position (called bottom center, BC) and both an inlet port and an exhaust port are open. Only two strokes of the piston, one inward to compress the fresh charge and the second outward to expand the burned gases, are needed to complete this two-stroke cycle, during which the crankshaft turns through one revolution. In the second scheme, called the *four-stroke cycle* (see Figure 9.2b–e), the outward power stroke is followed by a full inward stroke in which the piston displaces the combustion products, pushing them out through an open exhaust port located in the closed end of the cylinder (called the *cylinder head*). At the end of this inward stroke, called the exhaust stroke, the exhaust port is closed and an inlet port is opened, allowing the piston to suck in a fresh charge as it moves outward during the subsequent intake stroke. By adding these two extra strokes (an extra revolution of the crankshaft), a fresh charge is prepared in the cylinder, permitting the two subsequent strokes to compress, burn, and expand the charge, producing work. These four strokes (two crank revolutions) comprise the four-stroke cycle.

Almost all highway vehicles are powered by four-stroke cycle engines, predominantly gasoline-fueled SI engines for passenger vehicles and diesel-fueled CI engines for trucks. Two-stroke cycle SI engines are mostly used to power two-wheeled vehicles, such as motorcycles and mopeds. Engines in the latter vehicles have lower fuel efficiency and higher exhaust pollutant emissions, but are lighter and less expensive to manufacture (for a given power). The trend of increasingly stringent government regulation of pollutant emissions and vehicle fuel efficiency has spurred manufacturers to improve both two- and four-stroke cycle engines, but the predominant use of four-stroke cycle engines is unlikely to change in the foreseeable future, especially since two-stroke cycle engines have higher exhaust pollutant emissions.[6]

---

[6]For developments in the two-stroke cycle engine field, see Heywood, John B., and Eran Shaw. *The Two-Stroke Cycle Engine: Its Development, Operation, and Design.* Philadelphia: Taylor & Francis, 1999.

Although SI and CI engines are similar in that they both burn fuel with air in the cylinder when the piston is near TC, so as to create the pressure rise that eventually produces net mechanical work during the four-stroke cycle of events, they differ in three important ways. In the SI engine the fuel and air are mixed together outside the cylinder to form a uniform fuel–air mixture that is ingested during the intake stroke; in the CI engine the fuel is sprayed into air in the cylinder, near TC, forming a nonuniform fuel–air mixture. In the SI engine, the combustible mixture is burned when ignited by a spark at the appropriate time in the cycle near TC, whereas the CI fuel–air mixture ignites spontaneously, shortly after the fuel is sprayed into the engine combustion chamber at the requisite time near TC. A third difference is the method of controlling the power output of the engine, an important consideration since the vehicle engine must operate satisfactorily over a wide range of power in a continuously and frequently variable manner. This is accomplished in the SI engine by adjusting the air pressure in the intake manifold by means of a throttle valve that lowers that pressure below the atmospheric value, the more so as the engine power is reduced. In the CI engine, the power is lowered by reducing the amount of fuel injected into the cylinder at each cycle. These differences have important implications for the formation of pollutants and the fuel efficiency of each type of engine, SI or CI.

The power needed for a passenger vehicle engine is in the range of tens to hundreds of kilowatts. For many practical reasons, the cylinder size in light-duty passenger vehicles and trucks is approximately the same for all engines, with higher powered engines using more cylinders than small ones. Thus engines of increasing power may contain 2, 3, 4, 5, 6, 8, or 10 cylinders. Geometrically, the piston motions in these cylinders are phased so that the cylinders fire at equally timed intervals during the two revolutions of the crankshaft (720°) that constitute the four-stroke cycle. This also facilitates the mechanical balance of the moving parts, which would otherwise cause excessive engine vibrations. It is general practice to construct 2-, 3-, 4-, 5-, and 6-cylinder engines with all the cylinders aligned one behind the other, while 6-, 8-, and 10-cylinder engines have two banks of cylinders in a V-shape arrangement. This permits more efficient use of the engine space in the vehicle, especially for front-wheel-drive vehicles.

## 9.2.1  Combustion in SI and CI Engines

To perform properly, any reciprocating internal combustion engine must burn a mixture of air and fuel in a very short time, completing the combustion soon after the piston begins its outward power stroke. The duration of combustion should not much exceed about 50 degrees of rotation of the crankshaft, which would equal about 2 ms if the crankshaft were rotating at 3600 revolutions per minute (rpm), a typical value for a vehicle traveling at high speed. In order for a fuel molecule to burn, it must be mixed intimately with oxygen at the molecular level. Even then, the conversion of a hydrocarbon molecule to carbon dioxide and water vapor molecules will not occur rapidly enough unless the mixed air and fuel are sufficiently hot. This is assured by the rapid temperature rise of the cylinder charge during the inward compression stroke.

The combustion process in the SI engine is initiated by a timed electric spark that starts a flame front propagating through the air–fuel mixture. As it sweeps through the combustible mixture, heating the reactants to the point where they burn extremely rapidly, it converts fuel to products of combustion, with each fuel molecule being processed in about 1 $\mu$s. When the flame front reaches the walls of the combustion chamber it is extinguished, but not until all except a tiny proportion of the combustible mixture next to the wall will have been reacted to form combustion products.[7] However, to propagate a reliable, rapidly moving flame, the combustible mixture must not be too rich or too lean; that is, it must not have too much or too little fuel, compared with the amount of air, than is needed to completely consume all the fuel and oxygen present, called the stoichiometric mixture.[8] It is for this reason that the fuel and air are mixed in carefully controlled proportions prior to or while entering the engine cylinder. To reduce engine power, the amount of fuel burned per cycle is lowered by reducing the pressure, and hence density, of the incoming charge, its proportions of fuel and air, and its temperature, remaining unchanged. Thus the favorable high-speed flame propagation rate and rapid combustion are retained in SI engines, even down to idling conditions.

Fuel combustion in the CI engine proceeds quite differently, without flame propagation, albeit with comparable speed. Fuel injected into the very hot air within the cylinder is quickly evaporated, with the fuel vapor then mixing with the surrounding air and burning spontaneously without the necessity of spark ignition. The surrounding air swirls past the injector nozzle, providing a flow of oxygen needed to oxidize the evaporating fuel droplets as they emerge from the nozzle. When less power is needed, less fuel is injected into the fixed amount of air in the cylinder, consuming less oxygen and reducing the pressure rise in the cylinder. Thus the fuel–air ratio in the CI engine is variable and lean. At maximum power, some excess air is required to burn all the fuel because mixing conditions are not ideal.

At the molecular level, the combustion process is much more complex than might be inferred from the overall stoichiometry of the reaction. For example, the complete oxidation of one octane molecule ($C_8H_{18}$) requires $8 + 9/2$ oxygen molecules ($O_2$), forming 8 $CO_2$ and 9 $H_2O$ molecules. The rearrangement of the carbon, hydrogen, and oxygen atoms among the reacting molecules occurs one step at a time, requiring numerous individual changes as single H and C atoms are stripped away from the fuel molecule and oxidized. These changes are aided by very reactive intermediate species called radicals, such as H, O, OH, $C_2$, CH, $CH_2$, and so on, that act to accelerate the molecular rearrangement, but that disappear once the reaction is completed. Nevertheless, the combustion process is not perfect, so that small amounts of unreacted or imperfectly oxidized products may remain, molecules that reached dead-end paths and were unable to attain the complete thermochemical equilibrium of the bulk of the reactants. These molecules are dispersed among the principal combustion products and, unless removed, enter the atmosphere as air pollutants.

Nitric oxide (NO) is an important air pollutant that is a by-product of the combustion process. It is formed from molecular nitrogen and oxygen because it is thermochemically favored at the high temperature of the newly formed combustion products. It is produced rapidly by the two following reaction steps, facilitated by the presence of atomic oxygen in the combustion zone:

$$N_2 + O \rightarrow NO + N \tag{9.1}$$

$$N + O_2 \rightarrow NO + O. \tag{9.2}$$

---

[7]Under certain conditions it is possible for the combustible mixture to ignite spontaneously and burn explosively, in an uncontrolled and destructive manner. This is called combustion knock and is avoided by controlling the chemical properties of the fuel.

[8]See Section 4.3.

There would be no NO in the exhaust gas reaching the atmosphere if these reactions reversed their course as the product gas temperature declined during the expansion stroke and the NO maintained thermochemical equilibrium with the rest of the exhaust products. Unfortunately, this reverse process ceases during the expansion, leaving a residual amount of NO that can be reduced by subsequent treatment in the exhaust system outside the cylinder. This residual NO is sensitively dependent upon the air–fuel ratio and the flame temperature in the combustion process.

## 9.3   ENGINE POWER AND PERFORMANCE

The burning of fuel in the cylinder of a reciprocating internal combustion engine produces mechanical power by exerting a force on the moving piston. At any instant, if $V$ is the volume of gas in the cylinder and $p$ is the gas pressure, then the increment $dW$ in work done by the gas on the piston when it moves so as to increase the volume by an amount $dV$ is

$$dW = p\,dV. \tag{9.3}$$

The total work $W$ done by the gas in one cycle of the engine (two or four strokes) is found by integrating Equation (9.3) over a cycle,

$$W = \oint p\,dV \equiv \bar{p}V_c, \tag{9.4}$$

where $\bar{p}$ is an effective average gas pressure and $V_c$ is the cylinder displacement, that is, the volume of gas displaced when the piston moves from the innermost to outermost position. If the crankshaft rotates at a frequency $N$ and $n$ is the number of crank revolutions per cycle, then the time for a complete cycle is $n/N$ and the power input to the piston $\mathcal{P}$ is

$$\mathcal{P} = W\frac{N}{n} = \frac{\bar{p}V_cN}{n}. \tag{9.5}$$

Figure 9.3 sketches the variation of gas pressure in the cylinder of a four-stroke-cycle SI engine. The uppermost curve traces the pressure during the outward power stroke, which is followed by the inward exhaust stroke, for which the cylinder pressure is about equal to that in the exhaust manifold. The pressure in the following intake stroke is lower and equal to the intake manifold pressure so that the final compression stroke starts from a low pressure. The work per cycle, $\oint p\,dV$, can be seen to be the difference between the area enclosed by the upper loop of Figure 9.3 minus that of the lower loop.

Only a part of the work done by the gas on the moving pistons of an engine is delivered to the output shaft of the engine, where it can then be connected to the transmission and ultimately to the wheels to propel a vehicle. The friction of the pistons moving along the cylinder surfaces and of the piston, connecting rod and crankshaft bearings, and the power consumed in operating the valves, pumping lubricant and cooling water, reduce the power output of the engine below that delivered by the gas in the cylinders (Equation 9.5). This net output $\mathcal{P}_b$, called the *brake power*, is determined while running the engine in a laboratory at various rotational speeds $N$ and throttle settings. The power is computed as the product of the brake torque $\mathcal{T}_b$ (which is measured) and the

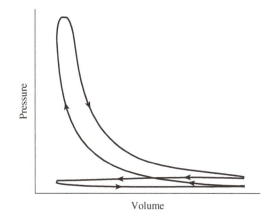

**FIGURE 9.3**    Sketch of the pressure, as a function of volume, in a four-stroke cycle SI engine cylinder.

shaft angular frequency $2\pi N$,

$$\mathcal{P}_b = \mathcal{T}_b(2\pi N). \tag{9.6}$$

Sometimes the power and torque are expressed in terms of an average cylinder pressure $p_{bmep}$, called the *brake mean effective pressure*, using Equations (9.5) and (9.6),

$$p_{mep} \equiv \frac{n\mathcal{P}_b}{V_e N} = \frac{2\pi n \mathcal{T}_b}{V_e}, \tag{9.7}$$

where $V_e$ is the *engine displacement*, the product of the cylinder displacement and the number of cylinders. The brake mean effective pressure is an indicator of the cycle work per unit displacement volume of the engine and is little affected by engine size. It directly reflects the amount of fuel burned per unit of cylinder volume.[9]

The size of engine installed in a vehicle is determined by the power required to propel the vehicle at the speed and acceleration needed to perform satisfactorily. An important characteristic of an engine is its power and torque when operated at maximum fuel input per cycle. For an SI engine, this occurs when the throttle is wide open. Figure 9.4 is a sketch of the brake power $\mathcal{P}_b$ and brake mean effective pressure $p_{bmep}$ (which is related to the brake torque $\mathcal{T}_b$ by Equation (9.7)) as a function of engine speed $N$. The brake power $\mathcal{P}_b$ rises monotonically with speed, reaching a maximum value max $\mathcal{P}_b$ at an engine speed $N_m$. On the other hand, the brake mean effective pressure $p_{bmep}$ varies only moderately over the range of speeds between normal idling speed $N_i$ (typically 800 rpm [13.3 Hz]) and the speed $N_m$ at maximum power. It reaches its maximum value max $p_{bmep}$ at a speed about two thirds of $N_m$.

At constant throttle setting or inlet pressure, the brake torque does not change much with engine speed. At low speeds, the torque declines somewhat because of heat losses from the cylinder. At high speed, the torque declines more precipitously, as pressure losses in the inlet and exhaust valves

---

[9]Pressure has the dimensions of energy per unit volume.

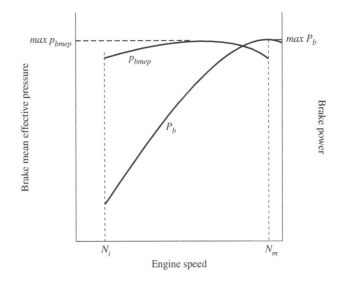

**FIGURE 9.4**   A sketch of the brake mean effective pressure $p_{\text{bmep}}$ and brake power $\mathcal{P}_{\text{p}}$ of an SI engine at wide-open throttle, as a function of engine speed $N$.

and increased piston friction reduce the brake mean effective pressure and cause the brake power to peak. To improve high-speed power, multiple inlet and exhaust valves may be used and the engine may be supercharged.

The characteristics of a sample of model year 2010 passenger cars and light-duty trucks equipped with SI engines are listed in Table 9.2, in order of ascending engine displacement and vehicle mass. The bigger, more powerful engines employ more cylinders, but the individual cylinder volume is more nearly the same among these engines than is the displacement. There are only small differences in the maximum brake mean effective pressure among these engines and comparably small differences in power per unit of engine displacement. As we shall see below, the differences in engine size reflect the differences in vehicle mass, ability to accelerate, and vehicle fuel economy.

### 9.3.1  Engine Efficiency

The fuel economy of the engine is usually expressed as the *brake specific fuel consumption* (bsfc), the ratio of the mass of fuel consumed per unit of mechanical work output by the engine shaft. The value of the brake specific fuel consumption depends upon the engine operating conditions. For SI engines, the most economical (i.e., the minimum) brake specific fuel consumption is about 0.27 kg/kWh, while for CI engines it is lower, about 0.20 kg/kWh.[10] An alternative measure of engine performance is the thermal efficiency $\eta_{\text{e}}$, the ratio of engine mechanical work to the heating value of the fuel mass consumed to produce that work. In terms of the LHV per unit mass of fuel,

---

[10]The brake specific fuel consumption is sometimes given in units of pounds (mass) per brake horsepower hour, where one pound (mass) per brake horsepower hour equals 1.644 kg/kWh.

**TABLE 9.2** 2010 Model Year Light-Duty Passenger Vehicle Characteristics (SI Engines)

| Manufacturer | Honda | Honda | Toyota | Honda | Volvo |
|---|---|---|---|---|---|
| Model | Civic | Civic Hybrid | Prius Hybrid | Accord | S40 |
| Size | Subcompact | Subcompact | Subcompact | Midsize | Midsize |
| Displacement (L) | 1.8 | 1.34 | 1.8 | 2.35 | 2.44 |
| Number of cylinders | 4 | 4 | 4 | 4 | 5 |
| Maximum power (kW) | 104.4 | 82.1 | 73.0 | 132 | 125 |
| Maximum bmep (bar) | 11.05 | 12.25 | 9.36 | 10.35 | 10.2 |
| Vehicle mass (t) | 1.22 | 1.31 | 1.33 | 1.47 | 1.48 |
| Frontal area ($m^2$) | 2.02 | 2.00 | 2.59 | 2.18 | 2.06 |
| Power/displacement (kW/L) | 58.0 | 61.3 | 40.6 | 56.1 | 51.4 |
| Power/mass (kW/t) | 85.6 | 62.7 | 54.9 | 89.8 | 84.4 |
| Urban/highway fuel efficiency (km/L) | 11.1/14.6 | 17.0/19.1 | 21.5/20.4 | 9.4/13.2 | 8.5/13.2 |

| Manufacturer | Lexus | Ford | Ford | Honda | Ford |
|---|---|---|---|---|---|
| Model | Is | Fusion | Fusion Hybrid | Pilot | F-150 |
| Size | Midsize | SUV | SUV | SUV | Truck |
| Displacement (L) | 2.5 | 2.5 | 2.5 | 3.47 | 4.6 |
| Number of cylinders | 6 | 4 | 4 | 6 | 8 |
| Maximum power (kW) | 152 | 131 | 142 | 187 | 185 |
| Maximum bmep (bar) | 11.4 | 10.5 | 14.9 | 11.32 | 10.2 |
| Vehicle mass (t) | 1.56 | 1.65 | 1.69 | 1.95 | 2.15 |
| Frontal area ($m^2$) | 2.05 | 2.12 | 2.12 | 2.89 | 3.04 |
| Power/displacement (kW/L) | 60.8 | 52.4 | 56.8 | 53.7 | 40.2 |
| Power/mass (kW/t) | 97.4 | 79.4 | 84 | 95.6 | 78.4 |
| Urban/highway fuel efficiency (km/L) | 8.9/12.3 | 8.5/14.5 | 17.4/15.3 | 7.2/9.8 | 6.0/8.1 |

these fuel economy measures are related by

$$\eta_e \equiv \frac{3.6}{\text{bsfc(kg/kWh)} \times \text{LHV(MJ/kg)}}. \tag{9.8}$$

The LHVs of vehicle fuels, per unit mass or volume, are empirically related to the fuel specific gravity SG by

$$\text{LHV} = \{28 + 11.2/(SG)\} \, \text{MJ/kg}$$

$$\text{LHV} = \{11.2 + 28(SG)\} \, \text{MJ/L}. \tag{9.9}$$

For gasoline of SG = 0.72, the LHV is 43.55 MJ/kg or 31.6 MJ/L, while for diesel fuel of SG = 0.85, the LHV is 41.18 MJ/kg or 35 MJ/L.[11] Converting to thermal efficiency by Equation (9.8), the best SI engine efficiency is about 31%, while that of a CI engine is 44%.

---

[11]Diesel fuel has more heating value per unit volume than gasoline, by about 11%. Where vehicle fuel efficiency is measured in terms of miles per gallon or kilometers per liter, a diesel-powered vehicle has an inherent 11% advantage, other things being equal.

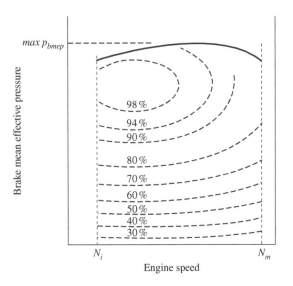

**FIGURE 9.5**    A sketch of contours of engine efficiency $\eta_e$ as a function of engine speed $N$ and brake mean effective pressure for an SI engine. Contour values are expressed as a percentage of the maximum value.

The engine efficiency $\eta_e$ is a function of the engine speed $N$ and brake mean effective pressure, as demonstrated in Figure 9.5. Its peak value occurs at about 35% of the maximum engine speed $N_m$ and 80% of the max $p_{bmep}$, where the engine power is about 35% of the maximum power. The efficiency declines rapidly with decreasing $p_{bmep}$ but less so with increasing speed. This decrease of efficiency is a consequence of the relative increase of engine friction and flow losses, compared with engine output, as $p_{bmep}$ and engine power are reduced below the optimum value. Because a road vehicle engine must provide the full range of its power over the speed range of the vehicle, it cannot operate at maximum efficiency all the time. Nevertheless, by proper matching of the engine to the vehicle it is possible to minimize the fuel consumption needed to meet a particular vehicle driving cycle.

## 9.4  VEHICLE POWER AND PERFORMANCE

Moving a vehicle along a highway requires the expenditure of mechanical power to turn the wheels. Part of this power is needed to overcome the drag force exerted by the air on the moving vehicle. Another part is needed to counter the resistance of the tires moving over the ground, called the rolling resistance. If the vehicle is climbing a hill, additional power is needed to lift it vertically in the earth's gravitational field. Finally, when the vehicle accelerates to a higher speed, power is needed to increase the kinetic energy of the vehicle.

The aerodynamic drag force acting on a vehicle is conveniently given as the product $\rho\, C_D A V^2/2$, where $\rho$ is the mass density of air (kg/m$^3$), $A$ is the frontal area (m$^2$) of the vehicle (about 80% of the height times the width), $V$ is the vehicle speed (m/s), and $C_D$ is the drag coefficient. The latter is empirically determined by testing vehicle models in a wind tunnel, and its value, generally in the range of 0.25–0.5, depends very much on the vehicle shape and external smoothness. Given the fact that a passenger vehicle must enclose the passengers in a safe structure employing good visibility and without excess material, it is difficult to reduce $C_D$ below about 0.25. The corresponding power $P_{air}$ needed to overcome the air resistance is the product of the force times the vehicle velocity,

$$P_{air} = (C_D A)\frac{\rho V^3}{2}. \tag{9.10}$$

The power required to overcome air resistance thus increases as the cube of the vehicle speed. At high vehicle speeds, air resistance is the major factor in determining the requirement for engine power.

It might seem that wheels should present little or no resistance to forward motion. While this resistance is small, it cannot be reduced to zero. The source of this resistance, called the rolling resistance, lies in the deflection of the tire where it comes into contact with the ground. This deflection is necessary to support the weight of the vehicle and to provide a contact area between the tire and the road that is needed to prevent the tire slipping along the road surface. The deflection is such that, when the wheel rotates, the road surface exerts a retarding torque on the wheel, which must be overcome by the engine drive system and a corresponding retarding force on the wheel axle. This force is usually specified as $C_R mg$, and the corresponding power $P_{roll}$ becomes

$$P_{roll} = C_R mgV, \tag{9.11}$$

where $m$ is the vehicular mass, g is the acceleration of gravity ($mg$ is the total vehicle gravity force supported by the tires), and $C_R$ is a small dimensionless constant whose value depends upon the tire construction and pressure. Stiff, highly pressurized tires will have smaller $C_R$, but will be more prone to slip and transmit road unevenness to the vehicle. The values of $C_R$ lie in the range 0.01–0.02. Since the rolling power grows only as the first power of the vehicle speed, it is generally smaller than $P_{air}$ at high speeds but can become larger at low speeds.

When climbing a hill of rise angle $\theta$, the force of gravity acting on the vehicle, $mg$, has a component $mg \sin \theta$ opposing the forward motion. The power required to maintain a steady climb rate is

$$P_{hill} = mgV \sin \theta. \tag{9.12}$$

Finally, the instantaneous power required to accelerate the vehicle, $P_{acc}$, is simply the time rate of increase of vehicular kinetic energy, $m(1 + \epsilon)V^2/2$, or

$$P_{acc} = \frac{d}{dt}\left(\frac{m(1 + \epsilon)V^2}{2}\right) = m(1 + \epsilon)V\frac{dV}{dt}, \tag{9.13}$$

where $\epsilon$ is a small dimensionless constant that accounts for the rotational inertia of the wheels and drive train, and $dV/dt$ is the acceleration of the vehicle. Of course, when the vehicle decelerates ($dV/dt < 0$) using the brakes or closing the throttle, the negative power of Equation (9.13) does not put power back into the engine, so that $P_{acc}$ is effectively zero during deceleration. In electric-drive vehicles, some of the deceleration energy can be recovered, stored in batteries, and used later in the operating cycle, saving fuel.

The total power thus becomes

$$P = P_{\text{acc}} + P_{\text{roll}} + P_{\text{hill}} + P_{\text{air}}$$

$$= mV \left( (1 + \epsilon) \frac{dV}{dt} + C_R g + g \sin \theta \right) + (C_D A) \frac{\rho V^3}{2}. \tag{9.14}$$

The first three terms on the right of Equation (9.14), the power required to accelerate the vehicle, overcome the rolling resistance, and climb a hill, are each proportional to the vehicle mass $m$. In contrast, the last term, the power needed to overcome aerodynamic drag, is independent of the vehicle mass, depending instead on the vehicle frontal area $A$, which is nearly the same for all light-duty passenger vehicles (see Table 9.2) but is noticeably larger for light-duty trucks. For a given driving cycle, heavy cars will require more powerful engines and will consume more fuel, more or less in proportion to vehicle mass (see Table 9.2), than will light vehicles. A very fuel-efficient vehicle is necessarily a light one.

For driving at a steady speed, only rolling and aerodynamic resistance must be overcome, the power required being

$$P_{\text{steady}} = C_R m g V + (C_D A) \frac{\rho V^3}{2}. \tag{9.15}$$

For the typical passenger vehicle, the rolling and aerodynamic resistance are equal at a speed of about 60 km/h. For highway cruising at 120 km/h, the aerodynamic drag would be four times the rolling resistance, with the power required being less dependent upon vehicle mass than when driving at low speed.

The parameters of vehicle design that lead to increased fuel economy include low values of vehicle mass $m$, drag area product $C_D A$, and rolling resistance coefficient $C_R$. In addition, recovery of vehicle kinetic energy during deceleration for reuse during other portions of the driving cycle will also improve vehicle fuel economy.

### 9.4.1 Connecting the Engine to the Wheels

Engine power is a maximum at the highest engine speed $N_m$, but less power is available at lower engine speeds (see Figure 9.4). If we wish to maximize the power available at all vehicle speeds, it is necessary to reduce the ratio of engine speed to wheel speed as the vehicle speed increases. This is accomplished in the transmission, a device attached to the engine that provides stepped speeds to the drive shaft that connects it to the wheels.

There are two forms of transmission, manual and automatic. In a vehicle with a normal transmission, the vehicle operator disengages a clutch and manually shifts to a different gear before engaging the clutch again. In an automatic transmission, a fluid coupling replaces the clutch and gear shifting is accomplished by computer-controlled hydraulic actuators. The more operator-friendly automatic transmission is less efficient than the manual one by about 10 percentage points. Most passenger vehicles employ four or five forward speeds and one reverse. The speed ratio between adjacent shifts is about 1.5.

The drive shaft connecting the engine/transmission to the wheels, either front or back (or both in the case of four-wheel drive), utilizes a differential gear to apply equal torque to both drive wheels while allowing different wheel speeds during maneuvering. The drive shaft provides for

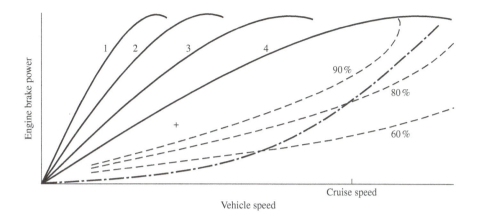

**FIGURE 9.6**  A sketch of the maximum engine power provided by a four-speed transmission as a function of vehicle speed (solid lines). The dash–dot line is the steady speed vehicle power requirement. Contours of relative engine thermal efficiency for lesser power operation in fourth gear are shown as dotted lines, with the 100% peak being marked by +.

the vertical motion of wheels with respect to the chassis and for the steering motion when using a front-wheel drive.

The maximum engine power available to a vehicle, as a function of vehicle speed and transmission gear ratio, is shown in Figure 9.6 (solid lines) for a four-speed manual transmission. Each of these four curves is identical to the brake power curve of Figure 9.4, but is plotted here against vehicle speed rather than engine speed. At any vehicle speed, the maximum engine power (which occurs at wide open throttle) is the ordinate of the corresponding gear shift level (1, 2, 3, or 4). At lower vehicle speeds, three or four levels are available, with the lowest yielding the greatest possible power. At the highest speed, only the highest gear can be used to deliver engine power.

For constant-speed travel, the engine power demand, as determined from Equation (9.15), is shown as a dash–dot curve in Figure 9.6. It rises rapidly with vehicle speed, as the aerodynamic drag outpaces rolling resistance. At any vehicle speed, the vertical distance between the "demand" (dash–dot) curve and the "maximum supply" (solid) curve is the maximum power available for vehicle acceleration and hill climbing. At low vehicular speeds, this difference is greatest for the lowest shift level, while at the highest speed this difference is least, shrinking to zero at the vehicle's top speed. The lower gear levels are needed to provide high acceleration at low vehicular speed.

It is clear from Figure 9.6 that only a small fraction of the maximum engine power is used at steady vehicle speeds, reaching about 40% at cruising speed. The remainder is available for acceleration and hill climbing. For average vehicle duty cycles, only a fraction of the time is used in acceleration, and the portion of that at which the engine power is a maximum is small. The vehicle capacity factor, the time-averaged fraction of installed power that is utilized by a moving vehicle, is less than 50% for the average duty cycle. Nevertheless, for satisfactory performance reserve power is required over the normal speed range of the vehicle.

Superimposed in Figure 9.6 are the relative efficiency curves of Figure 9.4 (dashed lines), for fourth gear. The 100% peak is marked by +. It is readily apparent that, for a steady vehicle speed, the engine relative efficiency is at best 50% for speeds less than half the cruise speed, rising to about 80% at cruise speed. The normal practice of downshifting at low speeds leads to even lower relative efficiencies. Taking into consideration the short periods of acceleration within a typical driving cycle, the time-averaged relative engine efficiency is certain to be less than 80%, perhaps closer to 60%. This performance could be improved a little by adding a fifth gear level, but that would leave less power reserve for acceleration, necessitating downshifting before accelerating at the higher speeds. An alternative is to employ a continuously variable transmission that can reach higher relative engine efficiencies over a range of steady speeds, but will downshift when acceleration is needed.

## 9.5    VEHICLE FUEL EFFICIENCY

Traditionally, the efficiency of use of fuel by a vehicle is measured by the distance it moves in a trip divided by the fuel consumed.[12] For road vehicles, this ratio is customarily reported in the units of kilometer per liter (or in the United States, mile per gallon). The vehicle's fuel efficiency determines both the fuel cost of the trip and the accompanying carbon emissions to the atmosphere, which are almost entirely in the form of carbon dioxide. The vehicle operator is concerned with the fuel cost while national authorities are concerned with the effects of aggregate vehicular fuel consumption on the problems of maintaining a reliable fuel supply and, most recently, on the contribution of vehicles to the national budget of greenhouse gas emissions. In the United States, where fuel retail prices are low compared with most other developed nations, fuel cost is a small fraction of total operating costs of a passenger car, yet it is still a factor in consumer choice of vehicle. Compared with European and Japanese owners, Americans on average drive larger, heavier, less efficient, and more expensive automobiles, but pay less for fuel. National tax policies that greatly affect the price of fuel play a large role in this difference.

The development of vehicle and engine technology that is more fuel efficient and less emitting of pollutants is primarily a response by manufacturers to national policies of regulation and economic tax disincentives that narrow the window of vehicle designs appealing most to consumers. In this section we consider the technological factors that directly affect vehicle fuel efficiency while meeting the restrictions on vehicle emissions.

### 9.5.1    U.S. Vehicle Fuel Efficiency Regulations and Test Cycles

In the United States, fuel efficiency of new passenger vehicles and light trucks is regulated by the National Highway Traffic Safety Administration of the U.S. Department of Transportation. This regulation had its origin in the oil shortages of the 1970s caused by an embargo on oil exports by OPEC nations. To ameliorate the nation's future dependence upon imported oil, Corporate Average Fuel Economy (CAFE) standards were promulgated, beginning with the 1978 model year. These standards required automobile manufacturers to design vehicles so that their sales-averaged vehicle

---

[12]We use the technical term, *fuel efficiency*, interchangeably with the common term, *fuel economy*, used in government regulations. A fuel-efficient vehicle would be economical to operate, having low fuel cost per distance traveled.

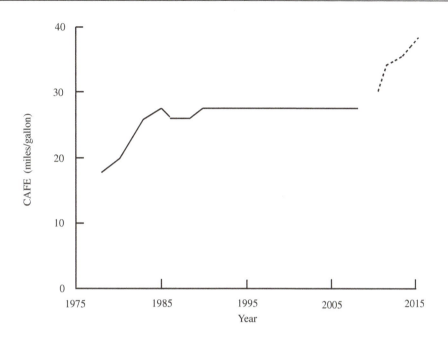

**FIGURE 9.7**  U.S. Corporate Average Fuel Economy (CAFE) standard for light-duty (passenger) vehicles for 1978–2008 (solid line) and proposed standard for 2011–2016 (dotted line).

fuel economy[13] did not exceed the level designated for each of two vehicle classes, passenger vehicles and light-duty trucks (pickup trucks, vans, sport utility vehicles).[14] In 2010, the CAFE standards for passenger cars and light-duty trucks are 27.5 miles/gallon (11.7 km/L) and 24.1 miles/gallon (10.2 km/L), respectively.

The history of the U.S. CAFE standard for automobiles, for the period 1978 until 2008, is shown in Figure 9.7 as a solid line. The first 10 years of the standard caused a remarkable increase of vehicle fuel economy made possible by advances in engine performance that accompanied the significant reductions in exhaust gas emissions. In subsequent years, when no further increase in fuel economy was required, improved engine performance made it possible for manufacturers to provide heavier, more powerful vehicles with the same fuel economy as their earlier predecessors

---

[13]While the average fuel economy is reported in units of miles per gallon, the averaging variable is the fuel consumption, measured in units of gallons per mile. Thus the average fuel consumption per mile traveled is regulated.

[14]Passenger cars are called light-duty vehicles (LDV) to distinguish them from heavy-duty passenger vehicles such as buses that carry more than 12 passengers. Light-duty trucks (LDT) are utility vehicles that can carry freight or passengers, whose gross vehicle weight rating (GVWR) is less than 8,500 pounds (3.856 t) and frontal area is less than 45 square feet or can operate off highways (e.g., sports utility vehicles). There are four categories of LDT (LDT1, LDT2, LDT3, and LDT4) depending upon the vehicle GVWR (<6,000 lb, >6,000 lb ) and the loaded vehicle weight (LVW), which is the sum of the vehicle curb weight plus 300 pounds (<3,750 lb, <5,750 lb). The vehicle curb weight (VCW) is the manufacturer's estimated weight of a fueled vehicle in operating condition.

and even lower exhaust emissions. But by 2010, the need to reduce $CO_2$ emissions had added greatly to the importance of reducing vehicle fossil fuel consumption, resulting in the proposed rapidly increasing CAFE standard shown in Figure 9.7 as a dotted line. This calls for a significant increase in fuel efficiency, most likely to be met by downsizing vehicle mass and engine power and use of electric hybrid or battery power.[15]

The measurement of vehicle fuel economy is based upon dynamometer simulations of typical driving cycles for urban and highway travel, originally devised by the U.S. EPA for evaluating vehicle emissions (Urban Dynamometer Driving Schedule). The test vehicle is operated in a stationary position while the drive wheels turn a dynamometer that is adjusted to provide the acceleration/deceleration and air resistance loads as described in Equations (9.10) and (9.13). While the dynamometer does not precisely simulate these forces at each point in the test, it suffices to provide a reasonable average for characterizing the emissions and fuel economy of the test cycle.

Two test cycles are used for measurements of vehicle fuel efficiency, one for urban driving and the other for highway travel. Figure 9.8 shows the vehicle speed versus time for each cycle. The urban cycle has many starts and stops (25 in a 17.8-km ride lasting 23 minutes) and a low average speed (while moving) of 31.4 km/h. For an average vehicle, about 34% of the energy needed to propel the vehicle through this cycle is dissipated in braking. In contrast, the highway driving cycle proceeds at a more uniform speed, averaging 78.5 km/h over a 16.5 km run, with only one stop.

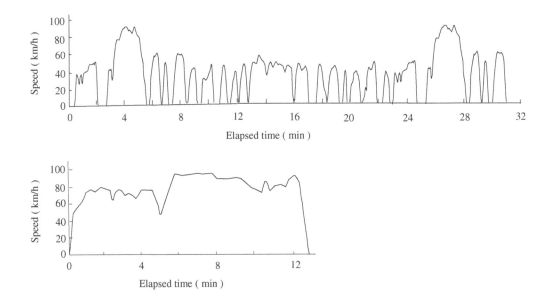

**FIGURE 9.8**  U.S. EPA 2010 driving cycles for evaluating emissions and fuel economy. Upper curve: Urban driving cycle (17.8-km length, 31.2-min duration, 34.1-km/h average speed, 91-km/h maximum speed). Lower curve: Highway driving cycle (16.5-km length, 12.7-min duration, 78.5-km/h average speed, 96.4-km/h maximum speed). (Data from U.S. Environmental Protection Agency.)

---

[15]Not shown in Figure 9.7 is the corresponding CAFE standard for light-duty trucks, which are equally significant contributors to fuel consumption and $CO_2$ emissions.

The two cycles represent two common types of vehicle use, stop-and-go driving in congested urban streets and uncongested freeway travel. For passenger cars, highway travel is between 12 and 70% more fuel efficient than urban travel (see Table 9.2). New vehicle purchasers are notified of the fuel economies for urban and highway travel determined from these tests.[16] For the purpose of enforcing CAFE standards, a weighted average is used (55% urban and 45% highway).[17]

## 9.5.2   Improving Vehicle Fuel Economy

In the United States in 1993, a consortium of U.S. vehicle manufacturers and the U.S. federal government undertook to develop a midsize passenger vehicle that would attain a fuel economy about three times the then-current value but would have performance properties equal to those of current vehicles, yet would be affordably priced. Known as the Partnership for a New Generation of Vehicles (PNVG), this ambitious program planned for introducing such a vehicle prototype by the model year 2004. The fuel economy goal was judged to be attainable using current or developing technologies. Although the PNGV program failed to reach its ultimate goal of a highly fuel efficient vehicle, the technological improvements investigated made it possible to greatly improve vehicle fuel efficiency in subsequent years.

There were two major thrusts of the PNGV program: improvements to the vehicle design and to the power source. Indeed, both vehicle and power source must be matched to obtain the maximum benefit. Nevertheless, in what follows, we consider these separately, with the understanding that they must not be considered in isolation.

### 9.5.2.1   Improving Vehicle Performance

In Section 9.4 we described the three components of power consumption in a moving vehicle: vehicle acceleration, rolling resistance, and aerodynamic drag. The first two are proportional to vehicle mass, while the latter depends upon vehicle size and shape, but not mass. The three major parameters determining these components are vehicle mass $m$, rolling resistance coefficient $C_R$, and the vehicle drag area $C_D A$. All of these are important.

**Vehicle Mass.** In current conventional vehicles, mass is the parameter that best correlates fuel economy (see Table 9.2). Large, heavy vehicles require big engines to perform well; they consequently consume more fuel. For a given vehicle size, reducing vehicle mass will permit reductions in engine and transmission mass, tire and wheel mass, braking system mass, fuel storage mass, steering system mass, engine radiator mass, and so on, compounding the gains in direct mass reduction of the vehicle frame. The principal means for reducing mass is the substitution of lighter materials of equal strength and stiffness, such as aluminum alloys or fiber-reinforced plastic for

---

[16]The consumer guide to vehicle fuel economy discounts the highway and urban test fuel economies by 10 and 22%, respectively, to account partially for disparities in the test procedure compared with average driving conditions. Studies of fuel consumption of vehicles in use indicate higher consumption than the consumer guide values.

[17]The weighting factors are applied to the fuel consumption, which is the reciprocal of the vehicle fuel economy.

steel and plastic for window glass, as well as the redesign of the vehicle structure to minimize structural mass. Reductions of up to 40% of vehicle mass seem possible.[18]

Reducing vehicle mass by material substitution may have implications for vehicle safety. Vehicle frames are designed to absorb the vehicle's kinetic energy in a crash while protecting the occupants from harm. Substitution of a lighter material of equal strength and energy-absorbing capacity in the body structure can maintain the same level of kinetic energy absorption and passenger protection while reducing overall vehicle mass. But in a two-vehicle collision of vehicles of unequal mass, the lighter vehicle absorbs more than its own kinetic energy and thereby suffers a safety disadvantage. In the current U.S. fleet of light-duty passenger vehicles and trucks, the mass ratio of the heaviest to lightest vehicle is about 3. As long as vehicles of different size persist in future fleets, this mass disparity will continue even if all vehicles are made lighter than current ones.

Vehicles with propulsion systems that utilize heavy energy storage systems (electric storage batteries, pressurized gaseous fuel such as natural gas or hydrogen stored in steel cylinders) incur vehicle mass penalties. For lead-acid storage batteries, the battery mass per unit of energy storage is 5,000 times that of gasoline, so that vehicle mass for electric battery vehicles is dominated by the battery mass needed to give the vehicle an adequate range. Lighter weight storage batteries, such as lithium-ion batteries, incur a lesser weight penalty, but their cost is much higher and is a significant obstacle to their widespread use in vehicles. For hydrogen- or natural-gas-fueled vehicles, the storage of the energy equivalent of 60 L of gasoline would add about 300 and 130 kg, respectively, to vehicle mass.

**Aerodynamic Resistance.** Reduction of aerodynamic resistance is limited to lowering the drag coefficient $C_D$ by careful streamlining of the entire body, since the frontal area $A$ is essentially fixed by the requirements of providing an enclosure for the passengers (see Table 9.2). The bulky form of the automobile limits what can be done to reduce aerodynamic drag, but attention to details can bring the drag coefficient into the range of 0.20–0.25 for passenger vehicles. There does not appear to be a weight penalty attached to low drag coefficients.

**Rolling Resistance.** The rolling resistance coefficient $C_R$ can be lowered to 0.05 from current values of 0.010–0.014 by improvements in tire design, but it is difficult to maintain durability, performance, and safety (traction) while reducing rolling resistance. Light alloy wheels reduce sprung mass, which is desirable, but more complex suspension systems will be needed to recover normal performance with low-resistance tires.

### 9.5.2.2 Improving Engine Performance

There are several paths to improving the efficiency of engines while supplying the requisite power needed by the vehicle. One is to improve the fuel (or thermal) efficiency of the engine, especially at

---

[18]The cost of direct weight reduction by material substitution is in the range of $1 to $3 per kilogram (Mark, Jason. *Greener SUVs. A Blueprint for Cleaner, More Efficient Light Trucks.* Cambridge, MA: Union of Concerned Scientists, 1999). But each kilogram of direct weight reduction by material substitution provides an opportunity for additional weight reduction if the engine and drive train are reduced in size in proportion to the vehicle mass reduction, thereby providing a cost savings. This would lower the net cost of the material substitution weight reduction.

off-optimum conditions where the engine is usually operated. A second is to utilize transmissions that permit the engine to maximize its efficiency for a required power output. A third is to reduce engine mass for a given power, so as to improve vehicle efficiency. However, there are serious constraints imposed on engine efficiency improvements by the requirements for limiting exhaust pollutant emissions. These may limit the possible efficiency improvements.

**Reducing Intake Stroke Losses in SI Engines.** As explained in Section 9.3, at partial load the cylinder pressure during the intake stroke is lowered to reduce the fuel amount in the cylinder, resulting in a loss of efficiency. This loss may be offset by varying the timing of the inlet and exhaust valves with the engine load. Variable valve timing (VVT) systems are currently used in four-valve engines, adding to the peak power output and engine mass reduction benefit. Another alternative is direct fuel injection (DI) into the cylinder during the intake stroke, forming a nonuniform fuel–air mixture at higher overall pressure and lower intake stroke power loss, albeit at some penalty in $NO_x$ emissions. The recirculation of exhaust gas into the cylinder during the intake stroke so as to reduce exhaust pollutant emissions can be arranged to reduce intake losses and improve engine efficiency at part load.

**Replacing SI Engines with CI Engines.** The indirect injection CI engine enjoys about a 25% advantage in fuel economy over the SI engine and does not suffer as much efficiency loss at part load as does the SI engine. Direct injection CI engines have even higher fuel economy advantage, about 30–40%.[19] On the other hand, the CI engine is heavier for a given power, thereby incurring a vehicle mass efficiency penalty, and is more expensive. Also, it is more difficult to reduce $NO_x$ and particulate emissions in the CI engine. If the latter can be overcome, the CI engine provides significantly better fuel economy.

**Supercharging.** Both SI and CI engines can be supercharged (or turbocharged) to increase maximum engine power for a given displacement and engine mass, with some improvement in engine efficiency at higher loads. This provides a vehicle mass efficiency benefit.

**Continuously Variable Transmission.** As explained in Section 9.4.1, the traditional multistep transmission does not permit the engine to operate at maximum thermal efficiency over the entire vehicle duty cycle. Indeed, the engine seldom operates at best efficiency. A continuously variable transmission (CVT) can be controlled to maximize the engine efficiency at any power level needed by the vehicle. Current CVT transmissions have shown that overall vehicle efficiency can be improved provided that the transmission efficiency is close to that of conventional transmissions.

**Engine Idle-Off.** In congested urban driving, considerable time is spent with the engine idling while the vehicle is stationary. If the engine is stopped during these intervals, fuel will be saved and exhaust emissions reduced. However, stopping the engine and restarting also consume fuel, so that the length of the idle period has to be sufficient for there to be a fuel reduction in idle-off control.

---

[19]In terms of thermal efficiency, the advantage is lower because of the difference in fuel heating value per unit volume of fuel between gasoline and diesel fuel.

## 9.6    ELECTRIC-DRIVE VEHICLES

In the earliest years of the development of the automobile, some were powered by electric-drive motors energized by lead-acid storage batteries. Despite their advantages of no noise and exhaust emissions, their limited range and difficulty of recharging confined their use to urban fleet delivery vehicles.

In recent years, electric-drive vehicles have been reconsidered. Advances in electric and mechanical energy storage systems and the energy savings to be made by storing and reusing vehicle braking power make such vehicles attractive from both the emissions and the vehicle efficiency point of view. Providing electric-drive power from an onboard fuel cell or ICE motor-generator set can provide more acceptable vehicle range. Various combinations of these drive systems might prove even more advantageous. In this section we review the progress made in incorporating technological improvements in vehicles of this type.

### 9.6.1    Vehicles Powered by Storage Batteries

The development of battery-powered electric drive vehicles was spurred in the United States when the state of California established a future requirement for some zero-emission vehicles in its new car fleet mix, beginning with the 2003 model year. Several manufacturers produced small passenger cars and pickup trucks that were battery powered. The specifications of one of these vehicles, the GM EV1, is listed in the last column of Table 9.3.

**TABLE 9.3**    2010 Model Year Electric Vehicle Characteristics

| Manufacturer | Nissan | GM | GM |
|---|---|---|---|
| Model | Leaf | Chevrolet Volt[a] | EV1[b] |
| Size | Five-passenger | Four-passenger | Two-passenger |
| Battery type | Lithium-ion | Lithium-ion | NiMH |
| Maximum electric power (kW) | 90 | 111 | 102 |
| Vehicle mass (t) | 1.13 | — | 1.32 |
| Frontal area ($m^2$) | 2.8 | 2.6 | 1.8 |
| Battery mass (kg) | 200 | 170 | 521 |
| Battery energy storage (kWh) | 24 | 16 | 26.4 |
| Vehicle range (km) | 160 | 64 | 121–209 |
| Charge time (h) | 8 | — | 6–8 |
| Highway energy efficiency (km/MJ) | 1.85 | 1.11 | 1.49 |
| Highway equivalent fuel efficiency (km/L)[c] | 17.5 | 10.5 | 14.1 |
| Battery/vehicle mass | 0.176 | — | 0.395 |
| Power/mass (kW/t) | 79.4 | — | 32.1 |
| Battery energy/mass (Wh/kg) | 120 | 94 | 51 |
| Battery energy/maximum power (h) | 0.27 | 0.144 | 0.26 |

[a] Plug-in hybrid. Data for electric battery operation only. Extended range is 1,030 km when using 53-kW engine/generator powered by 1.4-L, four-cylinder SI engine.
[b] Model year 2000 prototype.
[c] Assumes 30% equivalent engine thermal efficiency or 9.48 MJ/L.

Battery-powered vehicles have limited passenger- or freight-carrying capability compared with conventional vehicles and much smaller travel range between recharges of the onboard energy supply. Typically, they are powered by cooled AC induction motors drawing their power from battery banks via power conditioning equipment. The traction motor is geared to the drive wheels at a fixed speed ratio and can regenerate a partial battery charge during periods of vehicle deceleration.

The EV1 was equipped with either a lead-acid or nickel metal hydride battery bank, the latter storing 40% more energy with 12% less weight than the former and thereby achieving a 36% greater range. The time needed to charge the battery bank completely is 6–8 hours. After about 1,000 charge–discharge cycles the batteries must be replaced by new ones, but of course the depleted batteries can be recycled to recover battery materials.[20] Battery replacement will cost several thousand dollars every 50,000 miles.

The battery mass of the EV1 constituted about 40% of the vehicle mass, implying a very large vehicle mass penalty to the vehicle energy efficiency. Indeed, if we convert the vehicle electric energy efficiency of 1.3–1.5 km/MJ to an equivalent fuel efficiency of 12.5–14.1 km/L (see Table 9.3), then these vehicles are only slightly more efficient than current vehicles of comparable mass listed in Table 9.2. Despite the recovery of braking energy, the electric vehicles are unable to achieve higher fuel efficiency than comparable conventional ones.

Two model year 2010 battery-powered vehicles, the Nissan Leaf and the GM Chevrolet Volt,[21] are listed in Table 9.3. Both are powered by lithium-ion batteries especially developed for vehicle propulsion. In almost every respect the Leaf is superior to the EV1: power/vehicle mass, battery/vehicle mass, battery energy/mass, battery energy/maximum power, and vehicle range. When operated in battery-only mode, the Volt is less superior to its predecessor, but shows the superior advantages of lithium-ion battery properties.

The power/mass ratios of battery-powered vehicles, which lie in the range of 30 to 80 kW/ton, are less than those of conventional vehicles (Table 9.2), limiting their acceleration and grade-climbing abilities.

The limited range and fuel efficiency of battery-powered electric vehicles is tied to the low energy storage densities of currently available batteries, about 25–50 Wh/kg (90–180 kJ/kg). But other electrical storage systems being developed, such as electric capacitors and flywheels, have no higher energy storage densities (see Table 5.1), and other types of high density storage batteries are unsuitable for automobiles (see Table 5.2). Adding battery mass to the vehicle will not bring about a commensurate increase in range, which is better improved by increasing the efficiency with which stored energy is utilized in propelling the vehicle.

## 9.6.2 Hybrid Vehicles

Hybrid vehicles are those that combine conventionally fueled power sources (SI or CI engines) with electric motors to power the vehicle and include some measure of electric energy storage in batteries. In one extreme, the SI or CI engine can power a generator that delivers electric power

---

[20] In a lead-acid battery, the electrolyte sulfate ions are deposited on the surfaces of the electrodes during battery discharge. This process is not entirely reversed upon recharge, gradually leading to the loss of electrode active surface area and the energy storage capacity of the battery with each recharge cycle.

[21] The Volt is a plug-in hybrid, but its characteristics listed in Table 9.3 are for operation as a battery-powered vehicle only.

to an electric traction motor; this system is used in railway engines. For automobiles, it is more common to add a separate battery-powered traction motor/generator drive system, acting in parallel with the conventional engine/transmission drive system; either or both may operate independently or simultaneously. The motor/generator can store energy in a battery bank when excess power is available, during deceleration or when the power need is less than what the combustion engine can deliver, and can provide extra power to the wheels when it is temporarily needed for acceleration or hill climbing. Such a system can level the peaks and valleys of power demand in the driving cycle, enabling the combustion engine to run closer to average driving cycle power and potentially better average engine efficiency.

The characteristics of three hybrid vehicles of this type are listed in Table 9.2 for ready comparison with traditional vehicles of comparable vehicle mass. The Honda Civic Hybrid and the Toyota Prius Hybrid are intermediate in vehicle mass between the Honda Civic and Accord and have about 30% lower ratio of power to mass. However, their fuel efficiency is approximately double that of the comparable conventional vehicles, a consequence of the advantage of the hybrid system when tuned to the demands of the urban/highway driving cycle, including dynamic braking and engine-off operation at vehicle stops. On the other hand, the Ford Fusion Hybrid has nearly the same vehicle mass and engine power as the Ford Fusion, but substantially better urban fuel efficiency. It is clear that the hybrid drive system has better fuel efficiency than a conventional vehicle of comparable mass, especially if the hybrid vehicle power is lower. Offsetting this fuel economy advantage is the additional weight and cost of the electrical components of the hybrid system (traction motor/generator, batteries, and power conditioning equipment).

Evolution of standard vehicles in future years will move in the direction of hybrid drives. It is expected that the electric auxiliary system will change from 14 to 42 volts and triple in power. The current starter motor and belt-driven alternator will be replaced by a motor/generator directly connected to the engine drive shaft, permitting idle-off operation when the vehicle is stationary. Recovery of braking energy would be possible, depending upon the motor/generator power and electric storage capacity. Although the purpose of this development is to utilize electric drive for auxiliary power and thereby improve engine efficiency, it clearly can be extended to become a hybrid system.

### 9.6.3  Fuel Cell Vehicles

Prototypes of electric-drive vehicles whose electric power is supplied by fuel cells have been under development for several decades. Potentially, such vehicles could provide higher vehicle fuel efficiencies than conventional vehicles with little or no air pollutant emissions. Increasingly stringent exhaust pollutant emission standards, especially in California, and national policies in developed nations to secure both economic and environmental benefits of improving fuel economy have increased the incentives for manufacturers to develop vehicle fuel cell technologies.

As explained in Section 4.7.1, the oxidation of a fuel in a fuel cell has the potential to convert a higher percentage of the fuel's heating value to electrical work than does the typical combustion engine. The upper limit to this proportion is the ratio of the free energy change in the fuel oxidation reaction, $\Delta f$, to the enthalpy change, or fuel heating value FHV. For hydrogen, this ratio is 0.83, while for methane and methanol it is 0.92 (see Table 4.1); these upper limits are at least double what could be obtained by burning these fuels in a steam or gas turbine power plant. However, as Figure 4.2 illustrates, this high conversion efficiency is only reached at zero power output; at higher power the cell voltage declines nearly linearly with increasing cell current, resulting in only

50% of the upper value being recoverable at maximum cell power (41.5% for hydrogen and 46% for methane and methanol). Still, these are higher fuel efficiencies than are obtainable in vehicle SI engines at full power; the comparison is even more favorable to the fuel cell at part load because the fuel cell efficiency increases at reduced load.

But there are countervailing factors in vehicle fuel cell systems that lower this fuel efficiency. The only practical fuel cell for vehicle use requires gaseous hydrogen fuel. The most economical and energy efficient source of hydrogen, a synthetic fuel, comes from reforming from a fossil fuel such as natural gas, oil, or coal or from another synthetic fuel like methanol or ethanol. Whether this reforming is done at the fuel depot, where the hydrogen is then loaded and stored on the vehicle, or is accomplished on board the vehicle in a portable reformer, the reforming operation preserves at best only 80% of the parent fuel's heating value (see Table 4.2). This additional loss lowers the maximum power fuel efficiency of hydrogen to 33.2%. Furthermore, if hydrogen is liquified for storage on the vehicle, rather than being stored as a compressed gas in tanks, an additional energy penalty of about 30% is incurred because energy is needed to liquify hydrogen at the very low temperature of $-252.8°C$. Altogether, these synthetic fuel transformation penalties diminish the fuel efficiency advantage of fuel cells compared with conventional internal combustion engines in vehicles fueled by conventional hydrocarbon fuels.[22]

The hydrogen fuel cell used in vehicles utilizes a solid polymer electrolyte, called a proton exchange membrane (PEM). Only a fraction of a millimeter in thickness, the membrane is coated on both sides with a very thin layer of platinum catalyst material that is required to promote the electrode reactions producing the flow of electric current through the cell and its external circuit, as described in Section 4.7.1. Carbon electrodes, provided with grooves that ensure the gaseous reactants (hydrogen at the anode and oxygen (in air) at the cathode) are distributed uniformly across the electrode surfaces, are sandwiched on either side of the PEM, forming a single cell. As many as 100 or more cells are stacked in series, mechanically and electrically. The fuel and oxidant are supplied under a pressure of several bar, so as to increase the maximum power output per unit of electrode area, which is usually of the order of 1 $W/cm^2$. Water, formed at the cathode, must be removed, but the PEM must remain moist to function properly. In addition, heat is released in the cell reaction, so that the cell must be cooled, maintaining a temperature less than $100°C$.

Table 9.4 lists the salient features of several 2010 model fuel cell vehicles under development. These are comparable in vehicle mass to the midsize standard vehicles in Table 9.2, but with about 25% less maximum power. The Honda FCX Clarity and the GM HydroGen 4 include a lithium-ion battery storage system, so that these models function as hybrid electric vehicles with a fuel cell replacing the conventional engine. Earlier fuel cell vehicles developed during the 1990s in Europe and the United States included a compact car (NECAR 4) and a municipal bus (NEBUS). They both are based upon a conventional vehicle with the replacement of the engine and fuel tank by a fuel cell, electric motor, and fuel storage system, while maintaining the original vehicle's passenger capacity and range. The arrangement of the replacement fuel cell system in the vehicle is shown in Figure 9.9 for the NECAR 4 and NEBUS vehicles.

NECAR 4 stores its hydrogen fuel in liquid form in a cryogenic insulated fuel tank while NEBUS stores hydrogen gas in tanks pressurized to 300 bar and weighing about 75 times more

---

[22]Fuel cells used in electric power systems do not suffer such disadvantages. They do not require synthetic fuel and operate at high temperatures where coolant streams can be used to generate steam for extra electric power output.

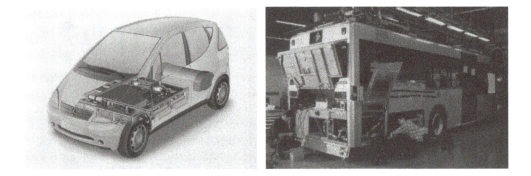

**FIGURE 9.9**   Left: Artist's sketch of the interior layout of the fuel cell (center) and liquid hydrogen fuel tank (rear) in the NECAR 4 vehicle. Right: The NEBUS municipal bus, showing the fuel cell in the lower rear and the hydrogen gas storage tanks on the roof.

**TABLE 9.4**   Characteristics of 2010 Fuel Cell Vehicles

| Manufacturer | Honda | GM | Daimler AG | Daimler AG |
|---|---|---|---|---|
| Model | FCX Clarity | HydroGen 4 | Mercedes-Benz F-Cell | NECAR 4[a] |
| Vehicle class | Midsize | SUV | B-Class | A(compact) |
| Passengers | 4 | 5 | — | 5 |
| Vehicle mass (t) | 1.6 | 1.65 | — | 1.48 |
| Frontal area (m$^2$) | 2.7 | 3.2 | 2.8 | 2.2 |
| Electric power (kW) | 100 | 93 | 100 | 55 |
| Electric power/vehicle mass (kW/t) | 62.5 | 56.4 | — | 37.2 |
| Cell voltage and battery type | 288, Lithium-ion | —, Lithium-ion | — | — |
| Cell mass (kg) | 67 | — | — | 5 |
| Fuel cell specific power (kW/kg) | 1.5 | — | — | 0.20 |
| Fuel mass (kg) and pressure (bar) | 4.1, 342 | 4.2, 700 | —, 700 | 5, — |
| Range (km) | 450 | 322 | 678 | 450 |
| Top speed (km/h) | — | 160 | — | 145 |
| Vehicle fuel economy (km/kg H) | 110 | 77 | — | 90 |

[a] 2000 prototype model.

than the fuel they contain, adding to the vehicle mass. In general, these vehicles weigh more than their conventional counterparts.

The fuel economy of these vehicles, expressed as kilometers per kilogram of hydrogen, are listed in Table 9.4. Compared with conventional vehicles, fuel cell vehicles have comparable fuel economies, but are clearly not distinctly superior to their conventional cousins. But if one considers the synthetic fuel penalties involved in preparing hydrogen, both gaseous and liquid, then the fuel cell vehicles, at least at this stage of their development, have little or no fuel economy advantage.

There is no question that fuel cell vehicles suffer from the limited availability and high cost of hydrogen fuel. Commercial-scale hydrogen reformers fueled by natural gas are complex systems, with the product stored at high pressure in tanks for transfer to vehicles. Hydrogen is especially prone to burn and explode, raising safety problems even for fleet vehicles that can be more carefully supervised than individually owned vehicles. But a major reduction in fuel cell cost and improvement in fuel economy will be required for fuel cell vehicles to be suitable substitutes for conventional ones.

## 9.7  VEHICLE EMISSIONS

By the middle of the 20th century vehicle exhaust emissions were recognized to be an important contributor to urban photochemical air pollution, especially in locations like southern California where high insolation and temperature and poor atmospheric ventilation combined with a rapidly growing automobile population to produce record levels of ground-level ozone concentrations. Not long thereafter, similar pollution problems appeared in other major cities around the world as urban vehicle populations blossomed. These problems are more acute in lower latitude locations, especially in developing countries where vehicle emission controls are not yet stringent.

Of course, vehicles provide only part of ozone precursor emissions. But they are mobile and more numerous than stationary sources and present different problems for abatement. Early in the pollutant regulatory history of the United States, it became obvious that it was more effective to require a few vehicle manufacturers to install control equipment on millions of new vehicles rather than to require hundreds of millions of vehicle owners to try to reduce their own vehicles' emissions. This scheme, adopted by all developed nations, replaces all vehicles with new and cleaner ones every 12–15 years, providing the opportunity to capitalize on the improvements in emission control technology. In the United States, it has resulted in significant reductions in air pollutant emissions from vehicles.

In this section we discuss the vehicle technologies that are used to reduce the emissions from ICE engines in vehicles in response to national regulations.

### 9.7.1  U.S. Vehicle Emission Standards

Vehicle emissions to the atmosphere are of two kinds: exhaust emissions and evaporative emissions. The first are the combustion gases emitted while the engine is running, whether or not the vehicle is moving. The second are emissions of fuel vapors from the fuel supply system and the engine when the vehicle is stationary with the engine not operating.[23] The federal government regulates both of these emissions by requiring the manufacturers of new vehicles sold in the United States to provide the technology needed to limit these emissions for the useful life of the vehicle and to warrant the performance of these control systems.

To certify a vehicle class for exhaust emissions, the manufacturer must test a prototype vehicle on a dynamometer following the Federal Test Procedure (FTP) (see Section 9.5.1), during which exhaust gases are collected and later analyzed for pollutant content. Regulated pollutants include nonmethane hydrocarbons (NMHC) or organic gases (NMOG), carbon monoxide (CO), nitrogen oxides ($NO_x$), particulate matter (PM), and formaldehyde (HCHO). The mass of each pollutant collected from the exhaust during the test is divided by the test mileage and reported as grams per

---

[23]Emission of fuel vapor while refueling may be separately regulated by state agencies under state implementation plans for conforming with federal air quality regulations.

mile. If the prototype vehicle's exhaust emissions do not exceed the standards set for its vehicle type, vehicles of its class and model year may then be sold by the manufacturer. The manufacturer is further responsible for ensuring that their vehicles' control systems continue to function properly during the life of the vehicles, currently set at 100,000 miles. Vehicles must also conform to the exhaust emission limitations of the Supplemental Federal Test Procedure (SFTP), designed to evaluate the effects of air conditioning load, high ambient temperature, and high vehicle speeds, not included in the FTP, on emissions.

Evaporative emissions are tested for two conditions, one where the vehicle is at rest after sufficient use to have brought it to operating temperature and the other for a prolonged period of nonuse. In these tests the vehicle is enclosed in an impermeable bag of known volume and the organic vapor mass is subsequently determined.

In the United States, vehicle emission standards are set by the U.S. Environmental Protection Agency in accordance with the provisions of federal air quality legislation. The regulation is based upon the recognition of the ubiquity and mobility of the automobile, its concentration in urban areas, its contribution to urban and regional air quality problems, and the ability of the manufacturer and not the owner to ameliorate its emissions. In the years since the early 1970s, when regulation was first introduced, emission standards have become more stringent as manufacturers devised better technologies and the difficulty of achieving desirable air quality throughout the United States became more apparent. Given the lead time required by manufacturers to develop new control technologies and incorporate them in a reliable consumer product, emission standards must be set years in advance of their attainment in new vehicles sold to the consumer.

U.S. exhaust emission standards for vehicles of model year 1996 and beyond are listed in Table 9.5. The standards apply for two time periods: 1996 to 2007 (Tier 1) and 2004 and beyond (Tier 2). The Tier 2 standards are phased in over the period 2004–2010, during which the Tier 1 standards are simultaneously being phased out.

Tier 1 standards limit four pollutants for five vehicle classes: light-duty vehicles (LDV, which are passenger vehicles for 12 passengers or less) and four types of light-duty trucks (LDT1, LDT2, LDT3, and LDT4, distinguished by the gross vehicle weight rating and the loaded vehicle weight [see Section 9.5.1]). The larger light-duty trucks are permitted greater emissions in recognition of their greater weight carrying capability. Like earlier standards, the Tier 1 standards apply to several vehicle classes, but within each class each vehicle model, small or large, must meet the same standard.

Tier 2 standards introduce a new method of limiting emissions. Like the fuel economy standard, each manufacturer must achieve a sales-averaged $NO_x$ emissions for all its vehicles of 0.07 g/mile, although individual vehicle models may emit more if they are offset by others that emit less. Each vehicle model is certified in one of seven emission categories (denoted by Bin 1, . . . , Bin 7 in Table 9.5), which ensures that the sales-averaged emission limits for pollutants other than $NO_x$ will also not be exceeded. This new method of limiting emissions will allow manufacturers to achieve the necessary overall reduction in their fleet's emissions as economically as possible by providing incentives to reduce emissions below the standard in light, low-powered vehicles that may then be credited to heavy, high-powered ones.

The Tier 2 standards, which begin with the 2004 model year, are quite stringent compared with those of the early 1970s, when standards were first applied (see Table 9.5). For the ozone precursor pollutants, $NO_x$ and NMOG, Tier 2 levels are about 2% of those for 1971 model year vehicles and approximately 0.2% of unregulated 1960's vehicles. This considerable reduction will be required

**TABLE 9.5**    U.S. Vehicle Exhaust Emission Standards

Tier 1[a]

| Veh. type | NMHC[b] (g/mile) | CO (g/mile) | NO$_x$ (g/mile) | PM (g/mile) | |
|---|---|---|---|---|---|
| LDV | 0.25 | 3.4 | 0.4 | 0.08 | |
| LDT1 | 0.25 | 3.4 | 0.4 | 0.08 | |
| LDT2 | 0.32 | 4.4 | 0.7 | 0.08 | |
| LDT3 | 0.32 | 4.4 | 0.7 | | |
| LDT4 | 0.39 | 5.0 | 1.1 | | |

Tier 2[c]

| Veh. type | NMOG[d] (g/mile) | CO (g/mile) | NO$_x$ (g/mile) | PM (g/mile) | HCHO[e] (g/mile) |
|---|---|---|---|---|---|
| All | 0.09 | 4.2 | 0.07 | 0.01 | 0.018 |
| Bin 7 | 0.125 | 4.2 | 0.20 | 0.02 | 0.018 |
| Bin 6 | 0.090 | 4.2 | 0.15 | 0.02 | 0.018 |
| Bin 5 | 0.090 | 4.2 | 0.07 | 0.01 | 0.018 |
| Bin 4 | 0.055 | 2.1 | 0.07 | 0.01 | 0.011 |
| Bin 3 | 0.070 | 2.1 | 0.04 | 0.01 | 0.011 |
| Bin 2 | 0.010 | 2.1 | 0.02 | 0.01 | 0.004 |
| Bin 1 | 0.000 | 0.00 | 0.00 | 0.00 | 0.000 |

[a] Model years 1996–2007. Five years, 50,000 miles.
[b] Nonmethane hydrocarbons.
[c] Model years 2004 and beyond, except 2008, for LDT3, LDT4. Full useful life (120,000 miles).
[d] Nonmethane organic gases.
[e] Formaldehyde.

to make it possible to achieve ambient ozone standards in U.S. metropolitan areas in the early part of the 21st century, despite increasing vehicle population and increased annual travel per vehicle.[24]

## 9.7.2    Reducing Vehicle Emissions

Vehicle exhaust pollutants are the remnants of an incomplete and nonequilibrium combustion process in the engine cylinder. Of the mixture of fuel and air introduced into the cylinder, all but a tiny fraction of the fuel is oxidized to $CO_2$ and $H_2O$, reaching a state of thermochemical equilibrium after releasing the chemical energy bound up in the fuel. But a small amount of the reactants do not reach this equilibrium state, instead remaining frozen into a metastable form. The principal molecules of this type are NO, CO, and various kinds of hydrocarbon (NMHC) fragments, all of

---

[24] In the United States, vehicle exhaust emissions of $CO_2$ are regulated indirectly via CAFE (fuel efficiency) standards, but climate change regulations pending in 2010 may alter this status.

which can contaminate the atmosphere into which the vehicle exhaust gas stream is introduced. The purpose of vehicle emission control technology is to reduce the amounts of these pollutants to such low values that the cumulative effects of many vehicles and other sources will not be great enough to cause any damage to living systems, including humans.

In SI engines, the amount of each of these principal pollutants is sensitive to the air/fuel ratio of the mixture inducted into the cylinder prior to ignition by the spark plug. The proportion of air to fuel must not be too far from the stoichiometric value for the engine to function properly and efficiently. If the mixture is fuel-rich (more fuel than can be completely oxidized by the available oxygen), some CO will be formed and not all of the fuel's heating value will be released. If the mixture is fuel-lean (excess, unused oxygen), the combustion product temperature and pressure will be less, resulting in less engine work per cycle. These differences in the chemical and thermodynamic state of the combustion gases influence the amounts of pollutants that leave the engine through the exhaust port.

Figure 9.10 shows how the mass of exhaust gas pollutants varies with the air/fuel ratio in an SI engine. In rich mixtures, there is insufficient oxygen to oxidize the fuel molecules, completely, leaving some unburned HC or incompletely oxidized CO and $H_2$, the more so as the oxygen deficiency becomes larger.[25] When there is surplus oxygen (lean mixture), CO and HC diminish to low values as the extra oxygen finds and oxidizes them. On the other hand, NO is formed by the reaction of $N_2$ with $O_2$ at the high temperature behind the flame front, but only in very small

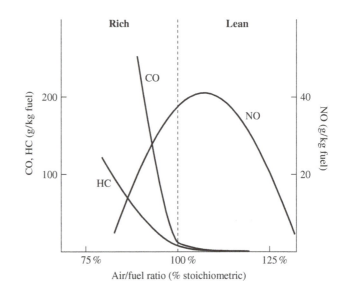

**FIGURE 9.10**  A sketch of the mass of exhaust gas pollutants, carbon monoxide (CO), nitrogen oxide (NO), and hydrocarbons (HC), in a typical SI engine, as a function of the air/fuel ratio.

---

[25]The hydrogen, not shown in Figure 9.10, is formed through the reduction of $H_2O$ by CO at the end of the combustion process. This hydrogen aids the catalytic reduction of NO in the exhaust converter.

quantities, and does not revert completely to $N_2$ and $O_2$ at the much lower exhaust temperature, as it should if thermochemical equilibrium prevailed. It is highest at or close to the stoichiometric mixture where the flame temperature is highest. The mass of HC and NO can be of the order of 1% of the fuel mass (about 0.1% of the exhaust gas mass), but the CO mass is 10 times larger. Such values are 10 or more times higher than allowed by U.S. exhaust emission standards.

For pollutants to reach the very low levels now being required of road vehicle exhaust streams, two steps must be undertaken simultaneously. The first is to reduce as much as possible the pollutant concentrations in the exhaust gas as it leaves the engine (engine-out emissions); the second is to reduce these emissions even further by exhaust gas treatment systems located between the engine and the tailpipe. Neither of these systems by itself can suffice to clean up the emissions to the required low levels.

### 9.7.2.1   Reducing Engine-Out Emissions

There are several features of modern SI engines that are nearly universally used to improve engine-out emissions.

**Precise Control of Air/Fuel Ratio.** Low values of the three principal pollutants, HC, CO, and NO, can be maintained if the air/fuel ratio is kept close to its stoichiometric value under all operating conditions. Fuel injection permits close control over fuel flow to each cylinder and can be computer controlled to be proportionate to the intake air flow. An oxygen detector placed downstream of the exhaust ports provides a sensitive signal used to correct the fuel flow so as to hone in on the desired air/fuel ratio. A further benefit of this control system is that it can provide optimum conditions for subsequent exhaust gas processing.

**Exhaust Gas Recirculation.** At the end of the exhaust stroke, when the exhaust valve has closed and the intake valve opens to admit a fresh charge of air–fuel mixture, the residual volume of the cylinder is filled with exhaust gas. This mixes with the incoming fresh charge, diluting it and reducing the temperature and pressure that is reached when that charge is fully burned at the beginning of the power stroke. Since the amount of NO formed is very sensitive to the peak temperature reached during combustion, we can reduce engine-out NO by diluting the fresh charge with even more exhaust gas than is normally encountered. This can be done by varying the exhaust and inlet valve timing or pumping exhaust gas from the exhaust system into the intake system. This is done at part load so the maximum engine torque and power are not compromised, but is acceptable since these maximum values are seldom utilized in standard driving cycles.

### 9.7.2.2   Catalytic Converters for Exhaust Gas Treatment

The exhaust gas pollutants HC, CO, and NO are not in thermochemical equilibrium with the rest of the exhaust gas. It should be possible to oxidize both HC and CO to $CO_2$ and $H_2O$ if enough oxygen is present and to reduce NO to $N_2$ and $O_2$, since these are thermodynamically favored. To make this happen quickly enough, these molecules must attach themselves to a solid surface coated with a catalyst, where they can react and their products evolve into the gas stream. Furthermore, this surface reaction will occur quickly only if the surface is hot enough and the proper catalyst is used. Current three-way oxidation-reduction catalysts utilize such catalysts as platinum and rhodium and must be heated to $250°C$ or more to be effective.

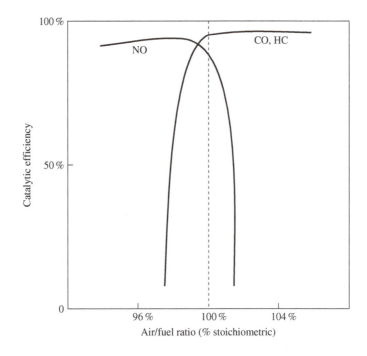

**FIGURE 9.11**    A sketch of the catalytic efficiency for CO and HC oxidation and NO reduction in a three-way catalytic converter as a function of the air/fuel ratio.

Simultaneous oxidation of CO and HC and reduction of NO in a catalytic converter requires very close control of the air/fuel ratio in the engine. Insufficient air will inhibit oxidation, while too much will prevent reduction. The window of air/fuel ratio that will remove equally all the pollutants is quite small, with only a few percent change being allowed. This is illustrated in Figure 9.11, which shows how the catalytic efficiency (percentage of pollutant removed in the converter) of the oxidation and reduction reactions depends critically upon the air/fuel ratio. Oxygen accumulation on the catalyst surface is built into these catalysts to widen the operating window.

For a catalytic converter to work properly, every pollutant molecule must have the chance to stick to the catalyst surface before it flows through the reactor. This requires that there be a large surface area coated with catalyst and that the flow passages surrounding this surface be finely divided. Either a honeycomb structure or a packed bed of catalyst-coated pebbles satisfies this requirement. Typically, the gas passage dimension is of the order of several millimeters, and the converter volume is about half the engine displacement. This allows the exhaust gas only about one engine cycle period to pass through the converter and be cleansed of most of the pollutants.

When an engine is first started at ambient temperature (called a cold start), the converter does not work until it has been warmed by the hot engine exhaust gas to its "light-off" temperature of about 250°C. During the minute or two that is required to reach converter light-off conditions, high engine-out pollutant levels are emitted from the tailpipe. Indeed, more than half of the emissions in the FTP may be emitted during this warm-up period. When engines are started in cold winter weather, excess fuel must be injected to achieve sufficient fuel evaporation to start the engine, giving greatly increased CO and HC emissions. Preconverters of low heat capacity, located close to the engine or electrically heated, are required to achieve low emission levels during cold starts.

Catalyst surfaces may be damaged by overheating if the exhaust gas contains excessive unburned fuel, which might occur if the air/fuel control system fails. Also, fuel impurities that leave surface deposits may destroy the catalytic function. Lead additives to gasoline have been phased out to protect converters, and current U.S. fuel regulations now require the removal of nearly all sulfur. Ultraclean fuels will ensure that converters will not deteriorate over the useful life of the vehicle.

### 9.7.2.3   Evaporative Emissions

Exhaust emissions of nitrogen oxides and incompletely burned fuel contribute to the formation of ground-level ozone in the atmosphere. But the inadvertent escape to the atmosphere of vehicle fuel vapor also can be an important contributor to the reactive organic compounds that participate in the formation of ozone. As a consequence, evaporative emission control systems are required on new U.S. vehicles.

There are several sources of vehicle evaporative emissions. Fuel stored in the fuel tank emits vapor into the air space above the fuel surface within the tank. This vapor can leak to the atmosphere during fuel refilling operations and during diurnal atmospheric temperature and pressure changes. At engine shutdown, unburned fuel remains in the engine and can subsequently leak to the atmosphere from the air intake or exhaust.

Fuel vapor has a different chemical composition than does the fuel because the vapor constituents are much richer in higher vapor pressure, lower molecular weight components than is the liquid fuel. Some of these components are also more chemically reactive in ozone formation, making it all the more important to prevent their release to the atmosphere. Fuel vapor pressure increases rapidly with temperature, so that uncontrolled evaporative emissions are higher in summer than winter. Since ozone formation is inherently higher in summer than winter, escape of fuel vapors exacerbates the summertime smog problem.

Fuel tank vapor emissions are controlled by placing a vapor-adsorbing filter in the vent line between the fuel tank air space and the atmosphere. If the fuel and tank air space warm up from solar heating, expelling some vapor/air mixture through the vent line, the filter will retain the vapor molecules on its adsorbing surface. To prevent the adsorbing surface from becoming saturated with fuel molecules, and thereby ineffective for further filtering, air is drawn inward through the filter when the engine is running, cleaning it of adsorbed vapor molecules. This inflow is ducted into the engine intake system so as to incinerate the desorbed vapor in the running engine.

When a fuel tank is filled, the vapor/air mixture in the tank air space is displaced by the incoming fuel. This mixture preferentially escapes through the fuel fill opening, and if not collected during the filling process by a vapor control system at the filling station it will be emitted into the atmosphere. U.S. states with ozone excedence problems usually require the installation of such equipment by service stations.

### 9.7.2.4   Reducing CI Engine Emissions

Diesel (CI) engines generally emit lesser amounts of CO and HC than SI engines, but more NO and particulate matter. The overall air/fuel ratio in a CI engine is always lean, more so at partial power levels, providing excess oxygen to oxidize HC and CO to $CO_2$ and $H_2O$, and $N_2$ to NO. But the evaporation and mixing of the fuel droplets in a CI engine with surrounding air is uneven, so that some of the fuel burns in an oxygen-deficient atmosphere, giving rise to tiny, solid elemental carbon (soot) particles. Most of this particulate matter eventually is oxidized, but some is not and remains unburned in the engine exhaust stream. Soot particles may be coated with low-volatility hydrocarbon molecules (polycyclic aromatic compounds), which are toxic to humans. Both NO and PM emissions are more difficult to control in CI engines than in SI engines.

Various modifications to direct injection diesel engine combustion are being tried to reduce both NO and PM engine-out emissions. One approach is to increase the fuel injection pressure and control its timing so as to provide more uniform fuel–air mixing and thereby better combustion conditions. Four-valve cylinders and exhaust gas recycling also help to control the combustion process so as to reduce emissions. If the mixture of fuel and air within the cylinder can be made nearly uniform (called homogeneous charge compression ignition), emissions are significantly reduced, at least at part load.

The use of catalytic converters to reduce NO molecules in the engine exhaust is less successful for CI engines than SI engines because there are fewer hydrogen-containing molecules needed for catalytic reduction of NO. Injecting small amounts of urea into the exhaust improves NO catalytic reduction. Nevertheless, substantial catalytic conversion is necessary to meet stringent $NO_x$ emissions standards for CI-powered vehicles.

Catalytic converters oxidize some of the engine-out PM, but not enough to meet stringent PM emissions standards. Particle filters can further reduce tailpipe PM emissions, but periodic cleaning of the filters by catalytic combustion or other means is necessary to ensure reliable PM reduction. Particle filters have not yet reached the level of development of catalytic converters, but are likely to be necessary for future diesel-powered light-duty vehicles in the United States.

CI engine emissions are affected somewhat by the composition of diesel fuel. The greatest effect is caused by fuel sulfur, which burns to $SO_2$ and hampers NO reduction in the catalytic converter. The sulfur content of vehicle fuels is currently regulated by the U.S. Environmental Protection Agency (EPA) to maintain good lifetime performance of catalytic converters.

The superior fuel efficiency of direct injection CI engines is a strong incentive for their use where consumer fuel cost is high and when vehicle $CO_2$ emissions are controlled. In the United States, where fuel prices are low, the fuel efficiency incentive resides with the manufacturer, who must meet CAFE standards, which is especially difficult for the current light-duty truck market. Should reduction of vehicle carbon dioxide emissions become a public policy goal in the United States, thereby encouraging greater use of CI engines in the passenger vehicle market, more intense development of CI engine emission control technology will be necessary to meet expected emission standards.

### 9.7.2.5  Fuel Quality and Its Regulation

We have already noted that fuel anti-knock lead additives and sulfur have been restricted to ensure the successful operation of exhaust gas catalytic converters. Other regulation of fuel properties have been directed at both exhaust and evaporative emissions.

To achieve desirable anti-knock properties, fuel refiners change the composition of the fuel, utilizing more volatile components that increase vapor pressure and thereby evaporative emissions and are more prone to generate ozone. It has been found that the addition of oxygenated fuel components, such as methanol or ethanol, improves fuel performance and reduces exhaust emissions, especially in older vehicles, so incentives to employ these additives have been utilized. Another fuel additive, MTBE (methyl tertiary butyl ether), has been required by some states with ozone problems, but has been found to be environmentally harmful in fuel leaks to ground water, in which it is very soluble. The use of ethanol additive may be required in such circumstances.

Natural gas is a clean vehicle fuel, yielding reduced exhaust emissions and no fuel vapor problem since it is very unreactive in photochemical ozone production. In addition, natural gas emits a lesser amount of $CO_2$, a greenhouse gas, than does gasoline or diesel fuel. But storing natural gas in a vehicle, either as a compressed gas in high-pressure tanks or as a refrigerated liquid at $-253°C$, is difficult and expensive and limits the vehicle range between fuel refills. At the present time, natural gas vehicles are restricted to fleet vehicles with limited daily range operating out of central fuel depots.

## 9.8  CONCLUSION

Among all transportation vehicles in the United States, light-duty passenger vehicles and trucks, in aggregate, are the predominant users of fuel and emitters of air pollutants. Transportation accounts for about a quarter of U.S. energy use, so substantial improvements in fuel efficiency and emissions of light-duty vehicles could contribute proportionally to reductions in national fuel consumption and pollutant emissions.

A substantial gain in vehicle fuel efficiency is a matter of improved vehicle design and engine efficiency. Current U.S. vehicle fleets offer a wide range of vehicle fuel efficiency, with the larger, more massive vehicles having poorer vehicle fuel efficiency than the smaller, lighter ones.

The vehicle design parameters affecting vehicle fuel efficiency are vehicle mass, aerodynamic drag, and rolling friction (in order of decreasing importance). For a given vehicle size, vehicle mass can be reduced below current designs by substitution of lighter materials of equal strength, particularly in the vehicle frame, without impairing vehicle safety in collisions. As vehicle mass is reduced, less power and mass are needed for the engine, transmission, wheels, tires, fuel tank, etc., compounding the gain in frame mass reduction. By careful attention to vehicle shape, aerodynamic resistance can be reduced. Efficient tires, in addition to mass reduction, lower the rolling resistance. Altogether, these technologies can improve vehicle fuel efficiencies independent of improvements to engine efficiency.

Improvements in engine fuel efficiency are closely constrained by the requirement to limit exhaust pollutant emissions. In the past, engine fuel efficiency has gradually improved while exhaust emissions were greatly reduced. There is still room for continued improvement in both respects for both SI and CI reciprocating engines.

The most fuel-efficient current vehicle, the hybrid electric vehicle, can achieve two times the vehicle fuel efficiency of current reciprocating engine vehicles of similar size and performance. Utilizing the same principles of vehicle design, vehicles powered by CI direct injection engine could achieve nearly comparable vehicle fuel efficiencies as current hybrids. Additional improvements seem likely in the future, given the long history of experience with these conventional technologies.

Although electric-drive vehicles powered by batteries or fuel cells are still under development, their vehicle fuel efficiencies are not as promising, being scarcely better than existing vehicles. Battery-powered vehicles suffer from an inherent weight problem that limits their equivalent vehicle fuel efficiency and vehicle range. Fuel cell vehicles have a lesser weight problem, but their dependence upon hydrogen fuel, whether generated on board or at fuel suppliers, complicates the vehicle technology. The thermodynamics of synthetic hydrogen production and utilization in the fuel cell does not yet provide a significant fuel efficiency advantage over the conventional utilization of fuel in vehicles to offset the economic and vehicle design advantages that improved conventional vehicles promise, especially considering the infrastructure needs of a hydrogen fuel economy. Furthermore, the very low emissions of these electric-drive systems become a less valuable offsetting benefit as the competing conventional vehicles become cleaner.

The technology for reducing exhaust emissions is well developed, especially for the SI engine. Improvements to the engine and catalytic converter could reduce emissions further, should it prove necessary to go beyond the U.S. national Tier 2 standards. Improvements to CI engine emissions are more difficult to achieve, and they may always have higher nitrogen oxide and particulate matter emissions than their SI counterparts.

The growing global movement for controlling greenhouse gas emissions will have significant effects on transportation vehicle design and operation. Both economic and technological regulation of transportation vehicles and their fuels will greatly affect the future development of the technologies discussed in this chapter.

# PROBLEMS

## Problem 9.1

Table 9.2 lists the characteristics of a selection of 2010 model year light-duty vehicles. The fuel efficiency (km/L) in highway mode and vehicle mass (t) are listed for 10 vehicles. (a) Plot the fuel consumption FC (L/km), the inverse of the fuel efficiency FE, as a function of vehicle mass M(t). Estimate or calculate by linear regression the value of the slope $m$, where $FC = mM$ is the best fit for a straight line through these points that passes through the origin. (b) Calculate the average value of FC times the vehicle mass M(t kg/L), for these vehicles, together with its standard deviation. (c) Discuss whether, and why, these figures support the analysis of Section 9.4.

## Problem 9.2

Tables 9.2–9.4 list characteristics of battery-powered electric, hybrid electric, and fuel cell vehicles. For each of these three vehicle types, calculate the average product of vehicle mass (t) times highway equivalent fuel economy (km/L).

## Problem 9.3

Table 9.3 lists characteristics of a selection of 2010 model year conventional vehicles. For these vehicles, calculate the average ratio of urban/highway equivalent fuel economy (km/L) and its standard deviation.

## Problem 9.4

A new sports utility vehicle averages 22 miles per gallon. It is expected to travel an average of 12,000 miles per year during a lifetime of 14 years. If fuel sells for $2.50 per gallon, calculate the lifetime expenditure on fuel.

## Problem 9.5

Using the data of Table 9.1 for the year 1995, calculate for each vehicle class its fraction of the annual fuel consumed by transportation vehicles.

## Problem 9.6

The engine power and size of a light-duty vehicle are related to vehicle mass. For the vehicles of Table 9.2, calculate the average values of power/displacement power/mass, and displacement per cylinder. Using these values, calculate the typical engine power, displacement, and number of cylinders for a 2-ton vehicle.

## Problem 9.7

A passenger vehicle diesel engine has a minimum brake specific fuel consumption of 0.22 kg/kWh. Calculate its maximum thermal efficiency.

## Problem 9.8

A 1.5-ton SI vehicle accelerates from rest to 100 km/h and then decelerates to a stop by braking. The average vehicle speed during this cycle is 50 km/h, and the cycle lasts 30 seconds. (a) Calculate the kinetic energy of the vehicle at its peak speed. (b) Calculate the time-average (W) and distance-average (J/km) of the energy dissipated in braking. (c) If 25% of the fuel heating value (31.6 MJ/L) is delivered to the wheels, calculate the average vehicle fuel economy in km/L for this start/stop mode, neglecting everything but the dissipation of braking. Compare this with the highway fuel economy of Table 9.2.

## Problem 9.9

Motor vehicle manufacturers list the maximum torque as well as the maximum power of the vehicle engine. Explain why the maximum torque has no direct influence on the vehicle performance.

## Problem 9.10

A 1.5-ton vehicle has a frontal area of 2 $m^2$, a rolling resistance coefficient of 0.1, and a drag coefficient of 0.3. Calculate the mechanical power delivered to the wheels at steady vehicle speeds of 50 and 100 km/h if the atmospheric density is 1.2 $kg/m^3$.

# BIBLIOGRAPHY

Appleby, A. J., and F. R. Foulkes. *Fuel Cell Handbook*. Malabar: Krieger, 1993.

Barnard, R. H. *Road Vehicle Aerodynamic Design. An Introduction*. Essex: Addison Wesley Longman, 1996.

Blackmore, D. R., and A. Thomas. *Fuel Economy of the Gasoline Engine*. London: MacMillan, 1977.

Heywood, John B. *Internal Combustion Engine Fundamentals*. New York: McGraw–Hill, 1988.

Heywood, John B., and Eran Shaw. *The Two-Stroke Cycle Engine: Its Development. Operation, and Design*. Philadelphia: Taylor & Francis, 1999.

Kordesch, Karl, and Gunter Simader. *Fuel Cells and Their Applications*. New York: VCH Publishers, 1996.

Mark, Jason, and Candace Morey. *Diesel Passenger Vehicles and the Environment.* Cambridge, MA: Union of Concerned Scientists, 1999.

National Research Council. *Toward a Sustainable Future. Addressing the Long-Term Effects of Motor Vehicle Transportation and a Sustainable Environment.* Washington, DC: National Academies Press, 1997.

Poulton, M. L. *Fuel Efficient Car Technology.* Southampton: Computational Mechanics, 1997.

Stone, Richard. *Introduction to Internal Combustion Engines.* London: MacMillan, 1985.

# Environmental Effects of Fossil Fuel Use

## 10.1 INTRODUCTION

The use of fossil fuels (coal, oil, and natural gas) almost always entails some environmental degradation and risk to human health. The negative impacts start at the mining phase, continue through transport and refining, and conclude with the fuel combustion and waste disposal processes.

Underground (shaft) mining of coal has claimed thousands of lives throughout the centuries because of explosions of methane gas in the mine shafts and also because of the inhalation of coal dust by the miners. At the mine mouth, mineral and crustal matter, called slag, is separated from the coal. The slag is deposited in heaps near the mine, thus despoiling the landscape. In recent times, most of the mined coal is crushed at the mine. The crushed coal is "washed" in a stream of water to separate by gravitational settling the adherent mineral matter, thereby beneficiating the coal prior to shipment. The "wash" usually contains heavy metals and acidic compounds, which, if not treated, contaminate streams and groundwater. Surface (strip) mining, which is much more economical than shaft mining, causes enormous scars to the landscape. Only recently were some regulations introduced in the United States and other countries to ensure the restoration of the wounds caused by the removal of overburden of the coal seams and recovery of the pits and trenches after the coal has been exhausted.

On- and off shore oil and gas drilling produces piles of drilling mud along with unsightly vistas of oil and gas derricks. Also, there is the risk of crude oil spills and explosions or fire at oil and natural gas wells.

Transport of coal, oil, and gas by railroad, pipelines, barges, and tankers carries the risk of spills, explosions, and collision accidents. In the refining process, especially of crude oil, toxic gases are emitted into the air or flared. Usually, liquid and solid by-products are produced that may be toxic. Strict regulations must be enforced to prevent the toxic wastes entering the environment, thereby threatening humans, animals, and vegetation.

The combustion of fossil fuels—coal, oil, or gas—inevitably produces a host of undesirable and often toxic by-products: (a) gaseous and particulate emissions into the atmosphere, (b) liquid effluents, and (c) solid waste. In many countries strict regulations are enacted as to the maximum level of pollutants that can be emitted into the air or discharged into surface waters or the ground from large combustion sources, such as power plants, industrial boilers, kilns, and furnaces. However, the smaller and dispersed combustion devices, such as residential and commercial furnaces and boilers, are not regulated, yet they do emit pollutants into the air. While great strides have been taken in many countries to control emissions from automobiles, trucks, and other vehicles, these mobile sources still contribute significantly to air pollution.

Perhaps the greatest long-term threat to the environment is the steadily increasing concentration of carbon dioxide in the atmosphere, which is for the most part a consequence of fossil fuel combustion. $CO_2$ and other so-called greenhouse gases may trap the outgoing thermal radiation from the earth, thereby causing global warming and associated climate changes.

In this chapter we deal with the environmental problems caused by the use of fossil fuels. The chapter is divided into the following sections: air pollution (with separate subsections on photo-oxidants and acid deposition), water pollution, and land pollution. The large looming problem of global climate change associated with $CO_2$ and other greenhouse gas emissions will be addressed in the following chapters.

## 10.2  AIR POLLUTION

Among the environmental effects of fossil fuel use, those that impair air quality are arguably the most problematic. Most emissions are a consequence of fossil fuel combustion. We are all familiar with the visible smoke that emanates from smokestacks, fireplaces, and diesel truck exhaust pipes. But in addition to the visible smoke, a plethora of pollutants is emitted from combustion "sources" in an invisible form. Emissions may occur also during the extraction, transport, refining, and storage phases of fossil fuel usage. Examples are fugitive coal dust emissions from coal piles at the mine mouth or storage areas at power plants; evaporative emissions from crude and refined oil storage tanks, as well as from oil and gasoline spills; evaporative emissions from gasoline tanks on board vehicles and during refueling; natural gas leaks from storage tanks and pipelines; fugitive dust from ash piles; and so on.

Air pollution is not a recent phenomenon. Air pollution episodes caused by open fire coal burning were observed already in medieval and renaissance England. In 1272, King Edward I issued a decree banning the use of "sea coal," coal that was mined from shallow sea beds and burned wet in open kilns and iron baskets. In 1661, John Evelyn, a founding member of the Royal Society wrote, "As I was walking in your Majesties Palace at Whitehall . . . a presumptuous Smoake . . . . did so invade the Court . . . [that] men could hardly discern one another from the Clowd . . . And what is all this, but that Hellish and dismall Clowd of Sea-Coal . . . [an] impure and thick Mist, accompanied with a fuliginous and filthy vapour"[1]

The association of air pollution episodes with human mortality and morbidity was recognized in the late 19th and early 20th centuries. In 1873, in London, during a typical "fog" episode, 268 deaths occurred in excess of what would be normally expected in that period. In 1930, in the heavily industrialized Meuse Valley, Belgium, during a 3-day pollution episode, 60 people died and hundreds were hospitalized. In 1948, during a 4-day episode, in Donora, Pennsylvania, where

---

[1]Quoted from Cooper, C. D., and F. C. Alley. *Air Pollution Control: A Design Approach*, 2d ed. Prospects Heights, IL: Waveland Press, 1994.

several steel mills and chemical factories are located, 20 people died and about one half of the 14,000 inhabitants got sick. A terrible fog episode occurred again in London from 5 to 8 December 1952. The excess deaths numbered 4,000! Most of the dead people had a history of bronchitis, emphysema, or heart disease. Apparently, individuals with a previous history of respiratory and cardiac diseases are predisposed to the impact of air pollution.

The 1952 London pollution episode induced the British Parliament to pass a Clean Air Act in 1956. This act focused on the manner and quality of coal burned in Great Britain. This act, and the fact that Great Britain shifted much of her fuel use from coal to oil and subsequently to natural gas, "cleaned" the air considerably over the British Isles. The ubiquitous London fog became a much rarer event; when it occurs, it may be truly a natural phenomenon, rather than of anthropogenic cause as it was often in the past.

In the United States the first Clean Air Act was passed by Congress in 1963. Subsequently, the Clean Air Act was amended in 1970, 1977, and 1990. Other countries followed suit by enacting their own clean air acts and various legislation and regulations pertaining to reducing air pollution. As a consequence, the air quality in most developed countries is improving steadily, although what is gained in reducing emissions from centralized sources is often negated by the ever-increasing number of dispersed sources, especially automobiles.

We classify air pollutants in two categories: primary and secondary. Primary pollutants are those that are emitted directly from the sources; secondary are those that are transformed by chemical reactions in the atmosphere from primary pollutants. Examples of primary pollutants are sulfur dioxide, nitric oxide, carbon monoxide, organic vapors, and particles. Particles may be composed of inorganic material, such as fuel ash, organic compounds originating in the fuel, and elemental carbon, commonly called soot. Examples of secondary pollutants are higher oxides of sulfur and nitrogen, ozone and other oxidants, and particles that are formed in the atmosphere by condensation of vapors or coalescence of primary particles. We shall later explain some of the processes that lead to the transformation of primary to secondary pollutants.

Most developed countries prescribe the maximum amount of pollutants that can be emitted from the sources. These are called *emission standards*. For large sources, the emission standards are usually set at a level that, after dispersion in the air within a reasonable distance, the pollutants will not cause significant human health or environmental effects. For small sources, such as automobiles, the emission standard may be set so as to prevent health effects from the cumulative emissions of all sources.

To protect human health and biota, most countries also prescribe maximum tolerable concentrations in the air. These are called *ambient air quality standards*. The emission and ambient standards are legal parameters, published in laws and decrees. If these standards are exceeded, the causative sources can be punished or penalized, and their licenses can be revoked.

## 10.2.1  U.S. Emission Standards

In the United States, emission standards have been promulgated for stationary and mobile sources. The Clean Air Act Amendments of 1970, 1977, and 1990 require the U.S. EPA to promulgate emission standards, called New Source Performance Standards (NSPS) and National Emission Standards for Hazardous Air Pollutants (NESHAP). Emission standards are specific to certain industrial categories, such as power plants, incinerators, steel plants, smelters, refineries, pulp and paper mills, chemical manufacturing, and so on, as well to mobile sources, that is, automobiles, trucks, aircraft, and ships. The maximum allowable emission rates are prescribed for a variety of

**TABLE 10.1**  U.S. NSPS Emission Standards for Fossil Fuel
Steam Generators with Heat Input > 73 MW (250 MBtu/hr)

| Pollutant | Fuel | g/GJ heat input |
|-----------|------|-----------------|
| $SO_2$ | Coal | 516 |
| | Oil | 86 |
| | Gas | 86 |
| $NO_x$ | Coal (bitum.) | 260 |
| $NO_x$ | Coal (sub-bitum.) | 210 |
| $NO_x$ | Oil | 130 |
| $NO_x$ | Gas | 86 |
| PM | All | 13 |

pollutants including $SO_2$, $NO_x$ (the sum of NO and $NO_2$), CO, PM, lead, mercury, arsenic, copper, manganese, nickel, vanadium, zinc, barium, boron, cadmium, chromium, selenium, chlorine, HCl, benzene, asbestos, vinyl chloride, pesticides, radioactive substances, and many other inorganic and organic pollutants. As an example, Table 10.1 lists the NSPS for fossil-fueled steam generators, including electric power stations, with a thermal power input of more than 73 MW (250 million Btu/hr). The thermal power input equals the fuel heating value times its mass rate of consumption. Thus, a power plant with a 500-MW output rating, operating at a thermodynamic efficiency of 33.3%, will have a 1,500-MW thermal input. Likewise, the NSPS are given in units of pollutant mass per fuel energy (heating value) input, grams/Joule or lb/Btu.

The estimation of the emission rate of $SO_2$ from a fossil fuel fired steam generator is quite simple. Because practically all sulfur atoms in the fuel burn up to form a $SO_2$ molecule, all we need to know is the weight percentage of sulfur in the fuel and its heating value. The mass emission rate of $SO_2$ is

$$E_{SO_2} = 2 \times (\%S \text{ by weight}) \, E(-2) \, FR \;\; g/s, \qquad (10.1)$$

where FR = firing rate of fuel in g/s. The specific emission rate per unit of fuel energy input, $e_{SO_2}$, is

$$e_{SO_2} = 2 \times (\%S \text{ by weight}) \, E(-2) \, / \, HV \;\; g/J, \qquad (10.2)$$

where HV = heating value of fuel in J/g. The factor of 2 arises because $SO_2$ has twice the molecular weight of S.

The emission rate of $NO_x$ or CO cannot be computed in that manner, because the formation of these pollutants is dependent on the combustion process and not on the weight percentage of the atoms in the fuel. The rate of emission of particulate matter is dependent on the content of incombustible mineral matter in the fuel and on the combustion process. The emission rates of $NO_x$, CO, and PM would have to be measured at the stack exit.

For mobile sources, the U.S. emission standards are given for four major pollutants that are emitted from these sources: carbon monoxide, hydrocarbons (HC), oxides of nitrogen, and particulate matter (PM). The U.S. emission standards for mobile sources are given in Chapter 9, Table 9.5. For light-duty vehicles, the standards are in g/mile: CO, 3.4; hydrocarbons (reckoned as molecular weight 13), 0.25; oxides of nitrogen (reckoned as molecular weight 46), 0.4; PM, 0.08. HC include all carbonaceous emissions, except carbon monoxide and dioxide, coming from the tailpipe and evaporative emissions from the fuel tank, fuel lines, fuel pump, and injection devices.

For certain situations, such as for heavily polluted areas, or if the source is near a national park, instead of prescribing emission rates, the U.S. EPA or state agency may specify the control technology that is presumed to achieve the desired emission standard. Thus, a multitude of acronyms came into being, such as BACT (Best Available Control Technology), MACT (Maximum Achievable Control Technology) and RACT (Reasonable Available Control Technology). The U.S. EPA determines, depending on the industrial category, pollutant, quality of fuel, and the severity of pollution in an "airshed," which control technology needs to be installed on new sources. The Clean Air Act Amendments of 1990 (CAAA 1990) went even further. For the control of emissions of precursors of acid deposition and precursors of photo-oxidants, the act requires that even existing sources install emission control technology. This means retrofitting control devices on older sources that in previous acts were "grandfathered." For example, on power plants that use coal in excess of 0.6% by weight of sulfur, the BACT for $SO_2$ control is a wet limestone scrubber (see Chapter 6); for coal with a lower sulfur content, it is a dry sorbent (usually limestone or lime) injection. For $NO_x$ control the present federal BACT is a low-$NO_x$ burner (LNB), but some states require SCR or NSCR for $NO_x$ control. For particle control on power plants, BACT is an ESP. For other industrial categories the requirement may be a fabric filter ("baghouse") or a spray scrubber.

Title IV of CAAA 1990 specifically addresses the acid deposition problem. By the year 1995 the average emission level of all major power plants and industrial boilers was to be limited to 2.5 lb $SO_2$ per million Btu heat input and by the year 2000 to 1.2 lb $SO_2$/MBtu. Indeed, currently the U.S. emissions of $SO_2$ are roughly one half of what they were in 1990, thanks to the installation of control technology on new and existing sources. The utilities and industry can choose any method they wish to achieve the emissions reduction, including installing FGD technology, fuel switching, seasonal fuel switching (e.g., using coal in the winter and natural gas in the summer), and marketable permits. The latter means that if one source reduces $SO_2$ emissions by more than the required quota, the excess can be sold to a source that does not wish to reduce that pollutant.

For hazardous air pollutants (HAP), in the United States, no longer are numerical emission rates prescribed, rather, the control technology is specified for each industrial category. The control technologies are designed so that the emissions of HAP are practically eliminated. This is called Maximum Achievable Control Technology (MACT). The control technologies for HAP include incineration, absorption (scrubbing) by a solvent (including water), and adsorption on a porous surface, such as activated carbon or zeolite.

The BACTs and MACTs are not supposed to be cast in concrete for all time. As new technologies are developed that prove to be more efficient and/or more economical, the U.S. EPA may adopt, after due process, new technologies in lieu of the old ones. The introduction of "technology forcing" as manifested by BACT and MACT is a revolutionary concept. It places the onus on government to develop and define appropriate emission control technologies, rather than merely setting an emission standard, and lets the sources find the devices that meet the standard. Also, the federally required control technologies are uniform across the nation. Individual states may impose even stricter control technologies, but never less efficient ones. This new way of controlling emissions places emphasis not only on measurement of emission rates, but also on monitoring and supervising the installation and proper functioning of the control devices.

## 10.2.2  U.S. Ambient Standards

The setting of emission standards has its purpose of ensuring that concentrations of air pollutants in the ambient air remain at a sufficient low level so that the population at large, and especially sensitive individuals, such as children and the elderly, will not suffer adverse health effects. The appropriate indices for exposure to harmful air pollutants are the ambient concentrations and the time periods for which these concentrations prevail. Therefore, the U.S. Congress mandated the U.S. EPA to promulgate ambient concentration standards, called the National Ambient Air Quality Standards (NAAQS). The NAAQS stipulate the concentrations and the average time periods for various air pollutants that should not be exceeded. In the United States, current ambient standards exist for six pollutants: particulate matter (PM), sulfur dioxide ($SO_2$), nitric oxides ($NO_x$, measured as $NO_2$), carbon monoxide, ozone ($O_3$), and lead (Pb). (Although ambient standards for lead are still listed, in actuality lead concentrations in the air are no longer monitored. Ever since leaded gasoline was phased out in the United States in the 1970s, air concentrations of lead fell precipitously. Of course, lead is still a health and environmental hazard, but its ingestion is not postulated to come from air inhalation.) These are the so-called *criteria pollutants*. This is not to say that concentrations of other pollutants need not be curtailed. Indeed, some toxic pollutants may be far more injurious to human health and biota than the aforementioned six criteria pollutants. The fact is that for the criteria pollutants a fairly well-known dose–response relationship has been established over years of clinical and epidemiological research, while such a relationship may not be known for other pollutants. For the other pollutants, classified as *toxic pollutants*, there are no specified ambient standards, but their emission into the atmosphere is to be prevented as much as possible by applying MACT at the sources.

The NAAQS for the six criteria pollutants are listed in Table 10.2. Primary standards set limits to protect public health, including the health of sensitive populations such as asthmatics, children, and the elderly. Secondary standards set limits to protect public welfare, including protection against decreased visibility and damage to animals, crops, vegetation, and buildings. The averaging times can be as short as 1 hour (for CO) to 1 year (for PM, $SO_2$, and $NO_2$). The reason is that the exposure of some pollutants at a high concentration for a short period may cause acute effects, whereas the exposure of others at a relatively low level for longer periods may cause chronic effects (see below). The ozone standard was revised in 1997 and then again in 2008. Before the revisions, the standard was 0.12 ppmv, 1-hour average, not to be exceeded more than once per year. Recent health and ecological studies indicated that ozone may be harmful at a lower level when exposed for a longer period, so the standard is now 0.075 ppmv, 8-hour average. Likewise, the PM standard has been revised at the same time. Before, only particles with an aerodynamic diameter of less than 10 $\mu$m were regulated (PM-10). However, recent epidemiological studies indicated that particles smaller than 2.5 $\mu$m are most detrimental to health because they lodge deeply in the lung's alveoli. Thus, in addition to PM-10 there is now a PM-2.5 standard.

The U.S. EPA is mandated to revise the NAAQS from time to time, as more results from health and environmental effects studies become available. For example, the U.S. EPA has been urged to promulgate a short time standard for $NO_2$ instead of the annual standard.

If the NAAQS are exceeded within an Air Quality Control Region, the state in which the region is located must develop a plan, called the State Implementation Plan (SIP), which lays out a strategy of how the region will attain compliance with the NAAQS in a reasonable time period. The SIP may include emission curtailments from emitting sources, traffic regulations, tightened inspection schedules and procedures, and other measures. However, research in the past few decades brought out clearly that air pollutants do not respect political or natural geographic boundaries. They travel over control regions, state lines, river valleys, mountains, and even oceans. Thus, no state can control its air pollution solely by its own means. A regional, national, and even international approach is necessary to control air pollution over a region, continent, and, for some air pollutants, the globe. In part, this is the reason why emissions of most air pollutants are regulated on the federal rather than on the state level in the United States.

**TABLE 10.2**   U.S. 2000 National Ambient Air Quality Standards

| Pollutant | Primary | | Secondary | |
|---|---|---|---|---|
| | ppm | $\mu g/m^3$ | ppm | $\mu g/m^3$ |
| Carbon monoxide (CO) | | | | |
| 8-hour average | 9 | 10 mg/m³ | | |
| 1-hour average | 35 | 40 mg/m³ | | |
| Nitrogen dioxide (NO₂) | | | | |
| Annual arithmetic mean | 0.053 | 100 | Same | Same |
| Ozone (O₃) | | | | |
| 3-yr average of annual 4th highest daily maximum 8-hr concentration | 0.075 | 147.5 | Same | Same |
| Particulate matter, diameter < 10 μm (PM-10) | | | | |
| Annual arithmetic mean | | 50 | | Same |
| Arithmetic mean of 24-hour 99th percentile, averaged over 3 years | | 150 | | Same |
| Particulate matter, diameter < 2.5 μm (PM-2.5) | | | | |
| Annual arithmetic mean | | 15 | | Same |
| Arithmetic mean of 24-hour 98th percentile, averaged over 3 years | | 65 | | Same |
| Sulfur dioxide (SO₂) | | | | |
| Annual arithmetic mean | 0.03 | 80 | | |
| 24-hour average | 0.14 | 365 | | |
| 3-hour | | | 0.5 | 1,300 |
| Lead | | | | |
| Rolling 3-month average | | 0.15 | | 0.15 |

## 10.2.3   Health and Environmental Effects of Fossil-Fuel-Related Air Pollutants

Air pollutants, when they exceed certain concentrations, can cause acute or chronic diseases in humans, animals, and plants. They can impair visibility, cause climatic changes, and damage materials and structures.

Table 10.3 lists some of the effects of air pollutants on human health, fauna and flora, structures, and materials. The order of listing does not follow any particular ranking—some individuals or plants are more sensitive to one kind of pollutant than to another. The five pollutants listed are the U.S. EPA-designated criteria pollutants. These are the pollutants for which dose–response relationships are fairly well known from clinical and epidemiological studies. Other pollutants not listed are suspected toxigens, mutagens, teratogens, carcinogens and possible animal and plant disease-causing agents.[2]

---

[2]Toxigen: a chemical agent that may cause an increase of mortality or of serious illness or that may pose a present or potential hazard to human health. Mutagen: any agent, including radioactive elements, that may cause biological mutation—that is, alteration of the genes or chromosomes. Teratogen: an agent that may cause defects or diseases of the embryo. Carcinogen: an agent that may cause cancer.

**TABLE 10.3**   Effects of Criteria Air Pollutants on Human Health, Fauna and Flora, and Structures and Materials

| Pollutant | Health effect | Fauna and flora effect | Structure and material |
|---|---|---|---|
| $SO_2$ | Bronchoconstriction, cough. | Cellular injury, chlorosis, withering of leaves and abscission. Precursor to acid rain: acidification of surface waters with community shifts and mortality of some aquatic organisms. Possible effect on uptake of Al and other toxic metals by plant roots. | Weathering and corrosion. Defacing of monuments. |
| $NO_x$ | Pulmonary congestion and edema, emphysema, nasal and eye irritation. | Chlorosis and necrosis of leaves. Precursor to acid rain. | Weathering and corrosion. |
| $O_3$ and photo-oxidants | Pulmonary edema, emphysema, asthma, eye, nose, and throat irritation, reduced lung capacity. | Vegetation damage, necrosis of leaves and pines, stunting of growth, photosynthesis inhibitor, probable cause of forest die-back, suspected cause of crop loss. | Attack and destruction of natural rubber and polymers, textiles, and materials. |
| CO | Neurological symptoms, impairment of reflexes and visual acuity, headache, dizziness, nausea, confusion. Fatal in high concentrations because of irreversible binding to hemoglobin. | n/a | n/a |
| Particulate matter | Nonspecific composition: bronchitis, asthma, emphysema. Composition dependent: brain and neurological effects (e.g., lead, mercury), toxigens (e.g., arsenic, selenium, cadmium), throat and lung cancer (e.g., coal dust, coke oven emissions, polycyclic aromatic hydrocarbons, chromium, nickel, arsenic). | n/a | Soiling of materials and cloth. Visibility impairment caused by light scattering of small particles. |

Adapted from Wark, K., C. F. Warner, and W. T. Davis. *Air Pollution: Its Origin and Control.* Reading: Addison-Wesley, 1998.

The definition of the deleterious effects of particulate matter is complicated and contentious. The ambient standards are given in units of mass per volume ($\mu g/m^3$). Surely the effects on health and biota are not dependent as much on mass concentrations of the inhaled particles, but on their characteristics, that is, their composition. While ordinary soil and road dust may not cause significant health effects, particles that contain acidic species, heavy metals, soot, and PAH may cause respiratory, neurological, and cancerous diseases. One reason that the U.S. EPA and other environmental protection agencies are using mass concentration as a standard for PM, rather than chemical composition, is that the determination of chemical composition requires complicated and expensive analytical instrumentation. Furthermore, the U.S. EPA maintains that there is evidence from epidemiological studies that excessive mortality and morbidity are correlated with mass and size of the particles, regardless of chemical composition.

### 10.2.4   Air Pollution Meteorology

The basic information necessary for air quality modeling are wind statistics for the modeling domain and the dispersion characteristics of the atmosphere. Winds blow from high- to low-pressure regions on the earth. Since the earth is a rotating body revolving around the sun, any spot on the earth receives constantly changing insolation over day and night and over the seasons. In addition, orographic effects—mountains and valleys—alter the course of winds, as does surface friction, sea–land interfaces, street canyons, etc. Thus, for air quality modeling, a multiyear wind statistic is necessary for predicting the advection by winds of pollutants from the sources to the receptor.

Meteorological data are available from numerous weather stations operating around the world, especially in the more developed countries. These weather stations measure and record surface and upper air winds, atmospheric pressure, humidity, precipitation, insolation, temperature on the ground, and the temperature gradient in the atmosphere—the temperature variation with altitude. The measurements are usually rendered twice daily at 0000 and 1200 Greenwich Mean Time, so that measurements are synchronized all over the world. Past and present weather data are available from national repositories, such as, in the United States, the National Weather Service in Asheville, North Carolina.

In the atmosphere, dispersion occurs mostly by turbulent or eddy diffusion. Such diffusion is orders of magnitude faster than molecular or laminar diffusion. The cause of turbulent diffusion is either mechanical or thermal shear. Mechanical turbulence is caused by wind shear in the free atmosphere (adjacent layers of the atmosphere move in different directions or speeds, or friction experienced by winds blowing over the ground surface and obstacles, such as tree canopies, mountains, and buildings). The other cause of turbulence is the thermal gradient in the atmosphere. In the lower troposphere, usually the temperature is higher near the ground and declines with altitude. In a dry atmosphere, the gradient amounts to approximately $-10°C/km$. This is called the dry adiabatic lapse rate. Occasionally, the gradient is steeper, meaning more negative than $-10°C/km$. In a moist atmosphere, the gradient is less steep because of the addition of the latent heat of condensation of water vapor. At night, because of radiative cooling of the surface, the gradient may become positive, with temperature increasing with altitude. This is called an *inversion*. Inversions can also occur aloft, when a negative gradient is interrupted by a positive one. The bottom layer up to the inversion is called the *mixing layer*, and the height to the inversion is called *mixing depth*. An inversion layer acts like a lid on the mixing layer. With an inversion, atmospheric conditions are especially prone to air pollution episodes, because pollutants emitted at the ground are concentrated in the shallow mixing layer. Later in the day, as the sun rises, the inversion layer may break up, allowing pollutants to escape aloft and thus alleviating the pollution episode. Valleys and urban areas surrounded by mountain chains experience frequent inversion layers and therefore are plagued with pollution episodes. Los Angeles, Denver, Salt Lake City, Mexico City, and industrial–urbanized river valleys across the continents are examples of places where pollutant concentrations frequently exceed air quality standards, and the population suffers from pollutant-caused respiratory and other diseases.

When the temperature in the upper layers is colder than in the lower layers, upper air parcels fall downward because of their larger density, and lower air parcels move upward. This movement generates turbulent or eddy diffusion. The steeper the temperature gradient, the greater the turbulent intensity. This is called *unstable* condition. A temperature gradient that is equal to the dry adiabatic lapse rate is called neutral condition and leads to moderate turbulence. A temperature gradient that is less steep than the dry adiabatic lapse rate, or even a positive gradient, is called a *stable* condition, in which there is minimal or no turbulence at all.

It is a convention to classify the turbulent conditions of the atmosphere into six stability categories, called Pasquill–Gifford stability categories. They range from A, very unstable (very turbulent) to F, very stable (little turbulence). Category D is called neutral, with moderate turbulence. The negative temperature gradient (lapse rate) of category D coincides with the dry adiabatic lapse, about $-10°C/km$. For categories A, B, and C, the magnitude of the negative gradient is greater than for D, for E the gradient is smaller than for D, and for F the gradient is positive.

While the lapse rate of the atmosphere is measured twice daily at the weather stations, the stability categories can be approximated by knowing the surface wind speed, insolation, and cloud cover. Table 10.4 lists the stability categories. It is seen that in daytime, low wind speeds and strong insolation lead to unstable categories A or B; high wind speeds and moderate to slight insolation lead to neutral categories C or D. At night, the stability categories are almost always neutral or stable, D, E, or F. Table 10.4 also lists the temperature gradients that correspond to the stability categories.

**TABLE 10.4**  Pasquill–Gifford Stability Categories

| Surface wind (m/s) | Day | | | Night | |
|---|---|---|---|---|---|
| | Incoming solar radiation | | | Cloud cover | |
| | Strong[a] | Moderate[a] | Slight[a] | Thinly overcast or ≥ 4/8 cloud | ≤ 3/8 cloud |
| <2 | A | A–B | B | [b] | [b] |
| 2–3 | A–B | B | C | E | F |
| 3–5 | B | B–C | C | D | E |
| 5–6 | C | C–D | D | D | D |
| >6 | C | D | D | D | D |

| | Stability category | Stability | dT/dz (°C/100 m) |
|---|---|---|---|
| | A | Extremely unstable | $\leq -1.9$ |
| | B | Moderately unstable | $> -1.9$ but $\leq -1.7$ |
| | C | Slightly unstable | $> -1.7$ but $\leq -1.5$ |
| | D[c] | Neutral | $> -1.5$ but $\leq -0.5$ |
| | E | Slightly stable | $> -0.5$ but $\leq 1.5$ |
| | F | Very stable | $> 1.5$ but $\leq 4.0$ |

[a]Zenith angle under clear skies > 60°; moderate = 35–60°; slight = <15°.
[b]Rural areas = F; large urban areas = D; small urban areas = E.
[c]Category D, neutral, applies to heavy overcast, day or night, all wind speeds.

## 10.2.5   Air Quality Modeling

After leaving the smokestack or exhaust pipe, the primary air pollutants disperse into the atmosphere by turbulent diffusion, advect by winds, and transform into secondary pollutants by chemical reactions among themselves and with other atmospheric species. The estimation of the concentration of pollutants in space and time is called air quality modeling. It is also called *source-receptor-modeling* or *dispersion modeling*, where the sources are emissions from point (e.g., a smokestack), line (e.g., a highway), or area (e.g., an urban area) sources, and the receptors are designated human habitats or ecologically sensitive areas.[3] In this section we shall deal only with nonreactive pollutants—pollutants that are not transformed from primary to secondary pollutants. In subsequent sections we shall incorporate transformation processes into air quality models.

### 10.2.5.1   Modeling of Steady-State Point Source

When a pollutant is released into the atmosphere at a constant rate, say from an elevated smokestack, while a steady wind is blowing and the atmospheric stability category is not changing for the modeling period, we have a time invariant (steady-state) situation. As the flue gas containing a pollutant leaves the top of the stack, it immediately begins to mix with the surrounding atmosphere, because the flue gas jet velocity is different from that of the surrounding atmosphere. The mixing process begins to dilute the concentration of the flue gas, the more so as more atmospheric air is entrained in the stack plume. Most of the time a wind is present so that the plume, as it mixes with the atmosphere, soon attains the horizontal speed of the wind, with the plume axis (or centerline) bending sharply and approaching the horizontal direction. At this point in the plume trajectory, mixing of the plume gas with the atmosphere continues at a rate determined by the turbulent motion of the atmosphere. As a consequence of this mixing process, the concentration of the plume gas, a mixture of the principal products of combustion and the air pollutants of interest, steadily declines with downwind distance from the stack exit. The physical evidence of this mixing is visible to the eye for smoky plumes where it can be seen that the plume width in both vertical and horizontal directions increases with downwind distance, marking the portion of the surrounding atmosphere into which the effluent gas has been mixed and thereby diluted.

Measurements of the pollutant gas concentration downwind of stacks show that, at any distance $x$ downwind, the concentration is maximum at the plume centerline and declines with vertical and lateral distance from the plume centerline. The centerline concentration and the effective vertical and horizontal widths of the plume change gradually with the distance $x$, in a way that is dependent on the atmospheric stability category. The concentration profile in both the $y$ and the $z$ directions can be approximated by a Gaussian (bell-shape) curve. A smokestack plume is depicted in Figure 10.1. The stack is placed at the origin of the coordinate system, with the $x$-axis pointing downwind, the $y$-axis crosswind, and the $z$-axis is in the vertical direction. The stack has a height $h$. Since the exhaust gas is usually warmer than the ambient air, it rises because of buoyancy by an incremental height $\Delta h$. This is the plume rise, which will be evaluated later. The final plume center line is at height $H = h + \Delta h$.

---

[3]Another type of modeling is called receptor modeling. These models attempt to identify and quantify the contribution of various sources to the amount and composition of the pollutant concentration at the receptor by using some characteristics of the sources. For example, receptor modeling may compare the distribution of trace elements at the receptor to the distribution of trace elements in the emissions of various sources.

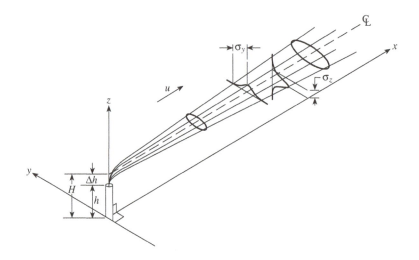

**FIGURE 10.1**    The Gaussian plume.

The pollutant time-averaged mass concentration $c(x, y, z)$ at a downwind distance $x$ from a source of pollutant emitted at a mass rate $Q_P$ g/s, at a height $H$, in the presence of wind of speed $u$, is given by the following equation, called the Gaussian plume equation (GPE).

$$c(x, y, z) = \frac{Q_P}{2\pi \sigma_y \sigma_z u} \exp\left[ -\frac{1}{2}\left( \frac{y^2}{\sigma_y^2} + \frac{(z - H)^2}{\sigma_z^2} \right) \right] \tag{10.3}$$

The $\sigma$'s represent the standard deviation (Gaussian width) of the bell-shape pollutant distribution in the horizontal and vertical plane, respectively, at the downwind distance $x$. In such a fashion, $\sigma$'s substitute for diffusion coefficients. Their dimension is in meters.

The variation of $\sigma_y$ (horizontal Gaussian width) and $\sigma_z$ (vertical Gaussian width) is depicted in Figure 10.2. The six curves correspond to the atmospheric stability categories A through F. Since the $\sigma$'s are largest for unstable conditions and smallest for stable conditions, stability category A leads to a rapid dispersion of the emitted pollutants and to highest ground concentration near the source. Category F leads to a slow dispersion and to highest ground concentration far from the source. It must be emphasized that the Gaussian plume model is only an approximation. It works best on level ground. Since the wind speed $u$ appears in the denominator, the GPE cannot be used for calms, when the wind speed is less than approximately 2 m/s. Also, the modeling distance should not be extended further than 20–30 km, because wind direction and speed, as well as the dispersion characteristics (atmospheric stability category), may change over longer distances. Since the GPE is a steady state model, the emission rate $Q_P$ and plume rise $\Delta h$ must also remain constant. Experience shows that within a limited distance, and on level terrain, the GPE gives concentrations on the ground that are within a factor of 2 of measurements. In valleys, hills, and urban areas, aerodynamic obstacle effects need to be considered. Numerous equations exist that work reasonably well when corrections for terrain complexities are incorporated into the Gaussian plume model.

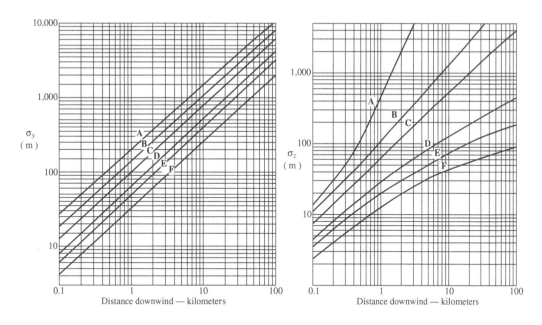

**FIGURE 10.2**   Horizontal and vertical dispersion coefficients.

## 10.2.5.2   Plume Rise

There are several empirical equations that allow the estimation of plume rise $\Delta h$ from known stack exit conditions. Here we give the Briggs plume rise equation, which is mostly used in U.S. EPA recommended dispersion models. For unstable and neutral conditions and $F \leq 55 m^4 s^{-3}$,

$$\Delta h = 21.4 F^{3/4} u^{-1}. \tag{10.4}$$

$F$ is the buoyancy flux parameter,

$$F = g v_s D_s^2 (T_s - T_a)/4T_s, \tag{10.5}$$

where

$g$ = acceleration of gravity (9.8 m s$^{-2}$)
$v_s$ = flue gas stack exit velocity, m s$^{-1}$
$D_s$ = stack diameter, m
$T_s$ = flue gas exit temperature, K
$T_a$ = ambient temperature at stack height, K
$u$ = wind speed at stack height, m s$^{-1}$.

For $F \geq 55$ m$^4$s$^{-3}$,

$$\Delta h = 38.7 F^{3/5} u^{-1}. \tag{10.6}$$

For stable conditions, regardless of the magnitude of $F$,

$$\Delta h = 2.6[F(uS)^{-1}]^{1/3}. \tag{10.7}$$

S is a stability parameter,

$$S = g(\delta\theta/\delta z)T_{\mathrm{a}}^{-1}. \tag{10.8}$$

The potential temperature gradient $\partial\theta/\partial z = 0.015$ K/m for category E and 0.025 K/m for category F.

### 10.2.5.3  Steady-State Line Source

When several point sources are aligned in a row, such as several smokestacks along a river bank or many automobiles and trucks traveling in both directions along a straight highway, the ground-level concentration is estimated by the following variant of the GPE,

$$c(x) = \frac{2Q_{\mathrm{L}}}{(2\pi)^{1/2}\sigma_z u \sin\phi} \exp\left[-\frac{1}{2}\left(\frac{H^2}{\sigma_z^2}\right)\right], \tag{10.9}$$

where

$Q_{\mathrm{L}}$ = average line source mass emission rate, g m$^{-1}$ s$^{-1}$

$H$ = average release height from sources

$\phi$ = angle between the source axis and the prevailing wind direction (which by convention always blows in the $x$ direction)

$x$ = distance of the receptor from the source axis in the direction of the wind.

The geometry of a line source is depicted in Figure 10.3. Note that the dispersion coefficient $\sigma_y$ does not appear in Equation (10.9) because the concentration along the $y$-direction is constant. Equation (10.9) should not be used when $\phi$ is less than 45°.

### 10.2.5.4  Steady-State Area Source

Emissions from an urban area can be considered as emanating from a bunch of parallel line sources, as depicted in Figure 10.4. The concentration of pollutant at a downwind distance $x$ is given by the equation

$$c(x) = \frac{2Q_{\mathrm{A}}l}{(2\pi)^{1/2}\sigma_z u} \exp\left[-\frac{1}{2}\left(\frac{H^2}{\sigma_z^2}\right)\right], \tag{10.10}$$

where

$Q_{\mathrm{A}}$ = average mass emission rate, g m$^{-2}$ s$^{-1}$

$l$ = width of the urban area along the wind axis, m.

Equation (10.10) can only be used at downwind distances $x \gg l$.

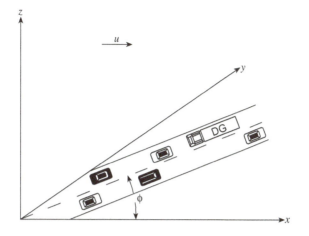

**FIGURE 10.3**  Line source.

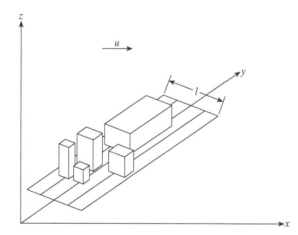

**FIGURE 10.4**  Area source.

## 10.2.6  Photo-Oxidants

Photo-oxidants are a class of secondary air pollutants formed from certain primary pollutants emitted by fossil fuel combustion. The name arises because these chemicals are formed under the influence of sunlight, and all of them have strong oxidizing capacity. They irritate and destroy (oxidize) the respiratory tract, eyes, skin, animal organs, vegetation tissues, materials, and structures. The major representative of this class of chemicals is ozone, $O_3$, but other compounds are included: ketones, aldehydes, alkoxy radicals ($RO\cdot$), peroxy radicals ($RO_2\cdot$), peroxyacetyl nitrate (PAN), and peroxybenzoyl nitrate (PBN). The symbol R denotes a hydrocarbon fragment, with one hydrogen missing; the dot denotes an unpaired electron.

Tropospheric ozone (the "bad" ozone) is to be distinguished from stratospheric ozone (the "good" ozone). Tropospheric ozone is mainly formed as a consequence of fossil fuel combustion, while stratospheric ozone is formed naturally by photochemical reactions under the influence of solar ultraviolet radiation.[4] A part of the tropospheric ozone is caused by intrusion of stratospheric ozone into the troposphere. This constitutes a background level of ozone. However, compared with concentrations of ozone in a polluted atmosphere, the background ozone is a small fraction of that concentration.

The current NAAQS (see Section 10.2.2) for $O_3$ is 75 parts per billion by volume (ppbv), 8-hour average. The previous standard was 120 ppbv, 1-hour average. These standards are being exceeded many times per year in most major metropolitan areas of the United States, especially in cities with high insolation and temperature and where topographic conditions are preventing good ventilation, such as Los Angeles, Salt Lake City, Phoenix, Houston, Dallas–Fort Worth, Denver, and Atlanta. High ozone levels are observed in most metropolitan areas of the world, especially areas with high insolation, such as Mexico City, Sao Paulo, Lima, Quito, Jakarta, Mumbai, Cairo, Istanbul, Rome, Athens, Madrid, Beijing, and Shanghai. Rural and remote areas that are downwind from metropolitan areas are also experiencing elevated ozone concentrations. The high concentrations in rural and remote areas may contribute to the degradation and die-back of plants and trees.

The only precursor that can initiate ozone formation in the troposphere is nitrogen dioxide, $NO_2$. As we shall see later, other gases, both manmade and natural, can abet ozone formation, but the initiation of the process is solely caused by $NO_2$. Nitrogen dioxide is a brown gas, formed by oxidation of nitric oxide, NO. $NO_x$ is the sum of NO and $NO_2$. Nitric oxides are formed primarily from the combustion of fossil fuels. A part of the $NO_x$ is formed on account of the inherent nitrogen content of fossil fuels, especially coal and petroleum. This is called *fuel-NO$_x$*. A greater part is formed during combustion. At the high flame temperature, a part of the air $O_2$ and $N_2$ combine to form NO. This is called *thermal-NO$_x$*.

Nitrogen dioxide can photodissociate in sunlight, forming nitric oxide and atomic oxygen. The resulting atomic oxygen combines with molecular oxygen to form ozone:

$$NO_2 \overset{h\nu}{\rightarrow} NO + O \qquad \textbf{(10.11)}$$

$$O + O_2 + M \rightarrow O_3 + M, \qquad \textbf{(10.12)}$$

where $h\nu$ denotes a photon of wavelength shorter than 420 nm, and M is an inert molecule that is necessary to bring about the combination of atomic and molecular oxygen. Since the formed $O_3$ can be destroyed by the same NO that is formed in Equation (10.12),

$$O_3 + NO \rightarrow O_2 + NO_2, \qquad \textbf{(10.13)}$$

this cycle of reactions cannot explain the large concentrations of ozone, which often exceed the concentration of the initiating chemical $NO_2$. It was assumed in the late 1950s and early 1960s by

---

[4] Stratospheric ozone has been steadily depleted over the past decades because of the penetration into the stratosphere of certain chemicals, the CFCs. This creates the so-called "ozone hole." Stratospheric ozone shields humans, animals, and vegetation from the penetration to the surface of the earth of harmful ultraviolet radiation. It is an irony that mankind is busy on one hand creating tropospheric ozone by emissions of combustion products of fossil fuel and on the other hand destroying stratospheric ozone by emissions of some other chemicals.

scientists at the California Institute of Technology that instead of Reaction (10.13), NO is reoxidized to $NO_2$ by some atmospheric oxidant other than $O_3$, thereby starting the photodissociation of $NO_2$ all over again, allowing the buildup of $O_3$ from a relatively small concentration of $NO_2$.[5] The "mystery" oxidant turned out later to be a peroxy radical, $RO_2\cdot$. This radical is formed in the following sequence of reactions:

$$RH + OH\cdot \rightarrow R\cdot + H_2O \tag{10.14}$$

$$R\cdot + O_2 \rightarrow RO_2\cdot \tag{10.15}$$

$$RO_2\cdot + NO \rightarrow NO_2 + RO\cdot \tag{10.16}$$

RH designates a hydrocarbon molecule, $OH\cdot$ is a hydroxyl radical, $RO\cdot$ is an alcoxy radical, and $RO_2\cdot$ is a peroxy radical. The naturally occurring peroxy radical, $HO_2\cdot$, can also oxidize NO to $NO_2$. The hydroxyl radical appears to be omnipresent in the atmosphere. It is formed by reaction of water vapor with an excited oxygen atom $O(^1D)$. The latter is formed in photodissociation of $O_3$ by radiation of wavelengths less than 319 nm.

Indeed, the peroxy radicals were subsequently found in the lower troposphere and in laboratory test chambers, called *smog chambers*. The picture of ozone and the other photo-oxidant formation processes became much more complicated. While $NO_2$ is necessary to initiate the ozone formation process, it is not consumed to a significant degree in Reactions (10.14)–(10.16). Hydrocarbons (RH) and other volatile organic compounds (VOCs), as well as naturally occurring oxidants (e.g., $HO_2$) all participate to boost the formation of $O_3$ well above the initial concentration of $NO_2$. Some of the RH and VOC are of anthropogenic origin, such as products of incomplete combustion, evaporation of fuels and solvents, emissions from refineries, and other chemical manufacturing. Others are of biogenic origin, such as evaporation and volatilization of organic molecules from vegetation, wetlands, and surface waters. For example, evergreen trees exude copious quantities of isoprene, terpene, pinene, and other compounds, which by virtue of double bonds in their molecules are very reactive with a hydroxyl radical. In essence, the ozone and other photo-oxidant formation processes have two classes of precursors—$NO_x$ and VOCs. The amount of photo-oxidants formed in the troposphere is a complicated (and nonlinear) function of precursor concentrations, as well as insolation and meteorological conditions.

### 10.2.6.1  Photo-Oxidant Modeling

Ozone and the other photo-oxidants are secondary pollutants; therefore, the regular dispersion models described in Section 10.2.5, pertaining to primary pollutants that do not transform while dispersing, are not applicable. For photo-oxidant modeling, in addition to meteorological parameters, the reactions of the primary pollutants among themselves, and those with atmospheric species, plus the interaction with sunlight need to be considered. A further complication is that some of the chemical kinetic processes are not linear, that is, they are not first-order rate reactions.

The rate of Reaction (10.14) depends on the kind of hydrocarbon compound. Molecules with a double or triple bond are very reactive, followed by aromatics, branched, and extended-chain aliphatics. The simple molecule methane, $CH_4$, reacts very slowly with OH; therefore, the hydrocarbons participating in (10.14) are called NMHC.

---

[5]Leighton, P. A. *Photochemistry of Air Pollution*. Academic Press, 1961.

The model that is frequently used is called the Empirical Kinetic Modeling Approach (EKMA). It is a Lagrangian model, that is, the coordinate system moves with the parcel of air in which chemical changes occur.

The architecture of EKMA is as follows. A column of air is transported along a wind trajectory. Starting time is 0800 local time (LT). The column height reaches to the bottom of the nocturnal inversion layer. The concentration of chemical species in the column is estimated from the 0800 LT emission rates of the following pollutants: NO, $NO_2$, CO, and eight classes of VOCs (olefins, parafins, toluene, xylene, formaldehyde, acetaldehyde, ethene, and nonreactives). Other input parameters are date, longitude, and latitude, which determine the insolation rate. As the column moves with the wind, fresh pollutants are emitted into the column. As time progresses, the solar angle increases and, the height of the column (mixing height) increases, with a consequent dilution effect. Inside the column the photochemical reactions occur in which ozone is generated. The rate constants for the chemical reactions are empirically determined based on smog chamber experiments. The model calculations stop when the ozone level reaches a maximum asymptotic value. This usually occurs between 1500 and 1700 LT.

Figure 10.5 presents an isopleth plot of maximum ozone concentrations versus 0800 LT concentrations of the sum of eight VOCs on the x-axis (in units of ppmv of carbon atoms) and the $NO_x$ concentration on the y-axis. The isopleths have the shape of hyperbolas. The diagonals drawn through the isopleths represent the morning ratios of VOC/$NO_x$. A ratio of 4:1 corresponds to a typical urban environment; a ratio 8:1 to a suburban environment; and 16:1 to a rural environment. Suppose the maximum ozone level reached is 200 ppbv. The "design" (i.e., NAAQS) value is 120 ppbv. To reach the design value one can go in two directions: reducing either the 0800 LT VOC or the $NO_x$ concentration, respectively, represented by the horizontal and vertical dashed lines in Figure 10.5. It can be seen that in an urban environment (4:1 diagonal), the 120-ppbv

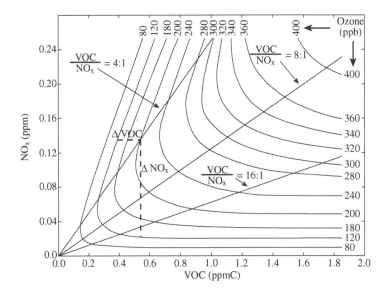

**FIGURE 10.5** Empirical Kinetic Modeling Approach plot of the maximum ozone concentration as a function of the initial $NO_x$ and VOC concentrations.

level can be reached with a much smaller reduction of VOC than $NO_x$ concentrations. In a rural environment (16:1 diagonal), a smaller reduction of $NO_x$ would be necessary. (A third way would be to go along the diagonal from the 200-ppbv isopleth to the 120-ppbv isopleth. This would require reductions of both VOC and $NO_x$, but in lesser quantities than by reducing only one kind of pollutant.)

The shortcomings of EKMA are (a) that it models the ozone concentrations only within a particular air column and (b) it pertains only for a single day. Because $NO_x$ is not appreciably destroyed in the ozone formation cycle, the $NO_x$ just keeps moving downwind with the column of air. The transported $NO_x$ will participate in ozone formation further downwind on the same day and even on subsequent days, because its lifetime can be several days, depending on atmospheric conditions and composition. This is the reason that rural and remote areas far downwind from metropolitan areas also experience high concentrations of ozone, even though the emissions of $NO_x$ in those areas may be quite small.

Because of the shortcomings of EKMA, regulatory agencies in the United States and other countries rely now on more sophisticated models that cover much larger areas than a single air column and a longer time period, usually selected to simulate an elevated pollution episode. Such models are of the Eulerian type, in which the coordinate system remains fixed, and the area covered is divided into grid cells. One such model is called the Urban Airshed Model and another is the Regional Oxidant Model. These models cover an area of several degrees latitude and longitude with variable size grid elements, typically 2 $km^2$. Vertically, the models are divided into several layers below and above the mixing height. The model inputs are the wind field in the modeled area, temperature, humidity, terrain roughness and vegetation cover (the latter affect deposition rates), emission inventory, and the background level of ozone. The governing chemical reactions are quite similar to EKMA, with eight VOC categories. These models can predict the time profile of ozone over a selected location or produce a contour map of maximum concentrations over the whole modeling domain. The time domain of the model is a meteorological episode lasting 3–5 days, usually terminating with a precipitation event. Figure 10.6 is a scatter plot obtained by the

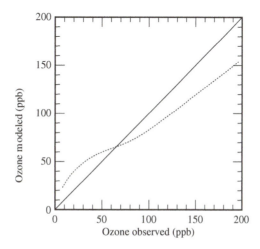

**FIGURE 10.6** Modeled versus observed $O_3$ concentrations (dotted line) from the Regional Oxidant Model.[6]

Regional Oxidant Model for about 15,000 data points. While there is a fairly good correlation between observations and predictions, there is a tendency to overpredict at low concentrations and underpredict at high concentrations.

The grid models bring out clearly that ozone is a regional problem that can only be solved with a regional emission reduction strategy. Because of transport of the precursors $NO_x$ and VOC and the formed ozone itself, high concentrations in the tens to hundreds of ppbv are observed not only in urban areas, but in rural and remote areas, at lower and higher elevations and even over the sea. Peak ozone concentrations are encountered not only in midafternoon hours as the EKMA model would predict, but also at all hours of the day and night.

In areas where the VOC/$NO_x$ ratio is high, $NO_x$ control provides better ozone reduction; in areas where the VOC/$NO_x$ ratio is low, VOC control appears more effective. In rural and remote areas where the availability of VOC is plentiful because of biogenic sources, $NO_x$ availability is the limiting factor to ozone production. In chemical kinetics, the reactant in short supply determines the rate of the reaction. For a subcontinental region, such as the northeastern or southeastern United States, the diurnal and seasonal average ozone concentrations can only be substantially reduced by region-wide $NO_x$ emission reductions.[6]

While in the United States and other developed countries great strides have been taken to reduce $NO_x$ emissions from both stationary and mobile sources, ambient $NO_x$ concentrations in the United States and worldwide are either on the increase or at best are level. What is gained in $NO_x$ control is lost in the ever-increasing number of $NO_x$ emitting sources, especially automobiles. In 2006, in the United States the number of registered automobiles reached 250 million with a population of close to 300 million. It remains to be seen whether in the future the old U.S. ozone standard of 120 ppbv, 1-hour average can be achieved, let alone the new 75-ppbv standard, 8-hour average.

## 10.2.7  Acid Deposition

Acid deposition is better known by the name acid rain. Acid deposition is a more appropriate term because acidic matter can be deposited on the ground not only as rain but also as other kinds of precipitation (e.g., snow, hail, and fog) and in dry form. The deposition by precipitation is called *wet deposition*; the direct impaction on land and water of acidic gaseous molecules and acidic aerosols (particles) is called *dry deposition*. Acid deposition is a secondary pollutant, because it is a result of transformation of primary emitted pollutants.

Already in the late 17th century, Robert Boyle recognized the presence of "nitrous and salino-sulfurous spirits" in the air and rain around industrial cities of England. In 1853, Robert Angus Smith, an English chemist, published a report on the chemistry of rain in and around the city of Manchester. He later coined the term "acid rain." It was first noticed in the 1960s and 1970s in Scandinavia and the northeastern United States and southeastern Canada that some lakes with a very low buffering capacity (low alkalinity) were slowly being acidified with a pH reaching as low as 5.5–6. Lakes act like a mildly alkaline solution in a beaker. When the solution is "titrated" over

---

[6]National Research Council. *Rethinking the Ozone Problem in Urban and Regional Air Pollution.* Washington, DC: National Academies Press, 1991.

the years with acid deposition, it becomes acidic.[7] As most aquatic organisms cannot survive in that kind of acid water, these lakes became devoid of life.

The acidity of rain precipitation in northwestern Europe and in northeastern America in the 1970s–1980s was measured to reach a pH value of 4.0–4.2 and occasionally as low as 3. Hydrogen ion deposition in those regions was measured in the range 0.4–0.6 kg ha$^{-1}$ y$^{-1}$ (Figure 10.7a and b).

In addition to lake and other surface water acidification, it was suspected that acid deposition causes damage to forests and vegetation and to materials and structures. Atmospheric deposition of nitrogenous species, such as $NO_3^-$ and $NH_4^+$, may in part be the cause of lake and coastal water eutrophication. Together with sewage seepage and storm runoff, atmospheric deposition of these nutrients may lead to algal blooms observed in many surface waters near urban–industrialized areas.

Power plants and industrial, commercial, residential, and mobile sources emit the precursors of acid deposition, sulfur and nitrogen oxides ($SO_x$ and $NO_x$). The precursors are advected by winds and dispersed by turbulent diffusion. During transport in the air, the precursors react with various oxidants present in the air and water molecules to form sulfuric and nitric acid, $H_2SO_4$ and $HNO_3$. The acids are deposited on land and water in the wet or dry form. A schematic of formation of acid deposition is presented in Figure 10.8.

The exact transformation mechanism of the precursors to acidic products is still being debated. It is surmised that there are two mechanisms of transformation, the gas phase and the aqueous phase mechanism. In the gas phase mechanism the following reaction sequence seems to occur:

$$SO_2 + OH \rightarrow HSO_3 \tag{10.17}$$

$$HSO_3 + OH \rightarrow H_2SO_4 \tag{10.18}$$

and

$$NO_2 + OH \rightarrow HNO_3. \tag{10.19}$$

The acids may directly deposit on land and water as gaseous molecules or adhere onto ambient aerosols and then deposit in the dry particulate form. The acid molecules and aerosols may be scavenged by falling hydrometeors and then deposited in the wet form. This is called *washout*. In the aqueous phase mechanism the precursors are first incorporated into cloud or raindrops, a process called *rainout*, followed by reactions with oxidants normally found in raindrops—hydrogen peroxide, $H_2O_2$, and ozone, $O_3$.

To exactly determine the hydrogen ion concentration, and thus the pH, it is necessary to analyze all of the anions and cations in precipitation. In the United States, weekly samples of precipitation are sent to central laboratories where all major ions are analyzed. The hydrogen ion concentration

---

[7]pH is the negative logarithm of hydrogen ion concentrations, pH = − log[H+]. Neutral water has a hydrogen concentration of E(−7) moles per liter, thus pH 7. A one-tenth molar concentration of hydrochloric acid has a hydrogen concentration of 0.1 moles per liter, thus pH 1. Lemon juice has an approximate pH of 3 and carbonic acid a pH of 5.6. Raindrops in contact with atmospheric carbon dioxide have a slightly acidic pH between 5 and 6, even without the addition of other acidic species, such as $H_2SO_4$ and $HNO_3$. On the other hand, raindrops in contact with atmospheric alkaline aerosols, such as $CaCO_3$, $CaO$, and $MgCO_3$, may actually have a pH greater than 7.

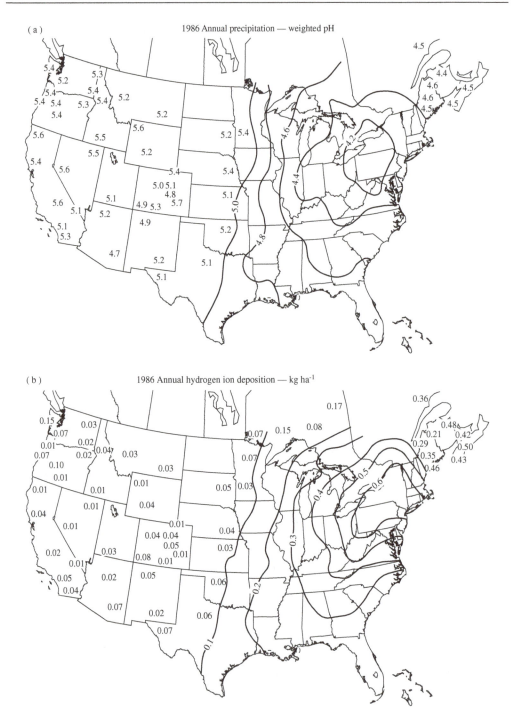

**FIGURE 10.7**    (a) 1986 annual precipitation-weighted pH and (b) annual hydrogen ion deposition, kg ha$^{-1}$. (Adapted from National Acid Precipitation Assessment Program. Washington, DC: Government Printing Office, 1991.)

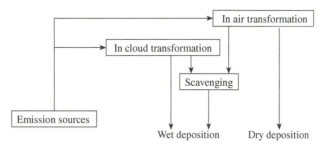

**FIGURE 10.8**  Acid deposition schematic, showing transformation and deposition paths.

is obtained from the ion balance equation,

$$H^+ = 2SO_4^{2-} + NO_3^- + Cl^- + HCO_3^- + 2CO_3^{2-} - NH4^+ - 2Ca^{2+} - 2Mg^{2+} - Na^+, \quad (10.20)$$

where the factor 2 pertains to bivalent ions and the ion concentrations are given in mol $L^{-1}$. Some of the ions are from manmade sources, whereas others are natural (e.g., sea salt, carbonic acid, and ions from crustal matter). In Eastern North America (ENA), an empirical ion balance equation appears to hold[8]:

$$H^+ = 1.63SO_4^{2-} + 0.95NO_3^-. \quad (10.21)$$

The empirical finding that the stoichiometric factor is not exactly 2 for $SO_4^{2-}$ and one for $NO_3^-$ indicates that other anions and cations participate in the hydrogen ion balance equation.

Figure 10.9 shows the trend of $SO_2$ and $NO_x$ (reckoned as $NO_2$) emission rates in the United States for the years 1970–2008. $SO_2$ emission rates reached more than 31 Mt $y^{-1}$ in 1970, but declined steadily to about 12 Mt $y^{-1}$ in 2008 as a consequence of installing $SO_2$ control technology on all new coal-fired power plants and retrofitting existing plants as required by the CAAA 1990. $NO_x$ emissions amounted to close to 27 Mt $y^{-1}$ in 1970 and declined steadily to close to 16 Mt $y^{-1}$ in 2008, thanks to installation of $NO_x$ emission control technologies on stationary and mobile sources. Concomitantly, the hydrogen ion deposition rates also decreased over this period. Current pH values in the northeastern United States are in the 4.5–4.6 range, and hydrogen ion depositions are in the 0.35–0.4 kg ha$^{-1}$ y$^{-1}$ range. This is a considerable improvement over the 1970–1980 values, although it can be argued that further emission reductions of acid deposition precursors are warranted.

### 10.2.7.1  Acid Deposition Modeling

Before enacting costly emission control strategies, national governments would like to know the expected environmental benefits of these strategies. For example, if $SO_2$ and $NO_x$ emissions were reduced by one half, respectively, will sulfate and nitrate ion deposition rates decrease proportionally everywhere? This relates to the linearity of transformation of primary emissions ($SO_2$ and $NO_x$) to secondary pollutants ($SO_4^{2-}$ and $NO_3^-$). Another question regards the geographic distribution of the secondary pollutants. Would it be possible to reduce emissions of primary pollutants to

[8]Golomb, D. *Atmos. Environ.*, 17 (1983): 1380–3.

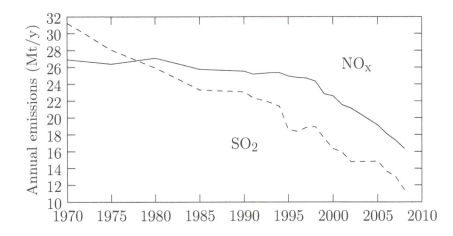

**FIGURE 10.9** Trend of $SO_2$ and $NO_2$ annual emissions in the United States, 1970–2008. (U.S. EPA. *National Emissions Inventory, Air Pollutant Emissions Trends, 1970–2008,* 2009.)

a greater degree in source areas that contribute most of the secondary pollutants to sensitive areas? To answer these questions one has to resort to atmospheric transport and transformation models, also called source-receptor models.

In the 1980s in the United States, Canada and Europe, literally dozens of models were developed, ranging from simple box models (no atmospheric dynamics, just chemical reactions in an enclosed box the size of a subcontinent) to complex Eulerian models stretching over a subcontinent, with individual emission rates in grids of a few square kilometers, and simulated wind, diffusion, precipitation, and other meteorological and topographic factors.[9] These models are called supermodels, and they require enormous computer capacity. The models have proved a linear relationship between emissions of primary pollutants and the concentration of secondary pollutants, including the deposition in wet or dry form of the secondary pollutants. The geographic extent of the secondary pollutants is very much influenced by meteorological conditions. Pollutants are advected in the direction of prevailing winds and are dispersed depending on the turbulent conditions of the atmosphere. The transformation of a primary to secondary pollutant is dependent on ambient temperature: higher temperatures in general promote faster transformation rates. The rate of wet deposition is dependent on the amount of precipitation at the receptor location.

As an example, we show the percentage contribution by various U.S. states and Canadian provinces to wet sulfate deposition at the Adirondack Mountains in New York State[10] (Figure 10.10). It can be seen that the high $SO_2$-emitting states, Pennsylvania, Ohio, West Virginia, Indiana, and New York, are the largest contributors to Adirondack sulfate deposition. These states lie upwind of the Adirondacks. A significant portion of lakes in the Adirondacks have been acidified, probably because of sulfuric and nitric acid deposition.

Acid deposition models could be used for "targeted" emission reduction strategies, wherein those source areas that contribute most to acid deposition to a sensitive area would curtail their

---

[9]National Acid Precipitation Assessment Program. Washington, DC: Government Printing Office, 1991.

[10]Fay, J. A., D. Golomb, and S. Kumar. *Atmos. Environ.* 19 (1985): 1773–82.

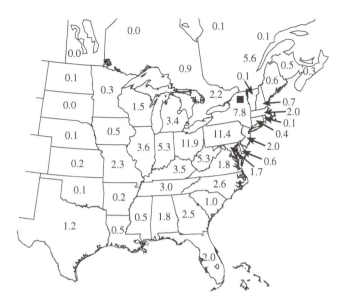

**FIGURE 10.10**   Contribution (%) by states and provinces to sulfate deposition at an Adirondack receptor.[10]

emissions to a greater extent than those source areas that contribute little or nothing. However, by virtue of Title IV of CAAA 1990, the U.S. Congress decided that $SO_2$ emissions will be cut by about one half of 1980 emissions by the year 2000, regardless of the location of a source vis-à-vis a sensitive area. From a political standpoint a uniform emission reduction strategy is more expedient than a targeted strategy, which would pit some states and regions against others.

## 10.2.8   Regional Haze and Visibility Impairment

Small particles (also called fine particles) less than 1–2 μm in diameter settle very slowly on the ground and can travel hundreds to thousands of kilometers from their emitting sources. Part of the fine particles are emitted directly from industrial, commercial–residential, and transportation sources. These are called *primary particles*. However, the majority of fine particles are a product of gas-to-particle transformation processes, including photochemical processes, where the precursor gases are emitted from the aforementioned sources. These are called *secondary particles*. The particles can envelope vast areas, such as the northeastern United States and southeastern Canada, California and adjoining states, western and/or central Europe, and southeastern Asia. Satellite photos often show continental areas covered with a blanket of particles, sometimes stretching out hundreds of kilometers over the ocean. This phenomenon is called *regional haze*. The haze is mostly associated with stagnating anticyclones, when a high barometric pressure cell remains stationary over a region, sometimes over a period of a week or more. The high-pressure cell causes air to circulate in a clockwise fashion around the cell, drawing in increasingly large amounts of emissions from the surrounding sources and forming the regional haze. Usually, the regional haze period is terminated by a cold front moving through the region with accompanying convective clouds, thunderstorms, and precipitation that washes out the particles.

The composition of the fine particles varies from region to region, depending on the precursor emissions. In the northeastern United States, central Europe, and southeastern Asia, more than half of the composition is made up of sulfuric acid and its ammonium and sodium salts, largely because of high sulfur coal and oil combustion. The rest is made up of nitric acid and its salts, carbonaceous material (elemental and organic carbon), and crustal matter (fine dust of soil, clay, and rocks). In the western and southwestern United States and in some other urbanized/industrial areas of the world, nitrate and carbonaceous matter makes up the majority of the composition of fine particles. This is because of heavy automobile traffic, chemical industries, refineries, gas-fired power plants, and other urban/industrial sources. A part of the haze may be caused by biogenic sources, such as the emissions of isoprene, terpene, pinene, and other isoprene derivatives from coniferous forests, but the predominant cause of regional haze is the emission of gaseous and particulate matter from anthropogenic sources, mostly from fossil fuel combustion. Condensed water is also an ingredient of fine particles, as the precursor gases and the formed particle nuclei attract water molecules from the air, especially during high humidity periods.

The small particles are efficient scatterers of light. The scattering efficiency is dependent upon the wavelength. Maximum scattering efficiency for visible light (400–750 nm) occurs with particles less than 1 μm in diameter, the so-called submicrometer particles. Light scattering prevents distant objects from being seen. This is called visibility impairment. During regional haze periods, one cannot distinguish distant mountains on the horizon, and occasionally one cannot see objects farther than hundreds of meters, such as the other wall of a river valley or buildings a few city blocks away.[11] Also, the increasing concentration of particles in urbanized parts of the continents causes the loss of visibility of the starlit nocturnal sky. These days, small stars, less than fifth order of magnitude, rarely can be seen from populated areas of the world.

The scattering of light causes a loss of contrast. This is depicted in Figure 10.11. Part of the incoming sunlight is reflected from the object and reaches the retina of the eye (ray 1). Part of the reflected light is scattered out by particles in the air and does not reach the retina (ray 2). Part of the incoming sunlight is scattered by particles between the object and the eye, reaching the retina, but this ray causes a loss of contrast (ray 3). Finally, part of the incoming sunlight is scattered by particles behind the object, reaching the retina, but with a loss of contrast (ray 4). The consequence of rays 2 to 4 is a loss of contrast of the object against the background. Empirically, it has been shown that an object is no longer visible when the contrast against the background is less than 2%. This is called the threshold contrast and can be expressed as

$$C_x = \frac{I_{obj} - I_{bkg}}{I_{obj}} \geq 2\%, \tag{10.22}$$

where $C_x$ is the contrast at distance $x$ meters, $I_{bkg}$ is the intensity of light rays from the background, and $I_{obj}$ is the intensity of light rays from the object. Alternatively, the contrast relationship can be written

$$C_x = C_0 \exp(b_{ext}x), \tag{10.23}$$

---

[11] Visibility impairment became a significant problem in some national parks in the United States. For example, on some hazy days the northern rim of the Grand Canyon in Arizona cannot be seen from the southern rim.

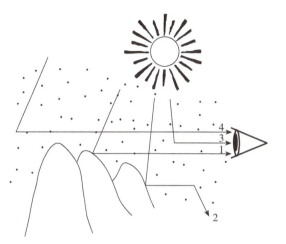

**FIGURE 10.11**  Reflection and scattering of light rays. Ray 1, reflected from object; Ray 2, reflected but scattered out of view; Ray 3, scattered by particles between object and observer into view; Ray 4, scattered by background particles. (Adapted from Seinfeld, J., and S. N. Pandis. *Atmospheric Chemistry and Physics.* New York: John Wiley & Sons, 2006.)

where $C_0$ is the contrast at zero distance, and $b_{ext}$ is the extinction coefficient in m$^{-1}$. Threshold contrast is when $C_x/C_0 = 0.02$; thus,

$$-b_{ext}x = \ln(0.02) = -3.912. \qquad \textbf{(10.24)}$$

Equation (10.24) is called the Koschmieder relationship. It gives the extinction coefficient $b_{ext}$ when an object loses its contrast against the background at a distance $x$ meters. In clear air $b_{ext} = \mathrm{E}(-5)$ to $5\mathrm{E}(-5)$ m$^{-1}$, and objects can be discerned as far as 80 to 400 km (of course, the latter distance is only visible from a height or to a height, so as to reach over the earth's curvature). In polluted air, $b_{ext}$ can be as large as $\mathrm{E}(-3)$ m$^{-1}$, when objects disappear at a few kilometers distance.

Visibility impairment is a significant and unpleasant side effect of fossil fuel use. Visibility improvement can be accomplished by reducing fine-particle and gaseous precursor emissions from fossil fuel combustion and other fossil fuel usage. In practice this means improved emission control devices for primary particles, $SO_2$, $NO_x$, and VOCs. In the United States, visibility improvement is addressed in Section 169A of the CAAA of 1977, which requires that visibility impairment at U.S. national parks must be lessened by reducing particles and their precursor emissions from nearby sources. As a consequence of better and wider use of emission control devices on stationary and mobile sources, visibility has improved steadily in the past decades, and regional haze occurrences in the United States are now less frequent than they were in the 1960s and 1970s.

## 10.3  WATER POLLUTION

The consumption of fossil fuel entails a significant impact on water quality and water usage. The contamination of water starts at the mining and extraction stage, through transport and refining, all the way to leaching into the ground water of ash and scrubber sludge left behind after combustion of fossil fuels. We shall limit this section to the effects of acid mine drainage, coal washing, leaching from coal and ash piles, and water pollution caused by atmospheric deposition of toxic by-products of fossil fuel combustion. We will not treat here the major environmental disasters following the collision or grounding of supertankers carrying crude oil. For example, in 1978 the *Amoco Cadiz* ran aground off the coast of France, spilling 223,000 metric tons of crude oil. In 1979, in the collision of the *Atlantic Empress* with the *Aegean Captain*, 50 miles northwest of Tobago, more than 287,000 tons were spilled. In 1983, the *Castillo de Bellver* spilled 252,000 tons off the coast of South Africa. In 1989, the marine tanker *Exxon Valdez* spilled 37,000 tons of crude oil in the environmentally sensitive Prince Williams Sound in Alaska. In 1991, the *ABT Summer* spilled 260,000 tons off the coast of Angola. On April 20, 2010, in the Gulf of Mexico, the *Deepwater Horizon* oil rig exploded because of a methane gas leak. Eleven workers on the rig evidently have died because their bodies were never found, and 17 workers were injured. Subsequently, oil gushed from the oil well at the sea floor, 1500 m deep, 66 km from the Lousiana coast. Until the gusher was finally capped on July 15, 2010, about 700,000 tons of oil spilled into the Gulf, the largest oil spill in the world's oceans ever. The spill caused yet to be completely assessed ecological and economic damage to the coastal areas of Luisiana, Mississipi, Alabama and Florida. The spill was a manifestation of the increasing environmental and human risks of oil exploration in the deep seas and inhospitable surroundings. It is estimated that between 3 and 4 million metric tons of oil are spilled annually into the world's rivers, lakes, seas, and oceans.

## 10.3.1   Acid Mine Drainage and Coal Washing

In terms of sheer quantity, the most serious water pollution problem associated with coal use is acid drainage from mines, especially surface mines, coal piles, and coal washing. Precipitation falling on open coal seams and on coal piles leaches out mineral matter. The leachate contains acids, toxic elements, and often radioactive isotopes. Leachates with a pH as low as 2.7 have been measured. The acid is a product of oxidation and hydrolysis of the pyritic sulfur in coal. The following toxic elements are found in coal mine and pile drainage in concentrations that exceed drinking water standards: arsenic, barium, beryllium, boron, chromium, fluorine, lead, mercury, nickel, selenium, vanadium, and zinc. Federal and state regulations now require that the leachate be either treated or disposed of in secure impoundments. Coal contains the radioactive isotopes $^{14}C$ and $^{40}K$. These elements were inherent in the original biologic tissue of the antecedents of coal. These radionuclides do not pose an environmental hazard. However, frequently uranium, thorium, and their radioactive daughter elements are also found in the mineral matter adhering to the coal seam. Thus, leachate from coal mines, piles, and slag needs to be monitored for its radioactivity before entering into the environment and for preventing mine worker exposure.

About 50% of all coal mined in the United States is now "washed" at the mine mouth prior to shipment to the user. Coal washing increases the heating value of coal by removing the incombustible mineral matter. More importantly, coal washing removes pyritic sulfur, which can amount to as much as 50% of the sulfur content of the raw coal. The most widely used technique for coal washing is mass separation. Mineral matter has a higher specific gravity than coal. By flushing crushed coal in a stream of water, the mineral matter settles out, while the lighter coal particles float in the stream. Eventually, the washed coal is dried by shaking it on a screen while the water is dripping off or drying the wet coal in a cyclone dryer. Thus, most of the water is recycled. However, the settled mineral matter, which is also wet, now contains a high concentration of acidic, toxic, and possibly radioactive compounds and elements. This slag must be analyzed for its content. If the content is toxic, the slag must be treated or safely disposed of in a hazardous waste impoundment. In the United States, discharges from mining activities are regulated under the Clean Water Act, specifically, the National Pollution Discharge Elimination System.

## 10.3.2   Solid Waste from Power Plants

Although much of the mineral matter of coal is removed at the mine mouth, coal delivered to a power plant or other facilities still contains adhering mineral matter, simply called ash. The ash content can amount anywhere between 1 and 15% of the coal weight. Even oil contains ash, amounting to 0.01–0.5% by weight. After combustion of the coal particle or oil droplet, the mineral matter remains uncombusted and either falls to the bottom of the boiler or is blown out with the flue gas as fly ash. In modern pulverized coal fired boilers, about 90% of the mineral matter forms fly ash and 10% bottom ash. Most of the fly ash is collected in an ESP (see Section 6.2.10.2), except for the very small particles (less than 1 $\mu$m in diameter), which escape into the air because they are not removed efficiently by the precipitator. The fly ash contains approximately the same acidic, toxic, and radioactive matter as the original mineral matter in coal. The fly ash has to be chemically analyzed and, if found hazardous to human health and the environment, properly disposed of. Nonhazardous bottom and fly ash is often used as aggregate material for concrete or asphalt.

As a consequence of acid deposition regulations in the United States and many other countries, coal-burning power plants and industrial boilers are required to be equipped with FGD devices. For high sulfur content coal ($\geq 0.6\%$ by weight), a wet limestone scrubber is necessary (see Section 6.2.10.3). After passing the ESP, the flue gas enters a scrubber tower. A slurry of limestone is injected from sprinklers at the top of the scrubber. Flue gas containing $SO_2$ and other sulfur compounds flows countercurrent to the limestone spray. A sludge is collected in the bottom of the scrubber, consisting of wet calcium sulfate (gypsum), calcium sulfite, and unreacted limestone. The sludge is dewatered as much as possible (unfortunately, gypsum is very difficult to dewater) and then disposed of. Because the sludge may also contain toxic elements, notably arsenic, cadmium, mercury, and selenium, its disposal may require a specially designed secure impoundment.

## 10.3.3   Water Use and Thermal Pollution from Power Plants

It is a consequence of the second law of thermodynamics that heat engines after performing useful work must reject some of the fuel heat input to a cold reservoir. Roughly, about one third of the inherent heating value of the coal is rejected to the cold reservoir. Another third is rejected into the atmosphere, and only one third is transformed into useful work, that is, electricity. The cold reservoir is usually a water body. Some power plants and industrial boilers are located near surface waters, such as a river, lake, or ocean, which can be used for rejecting the heat into them. Other facilities need to use a cooling tower or a cooling pond. The U.S. EPA requires that all new power plants use cooling towers for heat rejection, rather than once-through cooling water from adjacent surface waters. In cooling towers, cold municipal or well water percolates around the hot condenser tubes, taking up the rejected heat. The water is chilled in a draft of cold air and then recycled. Evaporated water rises from the cooling tower into the cold ambient air and condenses, thus forming the visible "steam" plume that emanates from the tower. A 1,000-MW power plant working at a thermodynamic efficiency of 33%, ambient temperature 15°C, loses about 1.7(E7) $m^3$/y (about 4.4 billion gallons/y) of cooling water to evaporation. In the United States cooling towers alone consume between 2 and 3% of total water withdrawals from artesian wells or surface waters. Some critics question the wisdom of using precious water resources for cooling purposes, rather than surface waters (where available), even though the latter entails some risk of thermal pollution of the surface waters and possible water contamination from leaching toxic elements from the heat exchanger components. It is seen here that there is no simple or unequivocal solution to environmental and natural resource problems engendered by fossil fuel usage.

## 10.3.4   Atmospheric Deposition of Toxic Pollutants onto Surface Waters

In Section 10.2.7 we dealt with one form of water pollution due to combustion of fossil fuels, that of acid deposition. But in addition to sulfur and nitrogen oxides, there are other combustion products that escape from the smokestacks and eventually are deposited on land and water, which may cause deleterious health and environmental effects. Two cases in point are the atmospheric deposition of toxic metals and PAH.

### 10.3.4.1   Toxic Metals

We noted previously that fly ash particles may contain toxic metals, such as arsenic, cadmium, mercury, lead, selenium, vanadium, and zinc. These metals are found in small particles, less than 1 $\mu$m in diameter. The small particles are not efficiently collected by the ESP and thus escape into the atmosphere. Because of their small size, these particles are little affected by gravity and can be transported over large distances, hundreds to thousands of kilometers. Eventually, they are deposited in dry or wet form on land and water. From the land, toxic metals may leach into groundwater or run off into streams, lakes, or ocean. Thus, they may enter the food chain, either directly in drinking water or via aquatic organisms that drink or feed in the water. For example, fish from the Great Lakes and other lakes in the northeastern United States and southeastern Canada often contain a concentration of mercury that is deemed unhealthy, especially for pregnant women. Children born to mothers who consumed mercury-laced fish may show birth defects or mental retardation. In addition to coal-burning power plants, municipal incinerators are sources of mercury, because batteries, switches, and fluorescent bulbs that are tossed into the garbage may contain mercury, which when incinerated escapes into the atmosphere. Lead may also cause mental retardation and other brain and central nervous system effects. Part of the lead enters surface and groundwater via atmospheric deposition. In the past, a compound of lead (tetraethyl-lead) was used to boost the octane number of gasoline. With the phasing out of leaded gasoline, lead deposition and lead-related diseases are on the decline.

### 10.3.4.2   Polycyclic Aromatic Hydrocarbons

PAH are organic compounds consisting of two or more fused benzene rings. Naphthalene has two rings, anthracene three, pyrene and chrysene four (in different arrangements), benzopyrenes (there are several) five, perylenes six, and coronene seven. Some of the PAH are known or suspected carcinogens, notably benzo(a)pyrene. PAH are a product of incomplete combustion. Whenever we see a sooty smoke, we can be sure that it contains PAH, because PAH adhere to soot particles. PAH in the vapor or particulate phase are carried by winds and subsequently deposited on land and water in the dry form, while those scavenged by precipitation are deposited in the wet form. Because PAH are unstable (they decompose in sunlight or by oxidation with atmospheric oxidants), they do not travel as far as metals, and they are deposited closer to the combustion sources. Because PAH are very hydrophobic, they are not readily incorporated into rain droplets, therefore, their dry deposition predominates over wet deposition. Rivers, lakes, and coastal waters surrounded by urban–industrial areas are especially affected by PAH deposition from the myriad combustion sources in those areas. Soot particles with adsorbed PAH settle in the sediments of these waters. Bottom-feeding fish and shellfish that live in the sediments may ingest PAH and are often found with cancerous lesions and other diseases. PAH dissolve in fatty tissues; therefore, these organisms may bioaccumulate PAH and transfer them to the food chain, including humans.

The irony is that PAH are not emitted as much from big centralized combustion sources as from small, dispersed sources. Large power plants, industrial boilers, and municipal incinerators are easily controlled for preventing emissions of products of incomplete combustion, of which PAH are a part. Basically, the control involves "good engineering practice," that is, combustion in excess air, thorough mixing of fuel and air, high flame temperature, and sufficient residence time in the combustion chamber. Under those conditions practically all carbonaceous matter burns up into $CO_2$ and $H_2O$, and no organic molecules or radicals are left that may recombine to form soot and PAH. Most of the soot and PAH emissions come from residential and commercial furnaces, wood stoves and fireplaces, open fires, barbecues, aircraft jet engines, gas turbines, and, most copiously, ICEs, such as diesel- and gasoline-fueled trucks and automobiles. We are all familiar with the cloud of soot emanating from a diesel truck under accelerating drive, when the fuel to air ratio becomes very rich, or the black trail left behind a jet aircraft during takeoff. If we want to minimize PAH emissions, we must find better ways of controlling these dispersed stationary and mobile sources. Cigarette smoke also contains PAH; the latter are probably the cause of lung and throat cancer in smokers.

## 10.4  LAND POLLUTION

In regard to fossil fuel use, the heaviest toll on land impact is caused by coal mining—in particular, surface mining, also called strip mining. In the United States more than 60% of about 750 million metric tons of coal mined annually is surface mined. Because the coal seam is rarely on the surface itself, this means removing the "overburden." The latter can amount to up to 100 m of soil, sand, silt, clay, and shale. Some coal seams appear in hill sides and river banks. Here, the coal is removed by auger mining, which uses a huge drill to dig the coal out of the seams.

Surface mining leaves behind enormous scars on the landscape, not to mention the disruption of the ecosystem that existed before mining started or other possible uses of land instead of mining. For that reason, the U.S. Congress passed in 1977 the Surface Mining Control and Reclamation Act. The law delegates to the states the authority to devise a permitting plan that must include an environmental assessment of the mining operations and procedures for reclamation of the land after cessation of operations. The permitting plan must describe the condition, productivity, uses, and potential uses of mine lands prior to mining and proposed postmining reconditioning of the mined lands, including revegetation. Stringent operational standards apply to mining in critical areas, including those with slopes greater than 20 degrees and alluvial valleys. The land must be restored to a condition capable of supporting any prior use and possible uses to which the land might have been put if it had not been mined at all. Needless to say, this law and the requirements imposed on safe and ecologically sound surface mining practices imposed a great effort and economic burden on surface mining companies and indirectly raised the price of surface-mined coal.

Deep shaft mining also places a burden on the land. We mentioned earlier that the mined coal is brought to the surface, where it is crushed and washed. The removed mineral matter accumulates in enormous slag piles that mar and despoil the landscape. Nowadays, strict regulations pertain in the United States and other countries relating to the disposal of coal mine slag and restoration of the landscape.

Finally, we should mention that the vast number of derricks associated with oil and gas exploration and exploitation is neither aesthetically pleasing nor helpful to the ecology that existed on the land prior to oil and gas mining.

## 10.5   CONCLUSION

The consumption of vast quantities of fossil fuel by mankind causes many deleterious environmental and health effects. These effects start from the mining phase of the fossil fuels through transportation, refining, combustion, and waste disposal. When coal is mined in deep shafts or strip mines, mineral matter is separated at the mines by milling and washing. The residual slag may contain toxic, acidic, and sometimes radioactive material, which needs to be properly disposed of without endangering the environment and humans. Oil and gas well derricks, on- and offshore, are aesthetic eyesores and are the cause of oil spillage. Oil and gas pipes may rupture and leak their contents. Oil tankers and barges spill on average 4 million tons per year of crude and refined petroleum on our waterways and oceans. Oil refineries are sources of (a) toxic emissions through vents, leaks, and flaring and (b) toxic liquid effluents and solid waste.

By far, the greatest environmental and health impact is caused by fossil fuel combustion in furnaces, stoves, kilns, boilers, gas turbines, and ICEs, which power our automobiles, trucks, tractors, locomotives, ships, and other mobile and stationary machinery. The combustion process emits pollutants through smokestacks, chimneys, vents, and exhaust pipes, such as PM, oxides of sulfur and nitrogen, carbon monoxide, products of incomplete combustion, and volatile toxic metals. Some of these pollutants are toxic to humans, animals, and vegetation per se, whereas others transform in the atmosphere to toxic pollutants, such as ozone, organic nitrates, and acids. The pollutants are advected by winds and dispersed by atmospheric turbulence over hundreds to thousands of kilometers, affecting sensitive population and biota far removed from the emission sources. Particles, besides containing toxic and potentially carcinogenic agents, often envelope whole subcontinental areas in a haze that reduces visibility and the enjoyment of the landscape and a starry sky.

Great strides have been taken, especially in the more affluent countries, to limit the emissions of air pollutants from combustion sources. For example, most particles can be filtered out of the flue gas by electrostatic precipitators or fabric filters. Sulfur oxide emissions can be reduced by wet or dry limestone scrubbers. Nitric oxide emissions can be reduced by catalytic and noncatalytic injections of chemicals into the flue gas. The catalytic converter, which is now applied to gasoline-fueled automobiles in many countries, reduced the emissions of automobiles significantly compared with the uncontrolled predecessors. Diesel engines, with proper tuning, are also emitting fewer pollutants, although a magic box like the catalytic converter has yet to be found for diesel engines. Alas, little or no emission control devices are applied to the myriad of dispersed sources, such as residential furnaces, stoves and fireplaces, and smaller industrial facilities.

The less affluent countries, because of more pressing economic needs, do not yet avail themselves of emission control devices. Consequently, air quality in those countries is much worse than in the more affluent ones. Because air pollutants do not stop at national boundaries, surrounding countries may feel the effect of emissions from neighboring states. It is incumbent on the more affluent societies to help the poorer ones in controlling air pollutant emissions, because the health of the whole human population and ecology of the planet is at stake.

Finally, this chapter has not addressed the great looming risk of global warming resulting from anthropogenic greenhouse gas emissions from fossil fuel usage. This will be the subject of the next chapter.

# PROBLEMS

### Problem 10.1

Calculate the emission rate of $SO_2$ (kg/s) from a large coal-fired plant that uses 2 million metric tons of coal per year with a sulfur content of 2% by weight.

### Problem 10.2

Calculate the emission rate of $SO_2$ per fuel heat input (g/GJ) of a large coal-fired power plant that uses coal having a heating value of 30 MJ/kg and a sulfur content of 2% by weight.

### Problem 10.3

A vehicle traveling along a highway at 90 km/h emits 80 g CO and 10 g NO per liter of fuel burned. The vehicle travels 8 km per liter of fuel burned. Calculate the emission rate of CO and NO in units of g/km.

### Problem 10.4

The table below shows ozone concentrations measured every hour on the hour on a high pollution day in Los Angeles. Plot the diurnal ozone profile. Determine the arithmetic average and geometric mean ozone concentrations for that day. Determine the maximum 8-h arithmetic average concentration and compare it with the new U.S. ambient $O_3$ standard of 75 ppbv.

| t (h) | 1 | 2 | 3 | 4 | 5 | 6 | 7 | 8 | 9 | 10 | 11 | 12 |
|---|---|---|---|---|---|---|---|---|---|---|---|---|
| $O_3$ (ppbv) | 12 | 8 | 4 | 4 | 6 | 10 | 23 | 42 | 63 | 88 | 112 | 183 |
| t (h) | 13 | 14 | 15 | 16 | 17 | 18 | 19 | 20 | 21 | 22 | 23 | 24 |
| $O_3$ (ppbv) | 212 | 254 | 302 | 312 | 315 | 312 | 204 | 162 | 88 | 34 | 23 | 18 |

### Problem 10.5

A coal-fired power plant has a smokestack 30 m high and 3 m in diameter. The flue gas exit velocity is 5 m/s, the exit temperature is 300°C, the ambient temperature is 30°C, the wind speed is 3 m/s, and the atmospheric stability category is B. Calculate the effective height above ground level (m) of the plume center line using the Briggs plume rise equation.

### Problem 10.6

Using the calculated emission rate of the plant with characteristics given in Problem 10.1, the GPE, stability category B, and the effective plume height calculated in Problem 10.5, calculate the $SO_2$ concentration ($\mu g\, m^{-3}$) at a downwind distance of 1 km, 100 m perpendicular to the wind direction at the top of a 100-m-high hill.

## Problem 10.7

Using the parameters of Problems 10.1 and 10.5, estimate the highest concentration of $SO_2$ ($\mu g$ $m^{-3}$) on level ground for stability category B. Estimate the downwind distance (m) at which the highest concentration ($x_{max}$) would occur.

## Problem 10.8

A highway with eight lanes has traffic of one vehicle per 30 m per lane. Each vehicle emits on average 10 g/s CO at an average height of 2 m. The wind speed is 6 m/s, wind direction is at an angle of 45° toward the highway, and stability category is C. Calculate the CO concentration at a receptor 100 m downwind from the highway.

## Problem 10.9

Using the EKMA plot of Figure 10.5, assuming the peak afternoon concentration of $O_3$ is 200 ppbv, estimate the percentage of the necessary reductions of initial concentrations of VOC and $NO_x$ to achieve an afternoon peak concentration of $O_3$ of 120 ppbv for an urban (VOC:$NO_x = 4:1$), suburban (VOC:$NO_x = 8:1$), and rural (VOC:$NO_x = 16:1$) environment. Estimate the concentration reductions in two ways: VOC only and $NO_x$ only. Explain why in one environment it is more effective to reduce VOC concentrations, whereas in another environment it is more effective to reduce $NO_x$ concentrations. Is it possible to achieve in a certain environment the 120-ppbv concentration by actually increasing morning $NO_x$ concentrations? In which environment is that possible, and what would that do to downwind locations?

## Problem 10.10

Measurements show that in an environmentally sensitive area the average sulfate ion concentration in precipitation is 25 $\mu$mol/L and the nitrate ion concentration is 15 $\mu$mol/L. Using the approximation Equation (10.21), calculate the pH of precipitation.

## Problem 10.11

(a) Estimate the maximum visibility distance (km) for a polluted atmosphere with $b_{ext} = 6E(-4)$ $m^{-1}$ and for a clean atmosphere with $b_{ext}$ 6E($-5$) $m^{-1}$. (b) If the maximum visibility is 6 km, what is the extinction coefficient ($m^{-1}$)?

# BIBLIOGRAPHY

Heinsohn, R. J., and R. L. Kabel. *Sources and Control of Air Pollution*. New York: Prentice Hall, 1999.

Kraushaar, J. J., and R. A. Ristinen. *Energy and Problems of a Technical Society*, 2d ed. New York: John Wiley & Sons, 1993.

Seinfeld, J., and S. N. Pandis. *Atmospheric Chemistry and Physics*. New York: Wiley Interscience, 2006.

Turner, D. B. *Workbook of Atmospheric Dispersion Estimates*, 2d ed. Chelsea, MI: Lewis, 1994.

Vallero, D. *Fundamentals of Air Pollution*, 4th ed. New York: Academic Press, 2008.

Wark, K., C. F. Warner, and W. T. Davis. *Air Pollution, Its Origin and Control*. Reading: Addison Wesley, 1998.

# Global Warming and Climate Change

## 11.1   INTRODUCTION

The earth's surface temperature and local climate vary with geographical position and season as a consequence of the energy balance between incoming solar radiation and outgoing thermal radiation to deep space, providing a thermal and nourishing environment that supports living organisms in the ocean and on land. The atmospheric and oceanic processes that sustain this balance comprise the earth's climate system.

All the earth's plants and animals are well adapted to present conditions of temperature and climate. Changes to these conditions could have profound effects on the current ecological balance and future viability of the diverse species of living systems that constitute the earth's ecosphere. Only a few degrees change in the earth's mean surface temperature could provoke a cascade of climatic effects that would irreversibly alter the ecological environment on which living systems depend. Regrettably, such changes have already started, primarily as a consequence of the industrialization of human economies on such global scale that the earth's atmosphere and hydrosphere have been affected, changing the solar–earth energy balance.

In the absence of the earth's atmosphere, its surface temperature would be lower by about $33°C$ and the climate totally different. The warmer surface temperature is caused principally by atmospheric water vapor and carbon dioxide, which absorb some of the earth's outgoing infrared radiation. This process is called the greenhouse effect and has operated to maintain a warmer earth ever since its atmosphere was formed hundreds of millions of years ago.

In recent decades scientists have become aware that the radiation-absorbing species in the atmosphere, called greenhouse gases (GHG), have been increasing in concentration. These increases might cause a rise in the average earth's surface temperature, called global warming, and consequent changes in earth's climate. Anthropogenic GHG are carbon dioxide ($CO_2$), methane ($CH_4$), chlorofluorocarbons (CFCs), nitrous oxide ($N_2O$), and tropospheric ozone ($O_3$). Anthropogenic aerosols may also contribute to global warming. Water vapor is a natural greenhouse gas, although its concentration in the atmosphere may be affected by human activities. Methane is also partially of natural origin.

The Intergovernmental Panel on Climate Change (IPCC) in its 2007 Summary for Policymakers[1] issued the following statements: "Warming of the climate system is unequivocal, as is now evident from observations of increases in global average air and ocean temperatures, widespread melting of snow and ice, and rising of global average sea level.... At continental, regional and ocean basin scales, numerous long-term changes in climate have been observed. These include changes in arctic temperatures and ice, widespread changes in precipitation amounts, ocean salinity, wind patterns and aspects of extreme weather including draughts, heavy precipitation, heat waves and the intensity of tropical cyclones.... Paleoclimatic information supports the interpretation that the warmth of the last half century is unusual in at least the previous 1300 years. The last time the polar regions were significantly warmer than present for an extended period (about 125,000 years ago), reductions in polar ice volume led to 4 to 6 m of sea level rise."

Of all environmental problems caused by human activities, global warming is the most perplexing, potentially most threatening, and arguably the most intractable. It is largely caused by the increased use of fossil energy. The prevention of global warming would require a very significant shift from our present energy use pattern toward one of lesser reliance on fossil fuels. Inevitably, such a shift would result in a price increase for energy commodities, employment disruptions (not necessarily job losses, but different employment patterns), development of alternative technologies, efficiency improvements, and conservation measures.

Because of the expected socioeconomic disruptions, there is considerable opposition to implementing preventative policies. Opposition comes from affected interest groups, including coal, oil, and gas suppliers, automobile manufacturers, steel, cement, and other heavy industries, their financiers and shareholders, and their political representatives. Opposition comes also from developing countries, which claim that their peoples' economic standard is so much lower compared with that in more developed countries that increased use of fossil fuel is necessary to elevate their standard. On the other hand, most scientists consider preventative measures a necessity to avoid significant changes in global climate and the consequent impact on human habitat and ecological systems.

In this chapter we explain the physical basis of global warming and its attendant climate change. In the following chapter we shall discuss possible measures that could be applied to mitigate manmade global warming.

## 11.2    WHAT IS THE GREENHOUSE EFFECT?

The term greenhouse effect is derived by analogy to a garden greenhouse. There, a glass-covered structure lets in the sun's radiation, warming the soil and plants that grow in it. The glass cover restricts the escape of heat into the ambient surroundings by convection and radiation, thereby raising the temperature inside the greenhouse to a higher value than would exist without the glass cover. Similarly, the earth's atmosphere lets through most of the sun's radiation, which warms the earth's surface, but the GHG restrict in part the outgoing terrestrial radiation, causing global warming and associated climate effects.

The warming effect on the earth's surface by certain gases in the atmosphere was first recognized in 1827 by Jean-Baptiste Fourier, the famous French mathematician. Around 1860, the British scientist John Tyndall measured the absorption of infrared radiation by $CO_2$ and water vapor, and he suggested that the cause of the ice ages may have been a decrease of atmospheric concentrations

---

[1]IPCC. *Climate Change 2007. Physical Science Basis*. Cambridge, UK: Cambridge University Press, 2007.

of $CO_2$. In 1896, the Swedish scientist Svante Arrhenius estimated that doubling the concentration of $CO_2$ in the atmosphere may lead to an increase of the earth's surface temperature by 5–6°C.[2] In 1938, G. S. Callendar pointed out that the combustion of fossil fuels is adding substantial amounts of $CO_2$ to the atmosphere, which may influence the average surface temperature of the earth. [3] In the 1950s, Gilbert Plass, the Canadian-born atmospheric scientist, published a series of insightful articles on the greenhouse effect caused by natural and anthropogenic polyatomic gases, in the atmosphere, notably $CO_2$. These articles were summarized in 1956 in a 6,000-word article in *American Scientist*.[4] Plass identified the 15-μm absorption band of $CO_2$ as a main cause for absorbing the outgoing terrestrial infrared radiation, thus providing a thermal blanket for the earth. He predicted that doubling of the $CO_2$ concentration in the atmosphere would increase the surface temperature by 3°C. He also predicted that if all the recoverable fossil fuels (coal and petroleum; natural gas was not yet a major fossil fuel in the 1950s) were burned up within 1,000 years, the earth surface temperature may rise by 7 to 13°C.

The major natural GHG are water vapor, carbon dioxide, and, in part, methane. If not for these gases, the earth would be quite inhospitable, with a temperature well below freezing. Why are we concerned about adding more manmade GHG to the atmosphere? Will not the earth become even more comfortable to living creatures? The answer is that humans and ecological systems have adapted to present climatic patterns. Perturbing those patterns may result in unpredictable consequences.

## 11.2.1   Solar and Terrestrial Radiation

The sun emits electromagnetic radiation ranging from very short wavelength gamma rays, to X-rays, to ultraviolet through visible radiation, to infrared radiation. A section of the solar spectrum reaching the earth, which contains the near-ultraviolet, visible, and near-infrared spectrum up to 3.2 μm, is shown in Figure 11.1. Three spectra are presented. The uppermost curve gives the solar irradiance outside the earth's atmosphere; the lowest curve is the spectrum of incident radiation at sea level. The dashed curve, which follows closely the upper curve, is the spectral radiance of a black body heated to a temperature of 5,900 K,[5] scaled to equal the total solar irradiance at the earth. This tells us that the sun's surface temperature is approximately 5,900 K. (The interior of the sun is much hotter, by many millions K, because of the nuclear fusion reactions occurring there.) A black body heated to 5,900 K emits its peak radiance in the visible portion of the electromagnetic spectrum, at about 500 nm (0.5 μm). The reason the solar radiance curve does not follow the black body radiance curve exactly is that at some wavelengths there is excess radiance and at others a deficit.[6]

---

[2]Arrhenius, S. *Philosophical Magazine*, 41 (1896): 237–76.

[3]Callendar, G. S. *Quarterly Journal of the Royal Meteorological Society*, 64 (1938): 223.

[4]Plass, G. N. *American Scientist*, 44 (1956): 302–16; reprinted in abbreviated version in *American Scientist*, 98 (2010): 58–67.

[5]A black body is an idealized radiator. It absorbs all the incident radiation of any wavelength and emits radiation at any wavelength commensurate with its temperature. A radiation field enclosed by a black body of temperature T has an energy density (and distribution of that energy with wavelength) that depends only upon the temperature and not on any other characteristics of the black body.

[6]Excess radiance results from excitation of atoms and ions in the solar corona (e.g., the sodium-D lines around 589 nm). Deficits result from self-absorption of the incident radiation by other gaseous atoms and ions, which are also present in the solar corona. These are called Fraunhofer lines.

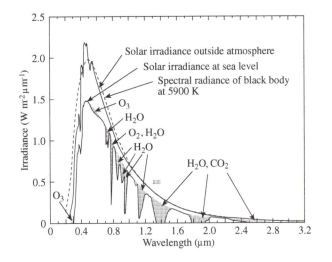

**FIGURE 11.1**   Solar spectrum at the top of the atmosphere (upper curve) and at sea level (lower curve). Shaded areas indicate absorption of radiation by atmospheric molecules. The dashed curve is the radiance spectrum of a black body heated to 5,900 K, scaled to the solar irradiance curve at the top of atmosphere. (Data from Valley, S. L., ed. *Handbook of Geophysics.* Bedford: U.S. Air Force, 1965.)

The solar irradiance spectrum at sea level (lower curve) is much different from the spectrum at the top of the atmosphere. This is because gases and aerosols in the earth's atmosphere absorb or scatter radiation. On the ultraviolet side, oxygen and ozone in the stratosphere are the major absorbing gases. In the visible portion, density fluctuations of atmospheric molecules scatter sunlight.[7] In the infrared, polyatomic molecules present in the lower atmosphere (troposphere), such as $H_2O$, $CO_2$, $O_3$, $CH_4$, and $N_2O$, absorb solar radiation.

The earth radiates outward to space a spectrum commensurate to her surface temperature. Figure 11.2 presents the earth's spectral radiance as observed by a spectrophotometer housed on an artificial satellite looking toward the earth's surface. The solid curves in Figure 11.2 represent black body radiances at various temperatures. The black body radiance curve that best represents the earth's radiating temperature at the spot where the spectrum was taken is about 280–285 K. The radiance curves span the far infrared region of the electromagnetic spectrum, from about 5 to 50 μm, peaking at about 18 μm. But the earth's radiance curve has several deficits. These deficits are the result of absorption of the outgoing infrared radiation by GHG. Each absorption band is composed of many individual absorption lines caused by vibrational–rotational transitions of the molecule. At some wavelengths there is overlap of absorption by individual species. Water vapor absorbs at 5–7 μm and at all wavelengths above 10 μm. The water vapor rotational lines are so crowded that they present a quasi-continuum. Carbon dioxide absorbs strongly at 12–18 μm,

---

[7]Scattered sunlight causes the sky to be blue. Atmospheric molecules (e.g., $N_2$, $O_2$) scatter radiation preferably in the blue region of the visible spectrum. With increasing particle concentration in the atmosphere the scattered radiation becomes colorless (i.e., white).

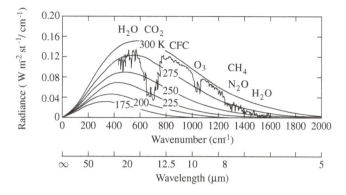

**FIGURE 11.2**   Earth spectrum ("earth-shine") as observed from the Nimbus-7 satellite. Major absorption bands by greenhouse gases are indicated. (Adapted from Liou, K. N. *Introduction to Atmospheric Radiation.* New York: Academic Press, 1980.)

methane and nitrous oxide at 7–8 μm, ozone at 9–10 μm, and the major chlorofluorocarbons ($CFCl_3$ and $CF_2Cl_2$) at 10–12 μm.

The deficits in the outgoing terrestrial radiation spectrum are a clear indication of the greenhouse effect. The absorption of radiation by GHG heats the surrounding atmosphere. By the laws of quantum mechanics, the molecules that absorbed the radiation reradiate some of the absorbed photons, called relaxation, in all directions. About one half of the radiated photons strike the earth's surface, warming it more than it would be for not receiving the additional radiation. That is the essence of global warming.

### 11.2.2   Sun–Earth–Space Radiative Equilibrium

The sun–earth–space radiative equilibrium is depicted in Figure 11.3. The numbers on the arrows are watts per square meter, average annual radiation received from the sun and emitted from the earth. (The intensities in Figures 11.1 and 11.2 are instantaneous radiation fluxes from the sun and earth, respectively.) About 30% of the incoming solar radiation is reflected into space by clouds and the earth's surface. This factor is called the *albedo*. Another part (25%) is absorbed by gases and aerosols, including clouds, in the earth's upper and lower atmosphere, exciting and splitting molecules, and ionizing molecules and atoms (for example, solar UV radiation causes the formation of ozone in the stratosphere and the formation of free electrons in the ionosphere). The rest (about 45%) reaches and is absorbed by the earth's surface, warming land and water. Much of the absorbed radiation goes into evaporating water from the oceans and other surface waters.

Quantitatively, the solar energy input to the earth on top of the atmosphere is

$$I_S = S_0 \pi R_E^2 \sigma T_E (1 - \alpha). \tag{11.1}$$

The factor $\pi R_E^2$ is the disk area presented by the earth toward the sun.

$S_0 = 1367\ \text{W m}^{-2}$ (solar energy impinging on top of the atmosphere, called the solar constant)

$R_E = 6371$ km (earth's radius)

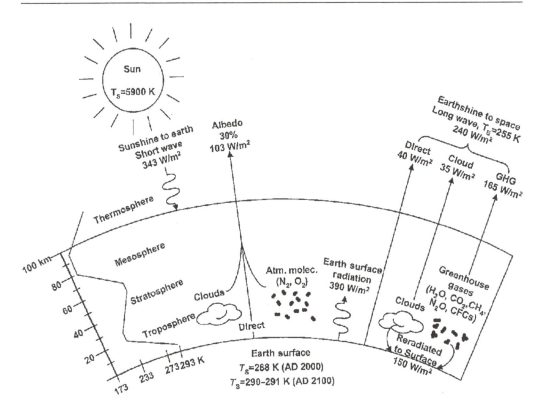

**FIGURE 11.3**    Schematic of the greenhouse effect.

$\alpha = 0.3 \pm 0.03$ (the present average albedo of the earth; it could change over time)

The earth's radiative output to space is

$$I_E = 4\pi R_E^2 \sigma T_E^4. \tag{11.2}$$

Here $4\pi R_E^2$ is the total area of the earth that radiates into space.

$\sigma = 5.67\text{E}(-8)$ W m$^{-2}$ K$^{-4}$ (Stefan–Boltzmann constant)

$T_E$ = equivalent black body radiative temperature of earth, K

From (11.1) and (11.2) we can calculate the earth's radiative temperature,

$$T_E = [S_0(1-\alpha)(4\sigma)^{-1}]^{1/4} = 255 \text{ K}(-18°C) \tag{11.3}$$

Here we used an albedo $\alpha = 0.3$. The temperature $T_E = 255$ K is the average radiative temperature of the earth and her atmosphere as it would appear to a space observer looking toward the earth. Currently, the earth's average surface temperature is $T_S = 288$ K (15 °C). The difference $T_S - T_E = 33$ K is a consequence of the greenhouse effect. This effect is present on other planets. Mars has a very thin atmosphere; therefore, the effective radiative temperature and the surface temperature are very close, 217 and 220 K, respectively. On the other hand, Venus, with a very dense $CO_2$ atmosphere, has an effective radiative temperature of 232 K and a surface temperature of about 700 K.

The earth's average surface temperature has varied over the geological ages, to wit the glacial and interglacial periods. In part, these temperature variations may have been caused by changing GHG concentrations in the atmosphere. Other factors for global climate change may have been related to the variations of the earth's orbit around the sun and the tilt of her axis. The first of these, the eccentricity of the earth's elliptical orbit, varies with a period of about 100,000 years, changing the average insolation of the earth. The second, the axis of rotation of the earth vis-á-vis the ecliptic, varies between 21.6° and 24.5° (currently 23.5°) with a period of about 41,000 years, affecting the amount of insolation of the hemispheres.

### 11.2.3  Modeling Global Warming

Modeling of global warming is based on calculations of radiative transfer between the radiating body (in this case the earth's surface) and atmospheric molecules, together with other energy transport processes in the atmosphere, ocean, cryosphere, and biosphere.[8] Modeling of the radiative transfer between the earth's surface and the atmosphere requires a detailed knowledge of the spectroscopic characteristics of GHG and the structure of the atmosphere. The exercise of the radiative transfer models requires very-high-capacity computers.

The absorption of energy at a particular wavelength by an individual molecule depends on line intensity, line half-width, and its quantum energy state. The energy absorption is a function of the temperature of the surrounding atmosphere with which the molecule is in equilibrium. The atmosphere is modeled by many layers, in each of which the absorbing gas concentration and temperature determine the radiative transport away from and toward the earth and from which the model predicts the surface temperature and atmospheric temperature profile.

The earth's atmosphere has a complicated temperature structure (see Figure 11.3). The bottom layer, extending to about 10 km altitude, is called the troposphere in which the temperature decreases with altitude. The next layer is the stratosphere, extending to about 50 km, in which the temperature increases. Following is the mesosphere, up to about 80 km, in which the temperature decreases. In the highest layer of the atmosphere, called the thermosphere, the temperature increases again.

The GHG molecules will have a net absorption of the outgoing terrestrial radiation only if they are at a colder temperature than the radiating earth. The terrestrial radiation "excites" the vibrational–rotational energy levels of the molecules. The excited molecules radiate this excess energy. We noted before that the earth's effective radiating temperature corresponds to a black body temperature $T_E = 255$ K. With an average surface temperature $T_S = 288$ K and an average temperature gradient (lapse rate) in the troposphere of about $-6.5$ K/km, the apparent height, called the mean radiating height, at which the temperature level of 255 K originates is 5.5 km. Since the effective radiating temperature of the earth will remain nearly at $T_E = 255$ K,[9] the addition of GHG to the atmosphere will change the surface temperature $T_S$, the temperature profile of the atmosphere,

---

[8]Solar heating of the land and ocean surface by radiation induces atmospheric and oceanic currents that distribute heat both horizontally and vertically. The atmospheric currents determine the amount of water vapor in the atmosphere, as well as precipitation to the surface. The currents also affect the temperature distribution in the atmosphere and thereby the degree of global warming.

[9]The radiative temperature of the earth can only change if the albedo changes (see Equation (11.3)).

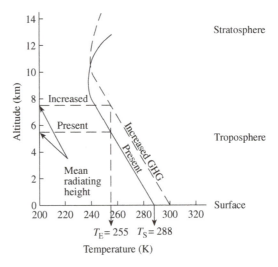

**FIGURE 11.4**   Effect of increased greenhouse gas concentration on surface temperature and vertical temperature profile of the atmosphere.

and the mean radiating height. The sun–earth–space system remains in radiative equilibrium; what changes with the addition of GHG to the atmosphere is the redistribution of energy, manifested by a redistribution of temperature between the earth's surface and her atmosphere. Increased GHG concentrations have the effect that in the higher troposphere, where the GHG molecules are colder, more outgoing terrestrial radiation is absorbed. That radiation is re-emitted in all directions. About one half is radiated downward to the earth's surface, raising $T_S$; the other half is radiated upward, warming the even colder upper layers of the troposphere. On the other hand, the lower stratosphere will be cooler because those layers receive less outgoing radiation from the surface, because of absorption of the intervening GHG. Taking the whole atmospheric temperature profile into account, the mean radiating height will be relatively higher than at present. These changes are depicted in Figure 11.4.

## 11.2.4   Global Warming Potential

Global warming potential (GWP) is a measure of how much a given amount of GHG released into the atmosphere is estimated to contribute to global warming. The GWP is based on a number of factors, including the radiative efficiency (radiation absorbing capacity) and the lifetime (decay rate) of a specific GHG relative to a reference gas, usually $CO_2$. Because the lifetime of each gas varies, a time horizon needs to be taken into account in estimating the global warming potential. Taking a 100-year time horizon, the GWP of several GHG are taken from the recently published report of the Intergovernmental Panel on Climate Change.[10] Thus, releasing into the atmosphere 1

---

[10]Intergovernmental Panel on Climate Change. *Climate Change 2007: Physical Science Basis.* Cambridge, UK: Cambridge University Press, 2007. Most subsequent data are taken from this report and will not be referenced separately.

**TABLE 11.1**  Greenhouse Gas (GHG) Concentrations, Global Warming Potential and Radiative Forcing of Several Anthropogenic GHG, and Other Radiative Forcing

| GHG | 2008 concentration (ppmv) | Global warming potential relative to $CO_2$ | Radiative forcing $(\triangle F_R \text{ W m}^{-2})$ | % of total |
|---|---|---|---|---|
| Carbon dioxide | 385 | 1 | $+1.66 \pm 0.17$ | 55.5 |
| Methane | 1.7 | 25 | $0.48 \pm 0.05$ | 16 |
| Nitrous oxide | 0.3 | 300 | $0.16 \pm 0.02$ | 5.5 |
| Halocarbons | 0.1 ppbv | 5,000–10,000 | $0.34 \pm 0.03$ | 11 |
| Ozone | 20–30 ppbv | ? | $0.35 \pm 0.1$ | 12 |
| | | | $+2.99 \pm 0.2$ | |
| Other radiative forcings | | | | |
| Clouds | | | $-0.7 \pm 0.3$ | |
| Aerosols | | | $0.5 \pm 0.4$ | |
| Land use | | | $0.2 \pm 0.2$ | |

kg of $CH_4$ has 25 times GWP compared with releasing 1 kg of $CO_2$ over a 100-year time horizon and $N_2O$ 300 times. While CFCs have an enormous GWP, it must be remembered that their production has been phased out according to the Montreal Protocol of 1987 and its amendments. Therefore, no further releases of CFCs are expected. However, because of the long decay time of CFCs in the atmosphere, these gases will contribute to global warming for decades and even centuries, albeit at a much smaller degree than at present. Currently CFCs are being replaced with hydrochlorofluorocarbons (HCFCs). These gases, when released into the atmosphere, also contribute to global warming, but because of their shorter decay time in the atmosphere their GWP is much less than that of fully halogenated CFCs. The GWP of anthropogenic GHG are listed in Table 11.1.

## 11.2.5  Radiative Forcing

Radiative forcing is defined as the change in net irradiance at the tropopause. Net irradiance is the difference between the incoming radiation energy and the outgoing radiation energy, measured in $\text{Wm}^{-2}$. Anthropogenic GHG and aerosols released into the atmosphere cause a change in the normal radiative forcing levels that occur at the tropopause. A positive value of radiative forcing indicates that more of the outgoing terrestrial radiation is absorbed in the troposphere, of which about one half is reradiated to the earth's surface and thus contributes to global warming. A negative value of radiative forcing indicates that more of the incoming solar radiation is reflected into space and manifested as increased earth's albedo, and thus contributes to global cooling. In the following we discuss the effect of individual anthropogenic GHG and aerosols on radiative forcing.

(a) *Carbon dioxide*. The present concentration of $CO_2$ is about 385 parts per million by volume (ppmv) and is growing approximately by 0.4% $\text{y}^{-1}$ (see Table 11.1). Before the preindustrial revolution, the atmospheric concentration of $CO_2$ was about 280 ppmv. The present contribution of $CO_2$ to radiative forcing is estimated at $+1.66 (\pm 0.17) \text{ Wm}^{-2}$, which is 55.5% of the radiative forcing caused by all anthropogenic GHG. Thus, $CO_2$ is the largest contributor to manmade global warming. The $CO_2$ absorption band at 12–18 $\mu$m is so strong that it is almost saturated at the center. Further additions of $CO_2$ to the atmosphere will result in more absorption at the wings of the bands, but little at the center. Thus, the increase of radiative forcing caused by increased $CO_2$ concentrations has a logarithmic relationship:

$$\triangle F_R = a(\ln C_t/C_0)\mathrm{Wm}^{-2}, \tag{11.4}$$

where $\triangle F_R$ is the radiative forcing increment, the constant $a = 5.35$ pertains to clear sky conditions, $C_t$ is the concentration of $CO_2$ (e.g., in ppmv) at time $t$, and $C_0$ is the reference concentration.[11] For example, a doubling of $CO_2$ concentrations from the preindustrial revolution level of 280 to 560 ppmv would cause a radiative forcing increment of 3.7 $\mathrm{Wm}^{-2}$, of which we already used up 1.66 ($\pm 0.17$) $\mathrm{Wm}^{-2}$. Similarly, a doubling of the present concentration of 385 to 770 ppmv would cause another increment of radiative forcing $3.7 - 1.66 = 2.04 \ \mathrm{Wm}^{-2}$.

(b) *Methane.* The present concentration of $CH_4$ is about 1.7 ppmv. In the 1970–1980s, $CH_4$ concentrations grew at a rate of 1% $\mathrm{y}^{-1}$. Recently the growth rate has leveled off, which implies that the removal rate of $CH_4$ is nearly as large as its emission rate (see Figure 11.10). Present radiative forcing $CH_4$ is estimated at +0.48 ($\pm 0.05$) $\mathrm{Wm}^{-2}$, 16% of the radiative forcing caused by anthropogenic GHG.

(c) *Nitrous oxide.* The present concentration of $N_2O$ is about 0.3 ppmv. The concentration is growing at a rate of about 0.25% $\mathrm{y}^{-1}$. Radiative forcing caused by $N_2O$ is estimated at +0.16 ($\pm 0.02$) $\mathrm{Wm}^{-2}$, 5.5% of the radiative forcing caused by anthropogenic GHG.

(d) *Chlorofluorocarbons and other halocarbons.* As mentioned above, fully halogenated carbons (CFCs) are no longer produced according to the Montreal Protocol of 1987 and its amendments. Because of the very long lifetimes of CFCs, they still contribute to radiative forcing. Furthermore, the substitute HCFCs also have strong IR absorption capacities. The combined radiative forcing of CFCs and HCFCs, called halocarbons, is estimated at +0.34 ($\pm 0.03$) $\mathrm{Wm}^{-2}$, 11% of the radiative forcing caused by anthropogenic GHG.

(e) *Tropospheric ozone.* Tropospheric ozone, also called the "bad ozone," is a consequence of photochemical reactions involving anthropogenic emissions of $NO_x$ and VOCs (see Chapter 10). Tropospheric ozone is to be distinguished from natural stratospheric ozone, the "good ozone," which shields the planet's biota and humans from harmful ultraviolet radiation. Tropospheric ozone concentrations vary widely over the planet (higher concentrations near urban–industrial areas), over seasons (higher concentrations during late spring, summer, and early fall), and diurnally (higher concentrations during daytime). The current global mean atmospheric concentrations of tropospheric ozone are estimated between 20 and 30 ppbv.[12] The mean radiative forcing caused by $O_3$ is estimated at +0.35 (+25 to +65) $\mathrm{Wm}^{-2}$. The wide margin of uncertainty is caused by the spatial and temporal variability of ozone concentrations over the planet. Tropospheric ozone currently contributes about 12% of the radiative forcing caused by anthropogenic GHG.

(f) *Water vapor (humidity).* Water vapor is a strong infrared absorbing gas at wavelengths between 5 and 7 $\mu$m and above 10 $\mu$m. In fact, water vapor is the strongest natural GHG.

---

[11]Myhre, G., et al. *Geophysics Research Letters*, 25 (1998): 2715–18, and private communication.

[12]Hough, A. M., and R. G. Derwent. *Nature*, 344 (1990): 645–8.

By the laws of thermodynamics, as the sea temperature increases (as well as land temperature), there will be more evaporation.[13] The total column water vapor content over the oceans increased by 1.2 ($\pm$0.3)% per decade over the period 1988–2004. The IPCC 2007 Assessment Report does not assign a specific value for radiative forcing caused by possible increasing tropospheric water vapor concentrations in the atmosphere because water vapor is not an anthropogenic GHG. The increase of its concentration in the atmosphere may be indirectly related to anthropogenic activities, that is, warming of the oceans, other surface waters, and evaporation from land surfaces. Therefore, the Working Group I of the IPCC report states that because of increased water vapor concentrations in the atmosphere, the increase of radiative forcing caused by increased water vapor concentrations may be twice as much as that caused by increased $CO_2$ concentrations.

(g) *Aerosols.* Manmade small particles suspended in air, called aerosols, can interfere with the incoming solar and outgoing terrestrial radiation. Manmade aerosols are mostly caused by fossil fuel combustion. They are composed of sulfates, nitrates, and carbonaceous matter. Their diameter ranges from submicrometers to a few micrometers. Especially, the smallest diameter aerosols are effective in scattering incoming solar radiation. Thus, the aerosol radiative forcing is thought to be negative, estimated at $-0.5$ ($\pm$0.4) $Wm^{-2}$.

(h) *Clouds.* Natural clouds have an enormous effect on global climate. Low-altitude cumulus-type clouds reflect solar radiation, increasing the earth's albedo. High-altitude cirrus-type clouds reflect the outgoing terrestrial IR radiation. In this context we have to consider the alteration of cloud patterns from manmade causes. Such alterations can include providing more condensation nuclei for cloud formation and changing cloud patterns. It is thought that manmade causes induce more cumulus-type clouds and hence a negative radiative forcing of $-0.7$ ($-0.3$ to $-1.8$) $Wm^{-2}$.

(i) *Ice albedo.* As the earth's surface warms because of increased GHG concentrations, the fringes of the Arctic and Antarctic ice caps may melt. Also, glaciers, which are already receding in this interglacial period, may recede even faster. Since ice has a higher albedo (reflects more sunlight) than water and land, the disappearance of ice will lead to a decrease in the earth's albedo (Equations 11.1 and 11.3). This will cause the earth's radiative temperature $T_E$, and concomitantly the surface temperature $T_S$, to increase slightly. However, the IPCC report does not include radiative forcing caused by changing ice cover.

(j) *Land use.* Changes in land surface (vegetation, soils, water) resulting from human activities can significantly affect local climate shifts in radiation, cloudiness, surface roughness, and surface temperatures. It is very difficult to assess the effect on land use changes on radiative forcing. The IPCC report assigns a negative value of radiative forcing caused by land use changes of $-0.2$ ($\pm$0.2) $Wm^{-2}$.

The individual contributions of GHG, aerosols, and clouds to radiative forcing are summarized in Table 11.1. The sum total of radiative forcing caused by anthropogenic GHG, excluding water vapor, is $+2.99$ ($\pm$0.2) $Wm^{-2}$. The estimated negative radiative forcings caused by cloud, aerosol, and land use effects are also listed in Table 11.1. The sum total of the negative radiative forcing is $-1.4$ ($\pm$0.54) $Wm^{-2}$. The sum total of radiative

---

[13]The Clausius–Clapeyron equation relates the rise of vapor pressure to temperature increase via the latent heat of evaporation: $d \ln p/dT \approx \triangle H_{vap}/RT^2$.

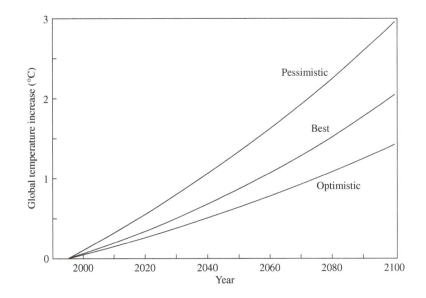

**FIGURE 11.5**  Projected trend of the earth's surface temperature increase. Upper curve: pessimistic scenario, based on "business-as-usual" fossil fuel consumption. Lower curve: optimistic scenario, based on slowing down of fossil fuel consumption. Middle curve: in between the worst and best scenarios. (Data from Houghton, J. T., G. J. Jenkins, and J. J. Ephraums, eds. *Climate Change, the IPCC Scientific Assessment.* Cambridge, UK: Cambridge University Press, 1995.)

forcing is positive, implying that a part of the outgoing terrestrial infrared radiation is absorbed in the troposphere. About one half of the absorbed radiation is reradiated to the earth's surface, warming the surface more than it would be in absence of GHG and other manmade causes.

## 11.2.6  Results of Global Warming Modeling

Based on radiative forcing models, the IPCC report projects the average earth's surface temperature increase in the 21st century as shown in Figure 11.5. This projection is based on estimated increases of $CO_2$ and other GHG emissions, as well on various other manmade perturbations. The middle, "best," estimate predicts a rise of the earth's surface temperature by the end of the 21st century of about 2°C; the "optimistic" estimate predicts about 1°C, and the "pessimistic" estimate predicts about 3°C. The optimistic estimate relies on slowing of $CO_2$ and other GHG emissions; and the pessimistic estimate on "business-as-usual" (i.e., on continuing rate of growth of $CO_2$ and other GHG emissions). The best estimate is somewhere in between.

## 11.2.7  Observed Trend of Global Warming

Since $CO_2$ concentrations have risen from about 280 ppmv from the start of the industrial era to about 385 ppmv to date, models show that to date there already should have been a global warming of about +0.5 to 1°C, depending on model and various assumptions. Has such warming actually occurred?

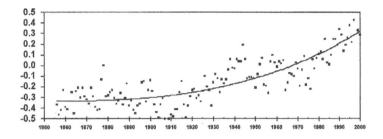

**FIGURE 11.6**  Observed average global surface temperature deviation, 1850–2000. The scale on the $y$-axis (°C) is indexed to the 1970 global average surface temperature. (Data points from Carbon Dioxide Information Analysis Center, Oak Ridge National Laboratory, 2000; best-fit curve by authors.)

Figure 11.6 shows the observed temperature trend from 1850 to 2005. Measured temperature data from hundreds of stations have been averaged in the northern hemisphere, southern hemisphere, and global average. As seen, there are wide fluctuations from year to year and over the hemispheres. Despite the fluctuations, there is an unmistakable upward trend. Averaging the fluctuations, the mean global temperature appears to have risen by about 0.5 to 1.0°C in accordance with predictions of radiative forcing models.

## 11.3  ASSOCIATED EFFECTS OF GLOBAL WARMING

As a consequence of increased GHG concentrations in the atmosphere and other manmade causes, the earth's surface temperature is likely to rise as discussed in previous sections. The surface temperature rise is likely to cause several ancillary effects on global climate and hydrogeology, which in turn will affect human habitat, welfare, and ecology.

### 11.3.1  Sea Level Rise

The sea level varied greatly over past geological periods. Before the onset of the last glacial period, about 120,000 years ago, the global average temperature was about 1°C higher than today. The average sea level was about 5–6 m higher than today. At the end of the last glacial period, about 18,000 years ago, the summer air temperatures were lower by 8–15°C over most of North America and Eurasia south of the ice sheets, and sea surface temperatures were about 2–2.5°C below present temperatures. The average sea level was over 100 m lower than at present. At that time the British Isles were joined with the European mainland, and the polar ice sheets extended in Europe as far as southern England and Switzerland and in North America to the Great Lakes and southern New England.

Several factors may contribute to future sea level rises: (a) thermal expansion of the surface layer of the ocean; (b) melting of glaciers and ice caps; (c) melting of the Greenland ice sheet; and (d) melting of the Antarctic ice sheet. Tide gauge observations over the period 1961–2003 indicated a sea level rise of 1.8 (±0.5) mm y$^{-1}$. Over the decade 1993–2003 satellite observations indicated a sea level rise of 3.1 (±0.7) mm y$^{-1}$. If that trend continues, by the end of the century, we can expect a sea level rise of 300 (±68) mm. This rise alone may inundate very low altitude coastal areas, such as southern Florida and the Florida Keys, the Netherlands, Bangladesh, and some Pacific atolls. However, considering the tidal rise and storm surges associated with sea level rises, other higher altitude coastal areas may be endangered.

### 11.3.2  Water Vapor and Precipitation Changes

The evaporation–precipitation cycle must be constant (what goes up must come down); therefore, global warming will result in increased evaporation and increased precipitation. However, it is extremely difficult to predict the changing patterns of precipitation. Regional precipitation patterns depend on temperature and pressure gradients in the atmosphere (which cause winds and storms) and orographic effects (mountains, valleys, vegetation cover, surface roughness). Some regions may become more arid and others more wet. Some areas may receive more snow and others more rain. It is likely that since about 1950 the number of heavy precipitation events in many land regions increased.

### 11.3.3  Hurricanes and Typhoons

Hurricanes and typhoons, also called tropical cyclones, spawn in waters that are warmer than 27°C, in a band from 5° to 20° north and south latitude. As the surface waters become warmer and the latitude bands expand, it is very likely that the frequency and intensity of tropical storms will increase. There is observational evidence for increased tropical cyclone activity in the North Atlantic since about 1970, correlated with increases of tropical sea surface temperatures. Estimates of the potential destructiveness of tropical cyclones (i.e., their intensity) suggest a substantial upward trend since the mid-1970s. The number of hurricanes in the North Atlantic has been above normal in the 9 years from 1995 to 2005.

### 11.3.4  Climate Changes

Predicting global and regional climatic changes as a consequence of average surface temperature rise is extremely difficult and fraught with uncertainties. It is expected that regional temperatures, prevailing winds, storm, and precipitation patterns will change, but where and when and to what extent changes will occur is a subject of intensive investigation and modeling on the largest available computers, the so-called supercomputers. Climate is influenced not only by surface temperature changes, but also by biological and hydrological processes and by the response of ocean circulation, which are all coupled to temperature changes.

It is expected that temperate climates will extend to higher latitudes, probably enabling the cultivation of grain crops further toward the north than at present. But crops need water. Precipitation patterns may alter, and the amount of rainfall in any episode may be larger than it is now. Consequently, the runoff (and soil erosion) may be enhanced, and areas of flooded watersheds may increase.

The well-known ocean currents, such as the Gulf Stream, Equatorial, Labrador, Humboldt, and Kuroshoi Currents, are driven by surface winds and density differences in the water. Density differences may arise because of melting ice caps. When ice melts, the surrounding water becomes less salty and less dense, replacing the saltier, denser water. The dense water sinks to a greater depth, setting up a thermohaline current that moves at the bottom of the ocean from polar regions equatorward. Near the equator, less dense water is upwelling and then moves poleward, completing the loop. The north–south and south–north currents are deflected by the Coriolis force, which is caused by the earth's rotation from west to east. The thermohaline currents have enormous influence on climate. For example, the relatively mild climate of Western Europe is attributed in part to the Gulf Stream. Another example is the El Niño event, also called El Niño Southern Oscillation. A warm current of nutrient poor tropical water replaces the cold nutrient-rich Humboldt Current, which supports great populations of fish. The El Niño creates ocean upwelling off the coasts of Peru and Ecuador, bringing nutrient rich waters to the coasts. The peak of El Niño typically occurs around Christmas, hence the name El Niño. The benefit of El Niño is a richer anchovy harvest. The downside is wide-reaching climatic effects, such as cool, wet summers in Europe, increased frequency of hurricanes and typhoons in the western Pacific, South China Sea, southern Atlantic, and Caribbean Sea, and heavy storms and precipitation battering the eastern Pacific coast from Mexico to British Columbia. The 1997–1998 El Niño caused winter storms along the Pacific coast of North America with heavy precipitation, mud slides, and resulting property damage.

Thus, the melting of the ice caps, in addition to changing the earth's albedo, may have far-reaching climatic consequences because of changes in the pattern of thermohaline currents.

## 11.4   GREENHOUSE GAS EMISSIONS

The increase of GHG concentrations in the atmosphere is a consequence of emissions of these gases from anthropogenic sources at a higher rate than their removal from the atmosphere. The most significant of these gases is $CO_2$, but $CH_4$, CFCs, and $N_2O$ emissions need also to be considered, as well as changes in tropospheric $O_3$ concentrations.

### 11.4.1   Carbon Dioxide Emissions and the Carbon Cycle

Because carbon constitutes a major mass fraction of all living matter, there is an enormous reservoir on earth of carbon in the living biosphere and its fossilized remnant. Sedimentary limestone, $CaCO_3$, contains about 12% of its mass as carbon. This limestone originates in part from shells and skeletons of past living creatures and in part from precipitation of supersaturated aqueous solution of $CaCO_3$.

There is a continuous exchange of $CO_2$ between the biosphere and the atmosphere. Carbon dioxide is absorbed from the atmosphere during photosynthesis of land vegetation and phytoplankton living in oceans and other surface waters. Carbon dioxide is returned to the atmosphere during respiration of animals and decomposition (slow combustion of carbonaceous matter) of dead plant matter and animals.

Figure 11.7 shows the rates of carbon exchange between the biosphere and atmosphere, ocean and atmosphere, and emissions from fossil fuel combustion and forest burning ($GtCy^{-1}$). Figure 11.7 also shows the carbon reservoirs residing in biota and soil litter on land (2000 GtC), dissolved in the ocean (40,000 GtC), in fossil fuels (5,000–10,000 GtC), and in the atmosphere (840 GtC).[14] The respiration and decomposition of land organisms emit about 60 $Gty^{-1}$ of carbon into the atmosphere, while photosynthesis absorbs about 62 $Gty^{-1}$. Thus, there is a net absorption of $CO_2$ from the atmosphere by biota. The oceans and other surface waters absorb 92 $Gty^{-1}$ of carbon by dissolution of $CO_2$ and by photosynthesis of phytoplankton. The oceans return about 90 $Gty^{-1}$ into the atmosphere by respiration and outgassing. Thus, the oceans are also net absorbers of

---

[14]Carbon and carbon dioxide will be used interchangeably in subsequent discussions. To convert from carbon to $CO_2$, multiply by $44/12 = 3.67$.

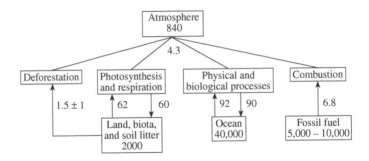

**FIGURE 11.7**    Rates of carbon exchange between biosphere and atmosphere and between ocean and atmosphere. Also shown are emissions from fossil fuel combustion and forest burning (Gt y$^{-1}$). Also listed are carbon reservoirs in soil, ocean, fossil fuel, and atmosphere (Gt).

$CO_2$. Currently, about 6.8 Gty$^{-1}$ of carbon (25 Gty$^{-1}$ $CO_2$) is emitted into the atmosphere by fossil fuel combustion. Another $1.5 \pm 1$ Gty$^{-1}$ is emitted because of deforestation and land use changes, mainly artificial burning of rain forests in the tropics, and logging of mature trees, which disrupts photosynthesis. Thus, the atmospheric carbon content should be increasing by about 4.3 Gty$^{-1}$. However, from atmospheric $CO_2$ concentration measurements (see Figure 11.9), the atmospheric carbon content is growing only by 3.6 Gty$^{-1}$. Most likely, the biosphere and oceans absorb more $CO_2$ than is indicated in Figure 11.7.

Figure 11.8 shows the growth of $CO_2$ concentrations in the global atmosphere since the year 1000. At the start of the industrial era in the middle of the 19th century, the concentration was about 280 ppmv. Figure 11.9 shows the growth of $CO_2$ concentrations since 1958, when Charles Keeling set up an accurate measurement device on top of Mauna Loa, Hawaii, using $CO_2$ infrared absorption. This instrument is still in operation, and the measurements thereof provide an accurate record of global average $CO_2$ concentrations ever since. The zigzag pattern of concentrations is attributed to seasonal variations in $CO_2$ concentrations. In the Northern Hemisphere, where most of the land masses are located, during the summer months, there is an uptake of $CO_2$ because of photosynthesis, resulting in a dip in $CO_2$ concentrations. During the winter months, there is a release of $CO_2$ because of respiration and decay of foliage. Also, in the winter months, there is increased combustion of fossil fuels. Currently, the concentration level reached about 385 ppmv. Drawing a smoothed exponential curve through the zigzag data points, we can estimate the rate of growth over the past decades to be about 0.4% per year. If that rate were to continue into the future, a doubling of the current $CO_2$ concentration would occur in about 175 years. However, if no measures are taken to reduce $CO_2$ emissions, because of the population increase and the concomitant enhancement of fossil fuel use, the rate of growth of $CO_2$ concentration will increase more than 0.4% per year, and the doubling time will be achieved sooner.

We have seen in Section 11.2.5 that among the anthropogenic GHG, the contribution of $CO_2$ to radiative forcing is about 55.5%.

## 11.4.2    Methane

Methane is emitted by natural processes, such as the anaerobic decomposition of organic matter in swamps and marshes and enteric fermentation of fodder in guts of animals. Anthropogenic emissions of $CH_4$ come from leakage of oil and gas wells, gas pipelines, and storage and transportation

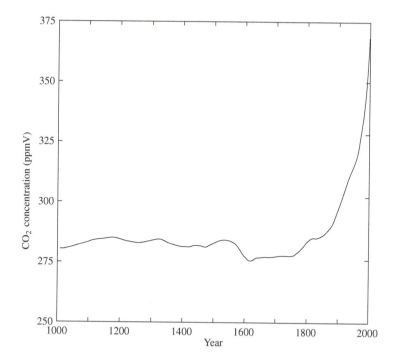

**FIGURE 11.8**    Trend of $CO_2$ concentrations in the atmosphere 1000–2000 AD (Data from Carbon Dioxide Information Analysis Center. *Trends Online: A Compendium of Data on Global Change*. Oak Ridge, CA: Oak Ridge National Laboratory, 2000.)

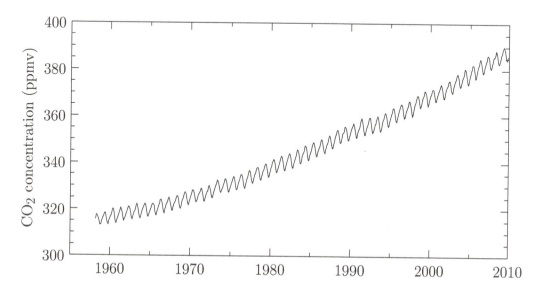

**FIGURE 11.9**    $CO_2$ concentration trend, 1958–2010. (Carbon Dioxide Information Analysis Center. Oak Ridge, CA: Oak Ridge National Laboratory, 2010.)

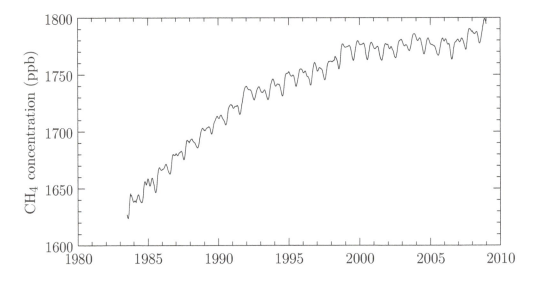

**FIGURE 11.10** Globally averaged $CH_4$ concentration trend, 1980–2009. (Carbon Dioxide Information Analysis Center. Oak Ridge, CA: Oak Ridge National Laboratory, 2010.)

vessels. Accumulated $CH_4$ in coal seams is mostly vented into the atmosphere when mine shafts are dug into the seams. In the 1970–1980s, $CH_4$ concentrations grew at a rate of 1% $y^{-1}$. Recently the growth rate has leveled off, as seen in Figure 11.10. Present $CH_4$ concentrations hover between 1.7 and 1.8 ppmv. Considering that $CH_4$ is a stronger absorber of IR radiation per weight than $CO_2$ (see Section 11.2.4), its contribution to radiative forcing of all GHG is 16%.

### 11.4.3 Nitrous Oxide

Nitrous oxide ($N_2O$) is to be distinguished from the air pollutants NO and $NO_2$ emitted during fossil fuel combustion.[15] $N_2O$ is emitted naturally during bacterial nitrogen fixation in soils and by lightning discharge. Minor quantities are emitted from fossil fuel combustion and in some chemical manufacturing processes (e.g., nitric acid, chemical fertilizer, and nylon production). Its current concentration in the atmosphere is about 0.3 ppmv and it is growing by about 0.25% per year. It has a relatively long lifetime in the atmosphere, about 120 years. Its contribution to radiative forcing of all GHG is 5.5%.

### 11.4.4 Chlorofluorocarbons

Chlorofluorocarbons (CFCs) are entirely manmade products, as they are produced in chemical factories for use as refrigerants, propellants in spray cans, foam-blowing agents, solvents, and other

---

[15]Nitrous oxide, called laughing gas, is used as an anesthetic in dental surgery and as a foaming agent in whipped cream.

uses. By the Montreal Convention of 1987 and its amendments, the manufacturing of CFCs is now banned. However, because of their slow venting from existing appliances and foam insulation materials, coupled with their long lifetime in the atmosphere (hundreds of years), they will contribute to global warming for a long time after they ceased to be produced. The fully halogenated CFCs have been replaced by hydrochlorofluorocarbons (HCFCs). The latter are also strong absorbers of the outgoing terrestrial infrared radiation. The contribution of the present concentrations of the combined halocarbons in the atmosphere to radiative forcing is 11%.

### 11.4.5   Ozone

In Chapter 10 we noted that there are two layers of $O_3$ in the atmosphere: one in the stratosphere (the "good" ozone) and the other in the troposphere (the "bad" ozone). Stratospheric $O_3$ is produced naturally from molecular oxygen under the influence of solar UV radiation. Tropospheric $O_3$ is produced from anthropogenic precursors: nitric oxides ($NO_x$) and volatile organic compounds (VOCs), under the influence of solar radiation. About 10% of the total column content of ozone resides in the troposphere. Since the upper troposphere and lower stratosphere are colder than the earth's surface, ozone molecules residing in those layers absorb part of the outgoing far infrared terrestrial radiation and then reradiate back to the surface, thus adding to global warming.

Ozone column amounts vary greatly as a function of latitude and altitude, seasons, and from urbanized–industrialized parts of the continents to remote areas. It has been estimated that the global mean concentration of tropospheric ozone is 20–30 ppbv. The estimated contribution of tropospheric ozone to radiative forcing of all GHG is 12%.

## 11.5   CONCLUSION

Global warming is caused by increasing concentrations of far infrared absorbing gases, called greenhouse gases (GHG), in the lower atmosphere, including $H_2O$, $CO_2$, $CH_4$, CFCs, $N_2O$, and $O_3$. Most of these gases are of anthropogenic origin, primarily associated with fossil fuel usage. Water vapor is also a GHG, but of course, it is of natural origin, although the global hydrogeologic cycle, and hence the instantaneous water vapor concentration in the troposphere, may be altered to some degree by anthropogenic activities. Methane is also in part of natural origin, such as emissions from swamps and bogs and from enteric fermentation of animals.

Anthropogenic aerosols may also alter the earth's surface temperature because of their scattering of incoming solar radiation or outgoing terrestrial radiation. Furthermore, aerosols may provide condensation nuclei for cloud formation. Possible cloud pattern changes, in turn, may have a profound effect on climate changes.

There is substantial evidence that the earth's average surface temperature already has risen by about 0.5–1°C since the middle of the 19th century. This temperature rise is ascribed to the historic increase of GHG concentrations from rapid industrialization and urbanization over the past century. By the end of the 21st century, the surface temperature may rise by another 2–3°C. This temperature rise may cause associated climatic effects, such as change in precipitation patterns, more frequent and intensive tropical storms, glacier and ice-sheet melting, ocean current variability, and sea level rises.

# PROBLEMS

### Problem 11.1

Calculate the earth's radiative temperature (K) for albedos $\alpha = 0.27, 0.3$, and $0.33$, assuming the solar constant is not changing.

### Problem 11.2

Calculate the radiative temperature (K) for the planets Mars, Earth, and Venus, given their solar constants $S_0 = 589, 1{,}367$, and $2{,}613$ $Wm^{-2}$, respectively, and their albedos $\alpha = 0.15, 0.3$, and $0.75$, respectively.

### Problem 11.3

Given the global surface temperature fluctuations shown in Figure 11.6, use a statistical program to draw a best-fit curve through the data. According to your best-fit curve, by how much did the surface temperature increase from 1850 to 2010?

### Problem 11.4

The present volume fraction of $CO_2$ in the atmosphere is 385 ppmv. What is the carbon content (Gt) of the atmosphere if $CO_2$ is the only carrier of carbon? The radius of the earth is 6,371 km, and the atmospheric mass per unit surface area is $1.033E(4)$ $kg$ $m^{-2}$. Hint: you do not need the height of the atmosphere.

### Problem 11.5

Given the $CO_2$ volume fraction (ppmv) in the atmosphere as shown in Figure 11.9, calculate the rate of increase (%/y) of the volume fraction in the years 1958 to 2008. Use exponential, not linear growth.

### Problem 11.6

The present volume fractions of $CO_2$ and $CH_4$ are 385 and 1.7 ppmv, respectively. The former grows by 0.4%/y, and the latter by 0.6%/y. What will be the volume fractions of these gases in 2100? Use exponential, not linear growth.

### Problem 11.7

Using Equation (11.4), calculate the radiative forcing increase $(\Delta F_R, W\,m^{-2})$ caused by the increase of $CO_2$ volume fraction by 10, 25, 50, 75, and 100% (the latter is a doubling of the volume fraction). Plot the result.

# BIBLIOGRAPHY

Brasseur, G. P., J. J. Orlando, and G. S. Tyndall, eds. *Atmospheric Chemistry and Global Change*. Oxford: Oxford University Press, 1999.

Houghton, J. T. *Global Warming*. Cambridge, UK: Cambridge University Press, 1997.

Intergovernmental Panel on Climate Change. *Radiative Forcing of Climate Change*. Cambridge, UK: Cambridge University Press, 1995.

Intergovernmental Panel on Climate Change. *The Physical Science Basis*. Cambridge, UK: Cambridge University Press, 2007.

Mackenzie, F. T. *Our Changing Planet, An Introduction to Earth System Science and Global Environmental Change*. Upper Saddle River, NJ: Prentice Hall, 1997.

Mitchell, J. F. B. *Review of Geophysics*, 27, (1989): 115–39.

Ramanathan, V., and J. A. Coakley. *Reviews of Geophysics and Space Physics*, 16 (1978): 465–90.

Schneider, S. H. *Science*, 243 (1989): 771–81.

# Mitigating Global Warming

## 12.1  INTRODUCTION

In Chapter 11 we concluded that present contributions to radiative forcing of global warming by anthropogenic GHG are caused by carbon dioxide (55.5%), methane (16%), nitrous oxide (5.5%), halocarbons (11%), and ozone (12%). (See Table 11.1.) These estimates are based on the radiative forcing that these gases exert on the energy balance of incoming solar radiation and outgoing far IR from the earth's surface. Other GHG also contribute to radiative forcing, notably water vapor and tropospheric ozone. Radiative forcing is a major cause of global warming, but not the sole cause. Clouds, aerosols (natural and anthropogenic), the variable earth albedo, and solar constant all may contribute to global warming or cooling. In the future, the absolute and relative contributions to radiative forcing by anthropogenic GHG may change, depending on their increasing or decreasing atmospheric concentrations.

In this chapter we describe the various technologies that could be applied to reduce emissions of anthropogenic GHG. We start with the minor contributors to radiative forcing: halocarbons, $N_2O$, and $CH_4$, and then tackle the more difficult problem of controlling anthropogenic emissions of $CO_2$, which is a consequence of fossil fuel combustion.

## 12.2  CONTROLLING HALOCARBON EMISSIONS

Halocarbons include the fully substituted carbon compounds, CFCs, such as CFC-11 and CFC-12, HCFCs, such as HCFC-22, perfluorocarbons (PFCs) and halons, such as methyl bromide, chloroform, and various brominated fire retardants. However, the contribution to radiative forcing of the PFCs and halons is relatively minor in comparison to that of the CFCs and HCFCs. Therefore, we shall address here only the controlling of CFCs and HCFCs.

Currently, the summed atmospheric concentration of several species of CFCs and HCFCs is about 1 ppbv. The rate of growth of CFCs is negative, because of their production phase-out. CFCs are carbon compounds in which the hydrogen atoms are fully substituted with fluorine and chlorine atoms. For most of these compounds, human activities are the sole source. They are used as refrigerants, solvents, degreasing agents, propellants in spray cans, and for foam blowing. They became implicated in the destruction of the stratospheric ozone layer (the ozone hole) with the potential for enhanced ultraviolet radiation reaching the earth's surface and the commensurate increase of incidences of skin cancer. It is for the risk of depleting the stratospheric ozone layer that, by international agreements (the Montreal Protocol of 1987 and its amendment of 1992) the production of CFCs has been phased out by the year 2000. Unfortunately, CFCs are very long-lived species, and they will persist in the atmosphere for decades and even centuries. Therefore, their contribution to radiative forcing will be significant for a long time to come. Meanwhile, their emissions into the atmosphere could be further restricted by disposing of old appliances carefully, so that the CFC content from compressors and insulating foams do not escape into the atmosphere.

The fully substituted CFCs have been replaced by partially substituted carbon compounds, called hydrochlorofluorocarbons, HCFCs. These compounds are not as long-lived in the atmosphere as the fully substituted CFCs because they can react with atmospheric oxidant species, such as hydroxyl radicals (OH), hydrogen peroxide ($H_2O_2$), and ozone ($O_3$). Nevertheless, while in the troposphere, HCFCs are strong far IR absorbents and thus can contribute to radiative forcing. However, the Montreal Protocol Amendment of 1992 mandates the phase-out of production of HCFCs by the year 2030. If that happens, and in view of their shorter lifetime in the atmosphere, these compounds eventually will not significantly contribute to radiative forcing.

## 12.3 CONTROLLING NITROUS OXIDE EMISSIONS

Currently, the atmospheric concentration of nitrous oxide ($N_2O$) is about 0.3 ppmv, with a growth rate of about 0.25% $y^{-1}$. Nitrous oxide, also called laughing gas (because it was used by dentists as a local anesthetics, which made patients laugh), is a relatively rare form of nitrogen oxides. The anthropogenic sources are from fossil fuel combustion, adipic acid manufacturing (used for nylon production), a by-product of nitrogen fertilizer manufacturing, and as an agent for whipping cream. Natural sources are lightning and photochemical reactions in the atmosphere and nitrogen-fixing bacteria. The installation of $NO_x$ control technologies on fossil-fueled power plants, such as selective catalytic reduction (SCR), may abate the emission of $N_2O$ as well. Abatement technologies for the reduction of $N_2O$ emissions from adipic acid manufacturing exist and are being used, consisting of thermal and catalytic destruction of the gas. Furthermore, it may be advisable to restrict the use of $N_2O$ for frivolous purposes, like whipping cream.

## 12.4 CONTROLLING METHANE EMISSIONS

The current concentration of methane ($CH_4$) in the atmosphere is about 1.7 ppmv and is growing at a rate of about 0.6 % $y^{-1}$.[1] Biogenic, anthropogenic, and anthropogenically related sources emit about 535 Mt $y^{-1}$ of $CH_4$. Methane is removed from the atmosphere by photochemical reactions with atmospheric oxidants, including hydroxyl radical (OH), hydrogen peroxide ($H_2O_2$), ozone

---

[1] Intergovernmental Panel on Climate Change. *Climate Change 2007: The Physical Science Basis*. Cambridge, UK: Cambridge University Press, 2007. Most subsequent data are taken from this report and will not be referenced separately.

($O_3$), and others. The ocean and soils absorb some of the $CH_4$, but the positive growth rate indicates that the emissions from the sources exceed the absorption capacity of the ocean and soils. The major controllable anthropogenic sources of $CH_4$ are coal mines, petroleum and natural gas operations, municipal landfills, and sewage treatment plants.

### 12.4.1   Controlling Methane Generated by Coal Mining

Worldwide emissions from coal mines amount to approximately 25–30 Mt $y^{-1}$. Methane is stored in the pores and in the cracks and fractures of the coalbed, as well as adsorbed on the surface of the coal granules. Methane and coal dust in coal mines cause explosions. Another gas that is stored and adsorbed on coal is carbon monoxide (CO). This gas causes asphyxiation (the proverbial cause of death of the coal mine canary). Both gases need to be thoroughly ventilated from the mines to prevent explosions and asphyxiation. The ventilation air contains too low a concentration of $CH_4$ for economic exploitation. Therefore, it is mostly vented to the atmosphere, although in principle, $CH_4$ from the ventilation gas could be captured by adsorption on suitable adsorbents, such as carbon black or pulverized coal. The adsorbed $CH_4$ could be desorbed by vacuum suction, called pressure swing adsorption/desorption, but that would be expensive.

High concentrations of $CH_4$ (greater than 90% by volume) in coalbed gases can be obtained in premining degasification. For this purpose, in-mine bore holes and vertical wells are installed to drain the $CH_4$ from the coalbeds. The $CH_4$ is compressed and injected into pipelines to be used for generating electricity or other uses. In 1996 in the United States, the CONSOL coal company reported a system-wide capture of 626,000 t $y^{-1}$ of $CH_4$. In several abandoned coal mines methane is displaced by $CO_2$ injection for the purpose of enhanced $CH_4$ exploitation and $CO_2$ sequestration (see Section 12.7).

### 12.4.2   Controlling Methane from Petroleum and Natural Gas Operations

Current methane emissions from the oil and gas industry are estimated between 50 and 75 Mt $y^{-1}$. Most of the associated gas at oil wells is flared (contributing to $CO_2$ and soot emissions), but some escapes into the atmosphere. At gas wells, after separation from undesirable gases ($CO_2$, $H_2S$, $N_2$), methane and some low-molecular-weight hydrocarbon gases are piped to the consumers or liquefied. Leakage of $CH_4$ may occur from gas compressors, pipelines, tanks, valves, and flanges. It is technically feasible (but not always practiced) to reduce $CH_4$ emissions from these sources by applying better sealants or replacing older, leaking equipment.

### 12.4.3   Controlling Landfill Methane

Another major source of anthropogenic methane is from municipal landfills and from sewage treatment plants. Worldwide these emissions are estimated at 30–35 Mt $y^{-1}$ and may grow to 60–65 Mt $y^{-1}$ by 2025. In open landfills, $CH_4$ escapes into the air. In secure, covered landfills, $CH_4$ is formed by anaerobic decomposition of cellulose, food, animal and garden waste, paper, and cardboard. Some of the $CH_4$ is flared (where it contributes to $CO_2$ emissions). Instead of flaring, the $CH_4$ could be economically captured and utilized for electricity generation and home heating or cooking purposes. As of 2006, approximately 395 landfill gas (LFG) energy projects were operational in the United States. These projects generate approximately 9 billion kilowatt hours of electricity per year and deliver just over 5.6 E(6) $m^3$ $d^{-1}$ of LFG for heating purposes.

The U.S. EPA estimates that approximately 600 additional landfills present attractive opportunities for landfill $CH_4$ utilization in the United States.

## 12.5  CONTROLLING CARBON DIOXIDE EMISSIONS

Currently, the atmospheric concentration of carbon dioxide ($CO_2$) is approaching 385 ppmv, and its rate of growth is approximately 0.4 % $y^{-1}$. Its contribution to radiative forcing is about 55.5% of all anthropogenic GHG. Anthropogenic $CO_2$ emissions amount to approximately 25 E(9) t $y^{-1}$ of $CO_2$.

In this section we discuss the possible technological practices for controlling emissions of $CO_2$ from fossil fuel combustion. By technological practices we mean reducing emissions at the source of combustion, not avoiding emissions by other means, such as conservation and efficiency improvements or substitution of renewable or nuclear energy for fossil fuels. The latter practices are discussed throughout this book (e.g., conservation and efficiency improvements in Chapter 2, nuclear power plants in Chapter 7, renewable (including biofuel) energy in Chapter 8, and automotive transportation in Chapter 9).

Technological practices will be effective and economic only at large stationary $CO_2$ emission sources. These sources contribute about 45% to worldwide $CO_2$ emissions (11.3 E(9) t $y^{-1}$). The electricity-generating plants contribute 78% of the total emissions from large stationary emission sources (Table 12.1). This is not to say that dispersed sources, such as transportation vehicles, commercial and residential boilers, and furnaces, do not contribute significantly to $CO_2$ emissions (transportation vehicles alone contribute roughly one third of anthropogenic $CO_2$ emissions), but for dispersed sources "end of the tailpipe," so to speak, emission controls are not deemed practical or economic.

In view of the predominant contribution of fossil-fueled electric power plants to $CO_2$ emissions from large stationary sources, the following sections are devoted to capturing and sequestering $CO_2$ from these plants.[2] However, industrial boilers, furnaces, and kilns employed in other large stationary sources may also be amenable to similar $CO_2$ emission controls, as are power plants.

**TABLE 12.1**  Large Stationary $CO_2$ Emission Sources

|  | Percentage of total |
|---|---|
| Fossil-fueled electricity-generating plants | 78 |
| Cement production | 7 |
| Petroleum refineries | 6 |
| Iron and steel industries | 5 |
| Petrochemical industries | 3 |
| Other industries | 1 |

Data from Intergovernmental Panel on Climate Change. *Special Report on CO2 Capture and Storage.* Cambridge, UK: Cambridge University Press, 2007.

---

[2]Intergovernmental Panel on Climate Change. *Special Report on CO2 Capture and Storage.* Cambridge, UK: Cambridge University Press, 2007.

## 12.5.1  Controlling $CO_2$ Emissions from Fossil-Fueled Electric Power Plants

In the United States, 36% of all $CO_2$ emissions come from electricity-generating plants, and a similar percentage is applicable worldwide (see Chapter 2). The electricity industry has many options to reduce $CO_2$ emissions while maintaining, and even increasing, electricity supply. They include the following:

- Shift from coal or oil to natural gas fuel
- Replacement of single steam cycle coal-fired power plants with natural gas-fired combined cycle plants
- Capture of $CO_2$ from flue gas by chemical absorption
- Oxyfuel combustion with $CO_2$ capture
- Integrated coal gasification combined cycle plants with $CO_2$ capture

### 12.5.1.1  Shift from Coal or Oil to Natural Gas Fuel

The simplest way of reducing $CO_2$ emissions from fossil-fueled power plants is to replace coal and oil with natural gas fuel. This is an option that can be retrofitted on existing coal- or oil-fired boilers. For equal heating value, natural gas emits less $CO_2$ than coal or oil. Taking the heating value of bituminous coal of 30 MJ $kg^{-1}$, residual oil 42.5 MJ $kg^{-1}$, and natural gas 50 MJ $kg^{-1}$ (Table 4.1), coal emits 1.67 times as much $CO_2$ as gas and oil 1.2 times as much as gas per heating value.

The substitution of gas for coal (usually pulverized coal) and oil requires minor modifications to the existing boilers. The modifications include addition of gas burners, proper tilting of the nozzles, and windbox changes. However, there will be a 3–5% boiler efficiency drop, expressed as the heat content of the steam generated per fuel heat input. Commensurately, the maximum continuous rating of the power plant will decrease by those percentages.[3]

### 12.5.1.2  Natural Gas-Fired Combined Cycle Plants

Replacing single (Rankine) steam cycle coal-fired power plants with natural gas-fired combined cycle plants (NGCC; see Section 6.3.2) would reduce $CO_2$ emissions to a greater extent than simply substituting gas for coal in existing boilers. As shown in the previous section, per unit heating value coal emits 1.67 times as much $CO_2$ as gas. The thermal efficiency of NGCC plants is in the range 45–50% compared with the range of modern single steam cycle pulverized coal-fired plants of 35–40%. (All thermal efficiencies are reckoned per lower heating value, LHV, see Chapter 4.) Therefore, the replacement of NGCC for a single cycle coal-fired power plant would emit approximately 50% less $CO_2$ per kWh electricity produced.

### 12.5.1.3  Capturing $CO_2$ from the Flue Gas by Chemical Absorption

Carbon dioxide has been captured from the flue gas of fossil-fueled power plants for a long time. It is the major source of commercial liquefied and solid (dry ice) $CO_2$ used for carbonated beverages, foam blowing, raw material for some chemical products (e.g., urea), and enhanced oil recovery. The

---

[3]Golomb, D., et al. MIT/E-Lab Technical Report, MIT-EL 86-009, 1986.

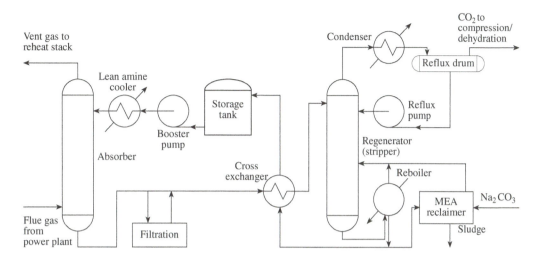

**FIGURE 12.1**    Flowsheet of the monoethanolamine process for $CO_2$ capture.[5]

time-honed capture technology is based on chemical absorption.[4] The most widely used absorbent for chemical absorption is a solution of monoethanol-amine (MEA) in water. The absorption reaction is

$$C_2H_4OHNH_2 + H_2O + CO_2(g) \Leftrightarrow C_2H_4OHNH_3^+ + HCO_3^- \qquad (12.1)$$

The reaction is reversible: the forward reaction proceeds at low temperatures and high $CO_2$ pressures, and the reverse reaction proceeds at elevated temperatures and low $CO_2$ pressures. The MEA solution absorbs $CO_2$, but not the rest of the flue gas, consisting mainly of $N_2$. The aqueous solution contains typically 30% by weight MEA, molecular weight 61. Assuming that each molecule of MEA absorbs one molecule of $CO_2$, molecular weight 44, for every ton of $CO_2$ we need 1.39 tons of MEA to pass through the absorption system and 3.23 tons of the aqueous solution. A 500-MW coal-fired power plant emits about 125 kg s$^{-1}$ of $CO_2$. Such a plant would require a flow rate of about 400 kg s$^{-1}$ of the aqueous solution. This gives an idea of the magnitude of materials handling for $CO_2$ capture from a medium-size (500-MW) coal-fired power plant.

A flow sheet of the MEA process is shown in Figure 12.1. The key process units are the absorption tower and the regeneration (stripper) tower. The absorption tower operates at 40–65 °C, which means that the incoming flue gas must be cooled before entering the tower. Some pressurization of the flue gas is necessary to overcome the pressure drop in the tower. The regeneration tower operates at 100–120°C, which requires siphoning steam from the power plant boiler. This is the major cause of efficiency loss for the power plant that uses this method. The reheater stage of the boiler (see Section 6.2.3) provides the heat for the regeneration tower. Assuming a heat capacity of the MEA solution of 0.9 cal g$^{-1}$ and a solution flow rate of 400 kg s$^{-1}$, the heating of the MEA solution from 40 to 100°C requires about 90 MW. Thus, 18% of the power of the 500-MW plant

---

[4]In chemical absorption the chemical composition of the absorbent is changed, as opposed to physical absorption where there is no such change.

will go to regenerate the MEA solution. We shall see later that the thermal efficiency of a coal-fired power plant with $CO_2$ capture by MEA solution degrades approximately by 30%. About two thirds of this degradation is caused by the heat requirement of the regeneration tower.

Coal and oil combustion produce copious quantities of $SO_2$, $NO_x$, and other contaminants that "poison" the solvent. When coal or oil is used as the source of $CO_2$, the flue gas must be thoroughly purified before entering the absorption tower. There are intensive research and development efforts worldwide for finding better chemical absorbents in terms of their stability, resistance to poisoning, and energy required for their regeneration.

In principle, the chemical absorption process could be retrofitted on a pulverized coal-fired power plant, provided that the plant is equipped with particle control, flue gas desulfurization, and flue gas denitrification. The thermal efficiency of a retrofitted plant would be even lower than that of a new plant where the absorption process is integrated in the plant design.

### 12.5.1.4  Oxyfuel Combustion with $CO_2$ Capture

When coal is burned in highly enriched or pure oxygen instead of air, the combustion gas consists mainly of $CO_2$ and $H_2O$. The water vapor can be easily condensed and the $CO_2$ captured. However, an oxyfuel combustion plant requires an air separation unit (ASU) in which oxygen is separated from the rest of air, that is, mainly nitrogen. When pulverized coal is burned in 20% excess air, the resulting flue gas contains about 17% by volume of $CO_2$. When burned in 95+% oxygen, the flue gas contains about 70% by volume of $CO_2$. However, the adiabatic flame temperature of oxyfuel combustion is much too high for ordinary boiler tube construction materials (e.g., steel) to tolerate. Therefore, the flame needs to be cooled. This can be accomplished by recycling a part of the flue gas into the burner. It is estimated that about 70–75% of the flue gas needs to be recycled into the burner to restore the flame temperature to tolerable values.[5] The $CO_2$ that is not recycled is condensed for eventual sequestration in deep geologic or ocean repositories.

Figure 12.2 presents a flow sheet for oxyfuel combustion of coal. Flue gas from the boiler first enters an electrostatic precipitator for fly ash removal. After exit from the ESP, the water vapor from the flue gas is condensed and discarded. The liquid oxygen from the ASU is evaporated and preheated with dry $CO_2$ from the flue gas and recycled into the burner. A $CO_2/O_2$ mass ratio of 2.42

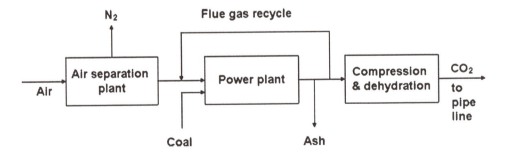

**FIGURE 12.2**  Air separation flue gas recycling power plant (schematic).

---

[5]Herzog, H., D. Golomb, and S. Zemba. *Environmental Progress*, 10 (1991): 64–74.

gives a similar flame temperature and heat transfer in the boiler as when air is used for combustion, where the $N_2/O_2$ mass ratio is about 3.65.

Air separation is a mature technology. Liquid and gaseous oxygen are used in many industrial processes (e.g., steel manufacturing, coal gasification, acetylene blow torches, as an oxidant in the space shuttle and other liquid fueled rockets, and for medical purposes).

The major thermal efficiency loss of an oxyfuel power plant is caused by the energy requirement of oxygen separation from air. On average, an ASU requires 250 kWh per ton of $O_2$. A 1000-MWel oxyfuel power plant may need upward of 10,000 t/d of oxygen. Few, if any, ASU exist today with that kind of capacity.

In principle, oxyfuel combustion can also be retrofitted to an existing air combustion plant, although it would require the addition of an ASU and other modifications to the plant, such as flue gas recycling to the burner and $CO_2$ condensation from the exhaust gas.

### 12.5.1.5  Integrated Coal Gasification Combined Cycle Plants with $CO_2$ Capture

The workings of an Integrated Coal Gasification Combined Cycle (IGCC) power plant were described in Section 6.3.2. An IGCC can be designed for carbon dioxide capture. Let us call such a plant IGCC$^2$. An IGCC$^2$ is proposed as one of the most promising ways to reduce significantly anthropogenic $CO_2$ emissions. First, IGCC$^2$ can use coal, which is a more abundant and cheaper fossil fuel than petroleum or natural gas. Second, it is predicted that IGCC$^2$ will achieve a thermal efficiency similar to a modern, single steam cycle pulverized coal-fired plant. Third, most, if not all, conventional pollutants (particulate matter, sulfur and nitrogen oxides, mercury, and other heavy metals) can be removed before combustion, vitiating expensive emission control systems. Last, but not least, an IGCC$^2$ emits very little or no $CO_2$. In a sense, an IGCC$^2$ is a zero-emission power plant. The captured $CO_2$ can be sequestered in deep geologic or ocean repositories (see below). In several countries prototypes of IGCC$^2$ are planned or already being constructed.

In the gasification process, synthetic gas (syngas) is produced from coal, consisting mainly of CO and $H_2$,

$$3C(s) + O_2 + H_2O(g) \rightarrow 3CO + H_2, \tag{12.2}$$

where $H_2O(g)$ designates steam. The hot, raw syngas is cold water quenched to about 300°C before entering a particle removal system, consisting of an electrostatic precipitator or fabric filter. After particle removal, the cooled syngas enters a reactor in which the water gas shift (WGS) reaction takes place (Figure 12.3).

$$CO + H_2O(g) \rightarrow CO_2 + H_2 \tag{12.3}$$

Reaction (12.2) proceeds at a temperature of 200 to 500°C in the presence of catalysts, such as $Fe_3O_4$ (magnetite), $Al_2O_3$, CuO, ZnO, or noble metals (palladium or platinum).

After the WGS step, all initial carbon in the coal is converted to $CO_2$, and most (but not all) of the initial heating value of the coal is transferred to hydrogen. The reason that not all the initial heating value of coal resides in hydrogen is that a part of the coal has to be burned to provide the work to run the air separation unit for supplying the oxygen in Reaction (12.2) and the steam for both Reactions (12.2) and (12.3). The heating value loss is dependent on the particular gasification and WGS processes. On the other hand, hydrogen can be used in more efficient thermodynamic

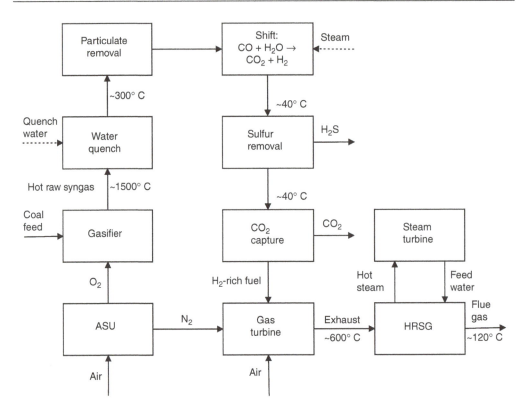

**FIGURE 12.3** Integrated coal gasification combined cycle power plant with $CO_2$ capture (IGCC$^2$).

cycles than coal itself. For example, when hydrogen is used in a combined cycle power plant, the thermal efficiency, reckoned as net electrical energy generated versus coal heat input, may be equal or larger in an IGCC plant than in a single steam cycle pulverized coal plant with pollution control. The WGS reaction is supposed to be one of the major suppliers of hydrogen in the so-called hydrogen economy in which hydrogen replaces natural gas or petroleum as a fuel (see Chapter 4). Other suppliers may include electrolysis of water, where the electricity is supplied by nuclear, solar, wind, or other renewable energies.

In an IGCC$^2$ plant the $CO_2$ produced in the water gas shift of Reaction (12.3) is separated from hydrogen by physical absorption or by membrane separation (see below). The captured $CO_2$ may contain acid gases: $H_2S$ and COS. When the captured gas mix is destined for deep underground sequestration, the presence of acid gases may not matter. Otherwise, the WGS gas must be cleaned up in a sulfur removal system, which is usually based on the Claus reaction (see Section 6.2.10.3). After the $CO_2$ is captured, the remaining $H_2$ can be used for running a gas turbine or hydrogen fuel cell to produce electricity or as a transportation fuel.

## 12.5.1.6 Capturing $CO_2$ after Gasification by Physical Absorption

After coal gasification and the water shift reaction, the resulting gas stream consists mainly of $CO_2$ and $H_2$. These gases can be separated by physical absorption.

Contrary to chemical absorption, in physical absorption the chemical composition of the solvent (absorbent) does not change when a gas (absorbate) is absorbed. Physical absorption is governed by Henry's law

$$c_A = K_A p_A, \tag{12.4}$$

where $c_A$ is the concentration of the absorbate (gas) in the absorbent, $p_A$ is the partial pressure of the absorbate (gas), and $K_A$ is Henry's constant for the particular a absorbent–absorbate system. For $c_A$ in $gL^{-1}$, $p_A$ in atm, $K_A$ is in units of mol $L^{-1}atm^{-1}$. Henry's constant is a function of temperature. Generally, Henry's constant decreases with temperature, hence, the concentration of the absorbate increases the colder the temperature of the absorbent. Henry's law applies only to dilute solutions. At high pressures of the absorbate, empirical relationships need to be devised for the particular absorbent–absorbate system.

The most often-used absorbents for $CO_2$ capture are based on methanol (e.g., Rectisol), dimethyl ethers of polyethylene glycol (e.g., Selexol), and $n$-methyl-2-pyrrolidone (e.g., Purisol). For example, Selexol absorbs 70–80 times as much $CO_2$ as $H_2$ and 50 times as much as $N_2$. However, Selexol may also absorb $H_2S$ and other sulfurous and nitrogenous gases. If $CO_2$ were used for food or other delicate products, these gases would need to be separated from $CO_2$ before absorption in Selexol. For geologic sequestration, the presence of these gases may not matter.

The gas mixture enters the bottom of an absorbing tower containing packing or trays. The cold, regenerated absorbent descends from the top of the tower, absorbing $CO_2$ in a counterflow fashion. The $CO_2$-rich absorbent leaves the bottom of the tower and flows into a regeneration tower. There the $CO_2$ is desorbed either by creating a vacuum over the absorbent (pressure swing absorption, PSA) or by reheating the absorbent (temperature swing absorption, TSA).

The Dakota Gasification Company has been gasifying lignite coal since 1984, producing syngas, methane, hydrogen, and other gases. In the process, $CO_2$ is captured using the Rectisol process, liquefied, and transported to the Weyburn oil field in Saskatchewan, Canada, where it is used for enhanced oil recovery (EOR).

## 12.5.1.7  Capturing $CO_2$ after Gasification by Membrane Separation

Another way of separating $CO_2$ from $H_2$ is by membrane separation. Membranes are porous materials allowing gaseous molecules to permeate through the pores. When a mixture of two gases with different molecular weight impinges on a porous membrane, the lighter molecule will diffuse faster through the pores than the heavier molecule. For a mix of $CO_2/H_2$, hydrogen will permeate faster through a porous membrane than carbon dioxide. Usually, more than one stage needs to be employed to increase the separation factor. For a $CO_2/H_2$ mix, ceramic, glass, metal, carbon, and polymer membranes can be used. Membranes can separate gases also on account of their chemical selectivity. A gas molecule is adsorbed on one side of the membrane, dissolves in the membrane, diffuses through the membrane, and desorbs on the other side. For example, polyimide membranes have a high selectivity for $CO_2$. On the other hand, palladium membranes have a high selectivity for $H_2$. The metal can adsorb large quantities of $H_2$ on its surface, followed by diffusion of the hydrogen through its pores. Palladium can be used in its metallic form or dispersed on ceramics.

## 12.6  THERMAL EFFICIENCY AND COST OF CONTROLLING $CO_2$ EMISSIONS FROM POWER PLANTS

In this section we compare the thermal efficiencies and the incremental cost of electricity production (COE) by introducing various technologies for the reduction of $CO_2$ emissions from power plants. The incremental COE is normalized to that of a conventional pulverized coal, single steam cycle power plant, the unit cost of which is set to 1. We do not consider here the incremental cost of fuel substitution (e.g., natural gas for coal) because the fuel price differential is highly variable with time and locale. The thermal efficiencies and normalized COE increments are listed in Table 12.2. Table 12.2 also lists the percentage of $CO_2$ emission avoidance compared with the emissions of an equivalent coal-fired plant. The $CO_2$ emission avoidance does not necessarily equal the amount of $CO_2$ captured, because more fuel may have to be burned to produce the same amount of electricity. Data are culled from various sources, mainly from the Intergovernmental Panel on Climate Change, *Special Report on CO2 Capture and Storage*, 2007.[2]

For simple natural gas substitution for coal in a single-cycle steam power plant, the thermal efficiency degradation is 3–5%; the incremental COE is negligible, not counting the incremental cost of the fuel. The $CO_2$ emission avoidance is about 35%. For a natural gas fired combined cycle (NGCC) plant, the thermal efficiency is in the range 45–50%. The capital cost of a new NGCC plant is actually less than that of a modern pulverized coal steam cycle plant when all the pollution control costs are included. The comparative cost of electricity production is largely dependent on the present coal/gas price differential. The $CO_2$ emission avoidance is about 50%. The thermal efficiency of a coal fired power plant with integrated $CO_2$ capture by MEA postcombustion chemical absorption is estimated on the order of 24–29%, the COE increment is around 50%, and the $CO_2$ emission avoidance is around 85%. An oxyfuel plant may degrade its thermal efficiency by 10–15 percentage points compared with an air combustion plant (e.g., from 40 to 30%), the COE increment is around 50%, and the $CO_2$ emission avoidance is about 95%. An IGCC[2] plant with precombustion $CO_2$ capture by physical solvent absorption will have a thermal efficiency of 35–40%, COE increment of 30–75%, and $CO_2$ emission avoidance around 90%. The same plant with membrane separation will have a thermal efficiency of 34–37%, COE increment of 85%, and $CO_2$ emission avoidance of 85%.

**TABLE 12.2**  Thermal Efficiency, Normalized Incremental Cost of Electricity Production, and $CO_2$ Emission Avoidance of Power Plants with $CO_2$ Control Technologies

|  | Efficiency (%) | Normalized COE increment (%) | $CO_2$ emission avoidance (%) |
|---|---|---|---|
| Pulverized coal (PC) with PM, $SO_x$, and $NO_x$ control, no $CO_2$ control | 35–40 | 0 | 0 |
| Coal substitution with natural gas, single cycle steam | 32–35 | $a$ | 35 |
| Natural gas combined cycle (NGCC) | 45–50 | $a$ | 50 |
| PC with MEA absorption | 24–29 | 50 | 85 |
| Oxyfuel combustion | 25–30 | 50 | 95 |
| IGCC with physical absorption | 35–40 | 30–75 | 90 |
| IGCC with membrane separation | 34–37 | 85 | 85 |

[a] Depends largely on the present coal/gas price differential.

Note that the estimates of thermal efficiencies, COE increments, and $CO_2$ emission avoidance of plants with $CO_2$ control technologies are tentative, because no large-scale commercial power plant with $CO_2$ emission abatement exists today to base the calculations upon. These estimates may change over time with experience and economy of scale.

## 12.7   CO$_2$ SEQUESTRATION

After being captured before or after combustion of the fossil fuel, $CO_2$ needs to be sequestered in a form that will not allow it to re-emerge into the atmosphere. This is called carbon capture and sequestration (CCS). The following repositories are considered, and in some cases already exploited, for $CO_2$ sequestration:

- Oil and gas reservoirs
- Coal seams
- Deep sedimentary basins
- Deep ocean

### 12.7.1   Sequestration in Oil and Gas Reservoirs

Oil and gas reservoirs are usually covered by an impenetrable layer of rock, which prevented the escape of the hydrocarbons over the eons. Carbon dioxide deposited into the reservoirs is not likely to re-emerge into the atmosphere. In respect to sequestering $CO_2$, oil and gas reservoirs behave differently. Whereas $CO_2$ can be injected into oil reservoirs while the oil is still being pumped out of it, it can be injected into gas reservoirs only after depletion of the gas (because of the miscibility of $CO_2$ and methane). In fact, injecting $CO_2$ into semidepleted oil reservoirs is a well-established technology. It is not done for sequestering $CO_2$, but rather for enhanced oil recovery (EOR). On average only 10–20% of a reservoir's original hydrocarbons in place is produced during primary recovery. Secondary recovery using pressurized water injection may increase the production to 20–30%. Tertiary recovery involving liquid or supercritical $CO_2$ injection may increase the recovery to 30–60%. The rest of the hydrocarbons are too viscous to be pumped out from the pores and throats of the reservoir, which consists mostly of sedimentary sandstone or limestone. Worldwide there are some 71 oil fields where carbon dioxide enhanced oil recovery ($CO_2$-EOR) is used. The majority of these fields are in the United States, in Texas and Colorado; others are in the North Sea. In the United States, total production from $CO_2$-EOR wells is not large compared with total oil production (about 4%). Other oil wells use water flooding. Water-EOR accounts for roughly 50% of U.S. oil production.

At the depth of the reservoir, the ambient temperature and pressure are usually above the critical constants for $CO_2$ ($T_c = 31°C$, $p_c = 7.3\,\text{MPa}$). The injected liquid $CO_2$ will transform at the release point into supercritical $CO_2$, which is less dense than liquid $CO_2$. A part of the injected supercritical $CO_2$ has a tendency to buoy upward in the reservoir, and may leak into overlying strata or potable aquifers. The rest of the supercritical $CO_2$ dissolves in the oil, making it less viscous, so it can be mobilized toward the production well. The $CO_2$ injection is usually followed by pressurized water injection to drive out the $CO_2$-diluted oil. This is called the water-alternating-gas (WAG) method. In the United States, most of the injected $CO_2$ comes from natural sources, rather than from flue gas capture. The natural sources are subterranean aquifers saturated with $CO_2$. In the Permian basin in West Texas, $CO_2$ from natural sources in Colorado is piped to several semidepleted oil fields in the area for $CO_2$-EOR. Most of the injected $CO_2$ is recovered and recycled at the well head. At the Weyburn, Saskatchewan, Canada, oil field, where liquid $CO_2$ is transported from the Dakota Gasification Company synfuel plant in a 40-cm pipe over a distance of 350 km, it is expected that approximately 18E(6) tons of liquid $CO_2$ will be injected over 25 years for the recovery of approximately 19E(6) tons of crude oil.

The global storage capacity in depleted or semidepleted oil and gas reservoirs is estimated at about 147 Gt $CO_2$. This compares with global emissions of about 25 Gt $y^{-1}$ of $CO_2$. Therefore, those oil and gas reservoirs that already have been depleted have a limited capacity for sequestering $CO_2$. Over time, more oil and gas fields will be depleted. When all the world's oil and gas fields have been depleted, it is estimated that about 500 Gt of $CO_2$ may be sequestered in them, about 20 years' worth of the current rate of emissions.

The problem with using oil and gas reservoirs for sequestering $CO_2$ is not only the limited capacity, but also the fact that very few existing large coal-fired power plants are within reasonable transport distance to the reservoirs. The transport cost of piping liquid $CO_2$ is significant. The transport costs are estimated at $8E(-3) to $3E(-2) per ton-km, depending on pipeline diameter and terrain over which the $CO_2$ is piped. Thus, a 1,000-MW coal-fueled power plant may have to spend in the range of $12–$56E(6) per year for transporting the $CO_2$ to a depleted oil or gas field that is 250 km away. This cost is in addition to the cost of capturing and compressing the $CO_2$ at the power plant.

Sequestering $CO_2$ in depleted and semidepleted oil and gas reservoirs can play a role in mitigating the greenhouse effect, albeit on a limited scale, and at an economic cost, except when it is used for enhanced oil recovery, in which case the cost may be defrayed by the price of additional oil that is recovered.

## 12.7.2  Sequestration in Coal Seams

Coal seams contain significant amounts of gases, including carbon monoxide (CO), methane ($CH_4$), and other volatile organic gases. The gases are physically adsorbed on the internal surfaces of the coal, incorporated in the coal molecular structure, or dissolved in the brine that permeates the coal seam.[6] Therefore, deep coal mines have to be constantly ventilated to prevent asphyxiation of the miners by toxic gases, or explosions. Coal seam methane, also called coal bed methane, (CBM) is an important source of $CH_4$. Methane can be extracted from deep coal seams by suction through bore wells. Methane in the coal seam diffuses from an area of high pressure within the coal seam into and up the borehole where the pressure is lower.[7] Global coal seam methane resources are estimated between 80 and 260E(12) standard cubic meters (scm),[8] equivalent to 3,000 to 10,000 EJ, which is equal or more than the global proven conventional gas reserves (see Chapter 2). In the San Juan Basin, New Mexico, 42E(3) scm $d^{-1}$ methane (1.6 GJ) are obtained per bore well by vacuum suction.

It has been demonstrated that the injection of gaseous or liquid $CO_2$ into coal seams liberates $CH_4$. Depending on coal characteristics (porosity, permeability, volatility, water content), two

---

[6] Juntgen, H., and J. Karwell. *Erdoel und Erdgas Petrochemie*, 19 (1966): 339–44.

[7] White, C. M. et al. *Journal of Air and Waste Management Association*, 53 (2003): 645–715.

[8] Kuuskra, V. A., et al. *Oil and Gas Journal*, Oct. 5 (1992): 49–54.

CO$_2$ molecules are adsorbed for each CH$_4$ molecule injected into high volatile bituminous coals, and up to 10 CO$_2$ molecules are adsorbed for each CH$_4$ molecule injected into subbituminous coals.[9] The injection of CO$_2$ would serve a dual purpose: CH$_4$ can be exploited and CO$_2$ can be sequestered. This is called CO$_2$ enhanced coal bed methane recovery, CO$_2$-ECBM. It is estimated that worldwide 300 to 1,000 Gt CO$_2$ could be sequestered by injecting it into "unminable" coal seams[10] compared with the global annual emissions from fossil-fueled power plants of 25 Gt CO$_2$. The problems with this scheme are manifold. First, most power plants are not located in the vicinity of coal seams. Transporting the captured CO$_2$ over long distances by pipeline, rail, or barges is an expensive proposition. Those coal-fueled power plants that sit at the mouth of a coal mine use the coal instead of disposing CO$_2$ into it. Second, how does one determine which coal seams are "unminable"? Coal seams that at present are deemed too expensive to exploit or are inaccessible by current technological methods may have to be exploited by future generations, when easily accessible coal seams become scarce or the price of coal surges. Third, there is some measured leakage of CH$_4$ from coal seams along faults and fractures. For example, the CH$_4$ leakage rate from the San Juan Basin in New Mexico is estimated at 13E(3) t y$^{-1}$ (26E(6) scm y$^{-1}$).[11] Injection of CO$_2$ into coal seams may cause a commensurate rate of leakage, which may be harmful to the immediate environment, let alone to the contribution to atmospheric CO$_2$, which we want to alleviate in the first place.

### 12.7.3  Sequestration in Deep Sedimentary Basins

Deep sedimentary basins underlie vast areas under the continents and oceans. The basins usually contain saline water (brine) and are separated from shallower aquifers—the source of much of the drinking water—by impermeable rock. The deep basins themselves consist of permeable, porous rock, such as sedimentary shale-, lime-, or sandstone, the pores of which are saturated with brine. Such basins are found at depths of 800 m or deeper. The injected CO$_2$ (in liquid or supercritical phase) would dissolve in the brine as carbonic acid. In the case of a limestone basin, some of the surrounding carbonate rock would dissolve into bicarbonate, furthering the absorption capacity of the reservoir and reducing the risk of leakage. The governing reaction is

$$CaCO_3 + CO_2 + H_2O \rightarrow Ca^{2+} + 2HCO_3^-. \tag{12.5}$$

There is limited information available on the extent and CO$_2$ absorption capacity of deep sedimentary basins. In the United States, the estimated storage capacity of subterranean basins ranges from 18 to 1,800 Gt of CO$_2$ (compared with annual emission rates from power plants of 6.2 Gt y$^{-1}$ of CO$_2$). Other estimates for worldwide capacities range from 400 to 11,000 Gt of CO$_2$. The problem with deep sedimentary basins is not so much their capacity, but their location vis-à-vis power plants. As discussed in Section 12.7, transport of liquid CO$_2$ in large diameter pipelines over large distances may prove to be expensive and energy intensive. Therefore, it will be advantageous to locate fossil-fueled power plants equipped with CO$_2$ capture as close to deep

[9]Wong, S., W. D. Gunter, and J. Gale. *Greenhouse Gas Control Technologies*. CSIRO Publ., pp. 531–537, 2001.

[10]Gunter, W. D., et al. *Applied Energy*, 61 (1988): 209–27.

[11]Clayton, J. L., et al. In *Organic Geochemistry: Developments and Applications to Energy, Climate, Environmental and Human History*. San Sebastian: AIGOA, 1995.

sedimentary basins as possible. Intensive research is ongoing to establish the location and capacity of the deep sedimentary basins worldwide.

Sequestration of fluids in underground porous sedimentary layers is a well-developed technology. It is very similar to the recovery of water, oil, or natural gas, but in the reverse direction. In the latter operations, a production well is drilled into the fluid layer, where the fluid pressure is equal to the gravitational stress in the surrounding sedimentary basin. As a consequence, the pressure at the wellhead is about 25 MPa per kilometer of well depth when there is no out-flow from the well. When the well is producing, the wellhead pressure falls as the outflow increases because of fluid friction in the well and the surrounding sedimentary layer. For a $CO_2$ injection well, the inflow stream must have a pressure exceeding the wellhead no-flow value (which is proportional to the well depth) by an amount proportional to the flow rate. This requires mechanical power, so the energy cost per unit mass of $CO_2$ injected increases with the well depth and injection flow rate. For high mass injection rates, multiple wells may be needed to minimize the economic cost of disposing of the $CO_2$.[12]

Since 1996, at the Sleipner gas fields in the North Sea, off the coast of Norway, liquid $CO_2$ is being injected into a subseabed sedimentary basin. The natural gas extracted from the Sleipner gas fields contains on average 9.5% by volume $CO_2$. The $CO_2$ is separated from the extracted gas by MEA absorption at the offshore gas production platform, compressed to a liquid, and injected into the Utsira basin at a depth of 800 m under the seabed at a rate of 1 Mt $y^{-1}$. The pipeline carrying the liquid $CO_2$ has a diameter of 18 cm. The release of the $CO_2$ occurs through holes in a horizontal section of the pipeline. The injection scheme is depicted in Figure 12.4.[13] The cost

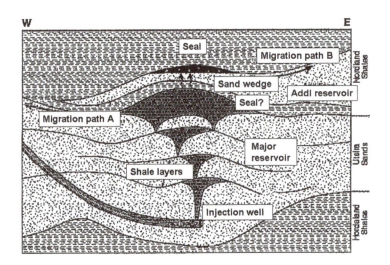

**FIGURE 12.4**   Schematic representation of the Sleipner storage system.[14]

[12]The economic cost is the sum of the operating cost per unit mass (mostly pumping power) and the fixed cost (capital cost of well and pump, annualized and averaged over the annual amount injected).

[13]Zweigel, P., et al. *Greenhouse Gas Control Technologies, 5th International Conference.* Collingwood: CSIRO Publishing, 2001.

of the injection is about \$15 per ton CO$_2$, which compares favorably with a tax of \$50 per ton of CO$_2$ that the government of Norway would levy if the CO$_2$ were emitted into the atmosphere.[14] In Salah, Algeria, at a liquefied natural gas plant, 1.2 Mty$^{-1}$ of CO$_2$ is injected in the sandstone formation at Krechba. Worldwide plans are underway to construct several IGCC$^2$ plants with deep underground sequestration of the captured and liquefied CO$_2$.

While CO$_2$ sequestration in deep underground sedimentary basins is touted to become the major route for disposing captured CO$_2$ from fossil-fueled power plants, industrial boilers, and other centralized emitters, it remains to be demonstrated that massive injection of liquid CO$_2$ into such basins is technically feasible, economically viable, and verifiable that the injected CO$_2$ will not leak into overlying fresh water aquifers or even back into the atmosphere. Taking the Sleipner example, the injection of 1 Mt y$^{-1}$ of liquid CO$_2$ requires an 18-cm-diameter pipe to be bored to a depth of 800 m under the seabed. A 500-MW coal-fueled power plant will produce about four times as much CO$_2$ per year as does Sleipner. Thus, a 500-MW plant may require four 18-cm pipes to be bored to the appropriate depth or one 36-cm pipe, assuming that the pipe capacity increases roughly as the diameter squared. This will add significantly to the cost of injection of CO$_2$ into deep saline sedimentary basins.

### 12.7.4  Sequestration in the Deep Ocean

The ocean is a natural repository for CO$_2$. The ocean is vast; it covers about 70% of the earth's surface, and the average depth is 3,800 m. There is a continuous exchange of CO$_2$ between the atmosphere and the ocean. We have seen in Chapter 11 that the ocean absorbs about 92 Gt y$^{-1}$ of carbon from the atmosphere, while it outgasses into the atmosphere about 90 Gt y$^{-1}$. Thus, the ocean is a net absorber of carbon, which probably is part of the reason that CO$_2$ concentrations in the atmosphere do not increase as fast as expected from anthropogenic emissions. Most of the ocean–atmosphere carbon exchange occurs within the surface layer of the ocean, which is on average about 100 m deep. This layer is more or less saturated with CO$_2$. Between 100 and 1,000 m depth the temperature of the ocean is steadily declining (the so-called thermocline). Beneath about 1,000 m depth, the temperature is nearly constant, between 2 and 4°C, and the density increases slightly because of hydrostatic pressure. This makes the deeper layers of the ocean very stable, and the turnover time (i.e., the time it takes for the deep layers to exchange waters with the surface layer is long) on the order of hundreds to thousands of years.

The deep layers of the ocean, 1,000 m or deeper, are highly unsaturated in regard to CO$_2$. The absorptive capacity of the deep ocean is estimated on the order of E(19) tons of CO$_2$.[15] Thus, conceivably all the carbon residing in fossil fuels on earth could be accommodated in the deep ocean without reaching even near the saturation limit. Since CO$_2$ emitted into the atmosphere eventually will wind up in the deep ocean, an artificial injection of CO$_2$ at depth would merely short-circuit the natural cycle that takes hundreds to thousands of years.

---

[14]This example shows how government intervention can foster greenhouse effect mitigation. Surely, if the Norwegian government had not imposed a carbon tax, the associated CO$_2$ from the Sleipner gas field would be allowed to escape into the atmosphere.

[15]Wilson, T. R. S. *Energy Conversion and Management*, 33 (1992): 627–35.

Models predict that the atmospheric concentrations of $CO_2$ will increase more or less in proportion to the rate of global fossil fuel combustion. Peak concentrations will occur sometime in the 23rd century, after which they will slowly decline because (a) mankind will run out of fossil fuels, and (b) the ocean will slowly absorb the excess $CO_2$ that built up in the atmosphere. The purpose of deep ocean injection of $CO_2$ is to shave off that peak that would be building up in the next 200–300 years.

An injection of $CO_2$ at 1,000 m or deeper is deemed necessary not only because the deep layers are unsaturated with regard to $CO_2$, but also on account of the physical properties of $CO_2$. If injected at 500 m or less, liquid $CO_2$ would immediately flash into gaseous $CO_2$ and bubble up to the surface. Between 500 to 1,000 m, liquid $CO_2$ injected from a diffuser at the end of a pipe would disperse into droplets of various diameters, depending on the release pressure of liquid $CO_2$ and the release orifices' diameter. Because at these depths the density of liquid $CO_2$ is less than that of seawater, the droplets would ascend because of buoyancy and may reach the 500-m level undissolved, where they would flash into gaseous $CO_2$. Various approaches are proposed to render the injected $CO_2$ plume heavier than ambient seawater, so that it will sink deeper from the injection point, rather than buoying upward. These approaches include precooling of the liquid $CO_2$, releasing it in a manner that assures hydrate formation,[16] or forming a dense emulsion consisting of tiny liquid $CO_2$ droplets coated with a layer of pulverized limestone dispersed in seawater.[17]

**$CO_2$-Hydrate.** At a hydrostatic pressure of about 5 MPa and temperatures below 10°C (corresponding to a depth of about 500 m), a solid $CO_2$ hydrate is formed, the crystalline structure of which contains one $CO_2$ molecule surrounded by 5–7 $H_2O$ molecules tied to $CO_2$ by hydrogen bonds. Laboratory and pilot-scale releases of $CO_2$ in the deep ocean confirm the formation of $CO_2$ hydrate.[17] At this time it is not clear whether hydrate formation is good or bad for $CO_2$ sequestration. If the hydrate crystals were heavier than seawater, they would sink to greater depth from the release point, thereby increasing the sequestration period. If the hydrate crystals were to occlude the liquid $CO_2$ droplets, they would ascend by buoyancy and hinder the dissolution in the surrounding seawater of the occluded $CO_2$. The hydrates may also clog the release pipe and diffuser.

**Transport of $CO_2$ to the Deep Ocean.** Liquid $CO_2$ can be transported to the deep ocean by pipeline or by tanker to a floating discharge platform. From the platform, an attached vertical pipe would deliver the liquid $CO_2$ to the appropriate depth. Because of the high vapor pressure of liquid $CO_2$ at ambient temperatures, tanker vessels would have to be refrigerated and pressurized. At the triple point temperature of −57°C, where solid, liquid, and gaseous $CO_2$ coexist, the vapor pressure of $CO_2$ is 0.52 MPa. Refrigerated and pressurized vessels are very expensive to build. Liquid natural gas (LNG) tankers cost on the order of $250 million per vessel of 1.5 E(5) $m^3$ capacity. LNG tankers are refrigerated but not pressurized. Because of the pressurization, tankers for liquid $CO_2$ may even be more expensive, not counting the investment cost for the discharge platform with the attached vertical discharge pipe. Therefore, pipeline transport of liquid $CO_2$ to the deep ocean is considered the most economic option. The laying of large diameter pipes on the continental shelf reaching to 1,000-m depth is a formidable task, but pipes have been laid to offshore oil wells to such depths in the Gulf of Mexico, the North Sea, and elsewhere. The cost of

[16]Tsouris, C., et al. *Environmental Science and Technology*, 38 (2004): 2470–75.

[17]Golomb, D., et al. *Environmental Science and Technology*, 38 (2004): 4445–50.

such pipelines is estimated between \$1 and 2 million per km length. Therefore, it is desirable to reach depths of 1,000-m or more as close to the shore as possible. Potential coastal sites where the deep ocean can be reached at distances of no more than 100–200 km are as follows:

### Americas

- Hudson, Delaware, and Hatteras Canyons
- Mississippi Canyon
- Baja California Trench, Monterey and Columbia River Canyon
- Several canyons along the coast of Mexico toward the Gulf of Mexico and the Pacific
- Along the west and east coasts of South America

### Europe

- Outflow of the Mediterranean Sea at Gibraltar
- Canyons along the coast of Portugal and the Bay of Biscayne
- Coast of Norway
- From western Scotland and Ireland over the Hebridian Shelf into several canyons

### Asia, Australia, and Oceania

- Several canyons from the coast of India into the Arabian Sea and Bay of Bengal
- Several canyons from the coast of China into the South China Sea
- Japan is surrounded by the deepest trenches in the world
- Canyons leading to the Japan Abyssal Plain from Korea and eastern Siberia
- Canyons along the southeastern coast of Australia
- Trenches along the Philippine and Indonesian islands

Altogether, access to the deeper layers of the ocean from industrial–urbanized continents is quite limited. For example, the industrial countries of central and eastern Europe are too far from the deep ocean. Since overland pipelines would be even more expensive to lay than underwater pipes, it is unlikely that power plants in those countries would avail themselves to ocean sequestration. The same is true for power plants in the midwestern United States and central Canada.

Apart from the technical and geographical problems, there are concerns about the environmental impact of deep ocean sequestration of $CO_2$. Even though $CO_2$ is not toxic, large-scale injection of $CO_2$ into the deep ocean would create localized regions of high carbonic acid concentrations. It is estimated that the discharge of $CO_2$ from ten 500-MW coal-fired power plants would create a volume of hundreds of $km^3$ in which the pH is less than 7.[18] In that volume fish and other creatures would either escape or die. However, the impacted regions would be minuscule in comparison to the world's ocean volume. Nevertheless, there is general opposition by marine ecologists to disposing of anything in the ocean. Furthermore, international treaties (e.g., the London Convention on Ocean Dumping of 1991) prohibit all dumping of radioactive and industrial waste. It is not clear whether $CO_2$ from power plants or other factories constitutes an industrial waste.

---

[18]Caulfield, J. A., et al. *Energy Convers. Mgmt.*, 33 (1997): S667–74.

## 12.7.5 CO$_2$ Removal from the Atmosphere

It is possible to remove a part of the CO$_2$ from the atmosphere. Atmospheric removal options include the following:

- Afforestation
- Ocean fertilization
- Mineral carbonation

### 12.7.5.1 Afforestation

Afforestation means planting trees on unused arable land without burning the wood after harvesting. It is estimated that typical coniferous and tropical forests absorb between 6 to 10 metric tons of carbon per hectare per year (t C ha$^{-1}$y$^{-1}$) because of photosynthesis. Indeed, the Kyoto Convention of 1997 and the subsequent Buenos Aires Conference of 1998 envisaged that part of the greenhouse effect mitigation efforts would come from afforestation. Large CO$_2$-emitting countries are encouraged to reduce global concentrations of CO$_2$ by investing in other countries (mostly tropical, less developed countries) to plant new trees. The emitting countries would obtain credit for CO$_2$ emission reductions equivalent to the amount of CO$_2$ absorbed by the new trees. This is called *emission trading*. The exact accounting procedures for such emission trading have not yet been established. We must note that deforestation has the opposite effect. Every hectare cleared of forests will reduce the absorption of CO$_2$ by 6–10 t C ha$^{-1}$y$^{-1}$. If the trees are burned, as is still widely practiced in tropical countries of the world, there is a double whammy effect: cessation of absorption of CO$_2$ and emission of CO$_2$ caused by wood burning. It is estimated that forest burning alone adds about 1.5 Gt y$^{-1}$ of carbon to the atmosphere.

### 12.7.5.2 Ocean Fertilization

In an ocean fertilization scheme iron and other fertilizers are spread over the ocean surface to enhance phytoplankton production. This scheme is strongly opposed by marine biologists because it may interfere with the natural ocean ecology. Also, it is not clear whether the plankton life cycle actually removes CO$_2$ from the atmosphere and sequesters it to great depth or to the bottom of the ocean from whence it will not re-emerge into the atmosphere, or whether it decomposes on the surface with re-emergence of CO$_2$ into the atmosphere.

### 12.7.5.3 Mineral Sequestration

When certain minerals, such as olivine, a magnesium iron silicate, are exposed to aqueous CO$_2$ (actually to bicarbonic and carbonic acid), some of the magnesium and iron silicates will convert to carbonate and precipitate from the solution. This happens naturally over geologic time scales. For anthropogenic mineral sequestration, the minerals would have to be pulverized to increase their exposed surface. That requires energy and cost. Also, the conversion from silicate to carbonate is a very slow process at ambient temperature and the relatively low CO$_2$ partial pressure in the atmosphere, currently about 3.8E(-4) atm (380 ppmv). Nevertheless, research efforts are underway in the United States and elsewhere to investigate the potential for mineral sequestration of CO$_2$.

### 12.7.5.4 CO$_2$ Utilization

Carbon dioxide is used on a commercial scale for dry ice manufacturing, carbonated drinks, and as a raw material for chemical products, such as urea, methanol, or other oxygenated fuels. The problems with this proposition are twofold: (a) most of the carbon in the product would eventually burn up or decompose back to $CO_2$ and wind up in the atmosphere; and (b) the conversion of $CO_2$ into useful products requires virtually the same amount of energy as was given off when carbon oxidized into $CO_2$. Also, the present market for chemical products that could be based on $CO_2$ is quite limited, amounting perhaps to less than 50–70 Mt $y^{-1}$, whereas the emission of $CO_2$ from a single 1,000-MW coal-fired power plant amounts to 6–8 M $ty^{-1}$.

Let us take the example of converting $CO_2$ to methanol:

$$CO_2 + 3H_2 \rightarrow CH_3OH + H_2O - 171 \text{ kJ mol}^{-1} \qquad (12.6)$$

The minus sign indicates that energy is liberated (i.e., the reaction is exothermic), with 171 kJ evolved per mole of $CO_2$ reacted. However, the reaction as written requires 3 moles of hydrogen for each mole of $CO_2$ to produce 1 mole of $CH_3OH$. The production of 1 mole of hydrogen from dissociation of water requires 286 kJ of energy; 3 moles require 858 kJ. Thus, the energy balance of Equation (12.5) is actually negative, requiring 687 kJ per mole $CH_3OH$ produced. The production of methanol from $CO_2$ and $H_2$ would only make sense if that hydrogen were derived from nonfossil energy, such as solar energy (e.g., photodissociation of water) or nuclear energy (e.g., electrolysis of water using nuclear electricity). Even if hydrogen were derived from nonfossil sources, the question is whether it should not be utilized directly, such as in fuel cells, rather than producing methanol. Furthermore, when methanol is burned in a heat engine, carbon is reoxidized to $CO_2$, and nothing has been gained in terms of global warming mitigation.

Another example is the production of urea from $CO_2$. Urea is an important industrial chemical because it is used in chemical fertilizers, polyurethane foam production, and as an intermediate in a host of other chemicals. The production of urea can be written as the following simplified reaction:

$$CO_2 + 3H_2 + N_2 \rightarrow NH_2CONH_2 + H_2O. \qquad (12.7)$$

This reaction is highly endothermic, with 632 kJ of energy required per mole of solid urea produced, including the 858 kJ of energy necessary for the production of 3 moles of hydrogen. This example shows again that unless hydrogen is produced from nonfossil energy sources, the utilization of $CO_2$ for converting to other chemicals makes no sense.

Nature uses $CO_2$ as a raw material for producing biomass by means of photosynthesis,

$$CO_2 + H_2O \xrightarrow{h\nu} CH_2O + O_2, \qquad (12.8)$$

where $h\nu$ represents a solar photon, and $CH_2O$ is a basic building block for plant sugar, starch, and cellulose. This reaction is endothermic; it requires 570.5 kJ of energy per mole of $CO_2$ consumed. That energy comes from the sun. All vegetation and phytoplankton that grow on the planet are based on Reaction (12.8). Because plants and planktons are at the bottom of the food chain, all animals and mankind are dependent on Reaction (12.8) for their sustenance. In fact, all fossil fuels were created over the eons by geochemical conversion of biomass to coal, oil, and natural gas.

The utilization of biomass for fuel in boilers or heat engines is $CO_2$ neutral. Every carbon atom burnt or utilized from biomass is reabsorbed in the next generation of plants or planktons. The utilization of biomass as a renewable energy source is discussed in Chapter 8.

## 12.8   CONCLUSION

Global warming could be mitigated by reducing emissions into the atmosphere of anthropogenic greenhouse gases (GHG). These gases include chlorofluorocarbons (CFCs), nitrous oxide ($N_2O$), methane ($CH_4$), and carbon dioxide ($CO_2$). In accordance with the Montreal Protocol of 1987 and its Amendment of 1992, CFC production is being phased out. Because CFCs are very long-lived species, they will still contribute to radiative forcing for decades and even centuries. Nitrous oxide is being emitted from fossil fuel combustion, adipic acid manufacturing (used for nylon production), nitrogen fertilizer manufacturing and decomposition, and as an agent for whipping cream. With more wide spread usage of nitric oxide emission control on fossil-fueled power plants and other large combustion sources, it is expected that $N_2O$ emissions from these sources will also be reduced. Similarly, abatement technologies for the reduction of $N_2O$ emissions from adipic acid manufacturing are being installed on an increasing number of plants. Major controllable emissions of $CH_4$ occur from coal mines, petroleum and natural gas operations, municipal landfills, and sewage treatment plants. (We consider $CH_4$ emissions from enteric fermentation of cattle and rice cultivation to be uncontrollable.) Methane in the ventilation gas from coal mines is too dilute to capture efficiently and economically, but premining bore holes into coal seams would allow a more concentrated stream of $CH_4$ to be captured and sequestered or exploited. Methane from municipal landfills and sewage treatment plants is being captured on an increasing scale.

Carbon dioxide concentrations in the atmosphere contribute about 55.5% to radiative forcing of all anthropogenic GHG. The control of $CO_2$ emissions constitutes a major technological, economic, social, and political challenge to the world. Obviously, a whole number of measures would need to be implemented to reduce significantly the emissions into the atmosphere of $CO_2$, ranging from conservation and efficiency improvements of $CO_2$-emitting sources, to substitution of fossil fuels by renewable and nuclear energy, and to $CO_2$ capture and sequestration. Among the $CO_2$ capture technologies, the approach with the lowest cost and energy penalty seems to be coal gasification with water shift reaction, followed by capturing the $CO_2$ by physical absorption. The remaining hydrogen can be combusted in a combined cycle power plant, called integrated coal gasification combined cycle power plant (IGCC). The hydrogen can also be used in fuel cells for electricity generation and as a transportation fuel.

The captured $CO_2$ must be sequestered in appropriate reservoirs from which it will not re-emerge into the atmosphere, at least not for the next few hundred years while fossil fuels are being exhausted. Depleted and semidepleted oil and gas reservoirs could accept about 25 years' worth of global $CO_2$ emissions. Larger-capacity reservoirs are deep subterranean sedimentary basins or the deep ocean. Intensive international efforts are ongoing to locate the underground reservoirs, and ensure that the $CO_2$—when deposited into them—will not leak back into the atmosphere. Similarly, further research is necessary to ensure that marine organisms will not be harmed if large quantities of $CO_2$ are injected into the deep ocean.

# PROBLEMS

### Problem 12.1

A 1,000-MW (el) power plant working at 35% thermal efficiency 100% of the time (base load) uses coal with formula CH and a heating value of 30 MJ/kg. How much $CO_2$ does this plant emit (metric tons/y)?

## Problem 12.2

This plant substitutes NG (formula $CH_4$) instead of coal, with a heating value of 50 MJ/kg, in a combined cycle mode having a thermal efficiency of 45%. (a) How much NG ($m^3$/y at STP ) is consumed? (b) How much $CO_2$ (t/y) is emitted?

## Problem 12.3

To ameliorate the $CO_2$-caused greenhouse effect, a power plant is built that enables the capture of $CO_2$ from the flue gas. The plant runs on pure methane and oxygen, so the flue gas consists only of $CO_2$ and $H_2O$,

$$CH_4 + 2O_2 \rightarrow CO_2 + 2H_2O - 0.244kWh(th)\ mol^{-1}CH_4,$$

where the minus sign means that the process is exothermic; that is, energy is evolved. The plant must produce its own $O_2$. From oxygen suppliers we know that energy requirement for air separation is 250 kWh (electric ) per metric ton of $O_2$. An equal amount of electric energy is required for liquefaction of 1 metric ton of $CO_2$. The plant has a gross thermal efficiency $\eta_{gross} = 0.45$ (that is, thermal efficiency calculated before electricity is syphoned off for $O_2$ production and $CO_2$ liquefaction). (a) What percentage of the plant's electricity output is available for dispatch to the grid? (b) What is the net thermal efficiency $\eta_{net}$ (electricity output per heat input after $O_2$ production and $CO_2$ liquefaction)?

## Problem 12.4

A nominally rated 1,000-MW coal-fired power plants emits 250 kg $s^{-1}$ of $CO_2$. Suppose all the $CO_2$ is absorbed by chemical absorption using MEA. The MEA aqueous solution contains 25% by volume MEA (MW = 61). Assume that each molecule of MEA absorbs one molecule of $CO_2$ (MW = 44). The absorption occurs in a tower where the MEA solution flows countercurrent to the flue gas. What is the flow rate ($m^3\ s^{-1}$) of the MEA solution through the tower to absorb all the $CO_2$?

## Problem 12.5

To boil off all the $CO_2$ absorbed in Problem 12.4, the saturated MEA solution is transferred to a desorption tower, where the solution is heated from 40 to 100°C. Assuming the solution has a heat capacity of 0.9 cal $g^{-1}$, how much heat (J) is required to boil off all the $CO_2$? What fraction of the nominally rated 1,000-MW plant is used for heating the solution?

## Problem 12.6

A 1,000-MW coal-fired power plant emits 250 kg $s^{-1}$ $CO_2$. How much water ($m^3\ s^{-1}$) at a temperature of 25°C has to be circulated through an absorption tower to absorb the $CO_2$? Use Henry's law with $K_{CO_2} = 3.38E(-2)$ mol $L^{-1}atm^{-1}$, $p_{CO_2} = 1$ atm.

## Problem 12.7

The Sleipner Project shows that 1 Mt $y^{-1}$ of liquid $CO_2$ requires a 18-cm-diameter pipe to carry the liquid to 800 m depth. What pipe diameter is necessary to carry 250 kg $s^{-1}$ liquid $CO_2$ to a depth of 800 m, assuming the throughput of the pipe scales with the square of the diameter?

# BIBLIOGRAPHY

Intergovernmental Panel on Climate Change. *Carbon Dioxide Capture and Storage.* Special Report, 2007.

International Energy Agency. *Capture of Carbon Dioxide from Fossil Fuel Fired Power Stations.* IEAGHG/SR2, 1993.

Pruschek, R., and G. Goettlicher. *Concepts of $CO_2$ Removal from Fossil Fuel Based Power Generation Systems.* Fachbereich: Universitaet Essen, 1996.

White, C. M., et al. *Journal of the Air and Waste Management Association*, 53 (2003): 645–716.

CHAPTER

13

# Concluding Remarks

## 13.1 ENERGY RESOURCES

At the beginning of this book, we reviewed in Chapter 2 the energy resources available to mankind and the uses of these resources in the various countries of the world; in subsequent chapters we outlined the environmental effects associated with energy use. The supply and use of energy has other important consequences, economic and political. Energy is a necessary and significant factor of national economies; energy expenditures amount to 5–10% of the GDP in industrialized nations. The availability of adequate energy to enterprises and individuals is a national goal and is thereby affected by governmental policies.

Among fossil energy resources, coal appears to be available in abundance for at least two to three centuries, while the fluid fossil fuels, petroleum and natural gas, may last for less than a century. The availability of gaseous and liquid fuel resources can be extended by manufacturing synthetic fuel from coal by coal gasification and liquefaction. Fluid fuels can also be obtained from unconventional resources, such as oil shale, tar sands, geopressurized methane, coal seam methane, and methane hydrates lying on the bottom of the oceans and under the icecaps. The manufacture of synfuels and the exploitation of unconventional fossil fuel resources will be more expensive than the exploitation of proven reserves, and the manufacturing and recovery processes will entail more severe environmental effects than those associated with exploitation of conventional reserves.[1]

Electricity is an essential energy component of modern industrialized societies; its use is increasing worldwide. Electricity is a secondary form of energy; it has to be generated from primary energy sources. Currently about two thirds of the world's electricity is generated from fossil fuels while the other third comes from hydroenergy and nuclear energy, with minor contributions from wind, biomass, and geothermal sources.[2]

---

[1] See Section 2.7.

[2] See Chapters 2 and 8.

In 2006, 15% of the world's electricity and 6.4% of its primary energy was supplied by nuclear power plants. The global resources of the raw material for nuclear power plants—uranium and thorium—would last centuries at current usage rates. These resources can be extended even further in the so-called breeder reactors, where artificial fissile isotopes can be generated from natural uranium and thorium. Nuclear power plants are much more complex and expensive to build and operate than fossil-fueled plants. Also, the real and perceived hazards of nuclear power plants, including the risks of the entire nuclear fuel cycle, from mining to refining to radioactive waste disposal, militates against building new nuclear power plants in many countries.[3] However, with depleting fossil fuel resources and the associated environmental risks of fossil-fueled power plants, notably global warming caused by $CO_2$ emissions, it is likely that in the future nuclear power will increase its share of the world's electricity generation.

Hydropower is a relatively clean source of electricity, but most of the high-gradient rivers and streams that are near population centers have already been dammed up for powering the turboelectric generators. Damming up more rivers is encountering increasing public resistance because of the risk to the watershed ecology and because it may entail massive population displacement. While several new hydroelectric power generators are being built or planned, notably the 18 GW hydroelectric stations on the Yangtze River in China, hydropower is not expected to increase substantially its share among other sources of electricity.

Other renewable energy sources than hydropower hold the promise of occupying an increasing share of electricity generation. Renewable sources are biomass, geothermal, wind, solar thermal and thermal electric, photovoltaic, and ocean tidal and wave energy. Biomass and geothermal plants are able to supply electric power dependably on a daily and annual basis. The other sources of renewable energy have diurnal and seasonal rhythms that do not necessarily match the demand for electric power. Because electricity cannot be stored easily, the renewable generators usually need to be backed up by conventional power sources. However, when producing electricity, the renewable sources can displace fossil fuel consumption and reduce air pollutant emissions, especially $CO_2$ emissions that affect long-term climate change. Renewable electricity generators require a greater capital investment than fossil power plants and are usually not economically competitive with these power plants except when $CO_2$ emission reduction is mandated to prevent climate change.

## 13.2  REGULATING THE ENVIRONMENTAL EFFECTS OF ENERGY USE

Mitigating the adverse environmental effects of energy use has been a chronic problem afflicting nations worldwide because national economies do not respond automatically to limit environmental degradation.

Economists label the release of pollutants into the environment as an *externality*, an activity that does not enter into the cost of production of a good or supply of a service. The capacity of the environment to tolerate the discharge of pollutants is considered a free good; the polluter pays no price for its use. But the environmental capacity to endure pollution is finite, and the cumulative effects of pollution from many sources degrades its quality, adversely affecting the interests of society as a whole to a much greater extent than the interests of one or even all of the

---

[3]Growth in global nuclear power plants may increase the risk of the spread of nuclear weapons.

polluters. Although the total social cost of bearing the ill effects of pollution may outweigh the cost of eliminating the pollution, the polluter's share of these social costs is usually too small to offset its abatement cost, so that abatement is uneconomic for each polluter, both individually and collectively.

A common solution to this social and economic dilemma is government regulation of pollutant-producing activities. Most often this takes the form of a performance requirement, such as a standard of maximum emissions per fuel input from energy using sources or a requirement that certain pollution reduction technologies be employed. Less often, economic incentives to abatement are used, such as emission taxes, pollutant fines, or tax deductions and credits for reducing emissions. Except for the case of deductions and credits, the cost of abatement is borne by the polluter, thereby internalizing the externality of pollution into the production process.

In almost all cases of pollution of air, water, or soil, the amounts of toxic pollutants emitted are but a tiny fraction of the fuel burned or material processed. In general, the cost of reducing ordinary pollutant emissions is only a small fraction of the economic value added in the production process, but the cost per unit of pollution removed is often very high, inevitably higher than any possible economic use of the sequestered pollutant. It is rare that any pollutant-reducing process pays for itself.

The goal of environmental regulation is to achieve social and environmental gains that accrue to society as a whole by regulating the activities of polluting enterprises without substantially vitiating their societal benefits. The role of technology in this effort is to provide the least costly way to achieve this goal. In the United States, for example, the responsibility for protecting the public health and welfare from harmful substances in the environment that are of anthropogenic origin is lodged with the U.S. EPA. The regulatory power of the U.S. EPA is embodied in a series of legislative acts setting forth the activities to be regulated, the requirement for promulgation of regulations, and their enforcement in federal courts. Frequently, legislation is quite specific concerning how activities are to be regulated and timetables set for achieving progress. In general, the legislation requires no balancing of costs and benefits; the economic costs of abating pollution are to be paid by the polluters, not the government, but funds have been appropriated to help local municipalities renovate municipal waste treatment plants and a superfund has been established to clean up abandoned toxic waste dumps. The U.S. EPA accomplishes its task by setting nationally uniform standards for air and water quality and for the processes that lead to their contamination. For example, the U.S. EPA has promulgated National Ambient Air Quality Standards for several prominent air pollutants associated with fuel combustion and emission or process standards for the sources of these pollutants. The U.S. EPA may require emission limits or the use of effective control technology for specific classes of sources, such as power plants and other stationary sources and mobile sources or by regulating the properties of fuel, especially motor vehicle fuel. These means have been effective in reducing the environmental effects of energy use despite the steady growth in the consumption of energy.

Surprisingly, increasing the efficiency of energy use plays no direct role in the environmental regulation of urban and regional pollutants because the degree of abatement needed is very much greater than can be garnered by the modest energy efficiency gains that are economical, while the cost of the requisite abatement technology is moderate. Nevertheless, there is some environmental benefit that accrues to energy efficiency improvement. Reducing electric power consumption by increasing the efficiency of its use would reduce the air pollutant emissions from power plants, given any level of control technology. Process modification could lead to less use of fossil fuels in the manufacturing of industrial goods. In the commercial and residential sector, fossil energy use, and thereby pollution abatement, could be achieved by better insulation in buildings, replacing incandescent with fluorescent lighting and using solar or geothermal space and water heating. In the transportation sector, great savings could be accomplished in fossil fuel usage and concomitant pollutant emissions by traveling in small, light vehicles or using hybrid internal combustion engines and electric motors for vehicle propulsion or more efficient fuel cell-powered electric motors.[4]

## 13.3   GLOBAL CLIMATE CHANGE

The accumulation in the atmosphere of GHG (mostly carbon dioxide, but with nonnegligible contributions from methane, nitrous oxide, and CFCs), which threatens to cause changes in the global climate and thus have adverse environmental effects, has generated international concern. Noncumulative urban and regional pollution, with but rare exceptions, has been seen as a national problem to be solved within the constraints of national economies. However, climate change is clearly a global environmental problem requiring the coordinated action of many nations over long time periods, on the order of a century. The regulatory regime used by a nation to cope with environmental degradation within its borders does not exist in the international community. Therefore, it will be necessary to seek multilateral international agreements for GHG emission control by the major emitting nations if this growing problem is to be ameliorated.[5]

Any effective program for limiting the amount of climate change to be experienced in the 21st and subsequent centuries must necessarily have significant impacts on the supply and consumption of energy by nations and on their economies. The magnitude and comprehensiveness of the needed adjustments to the energy economy will be much greater than what has been required for dealing with urban and regional pollution. The first steps toward seeking international agreement for climate change control were taken in the 1990s. These included seeking an international scientific consensus on the understanding of climate change and the prospective effects of remedial measures, along with a plan for allocation of national annual emission caps to be implemented within the first two decades of the 21st century. This consensus resulted in the Kyoto Protocol of 1997, which aims to reduce annual global GHG emissions by an average of 55% below 1990 levels. However, as of the year 2010, not all of the industrialized nations have ratified the Kyoto Protocol. Nevertheless, individual countries are assessing (a) the measures they may need to secure some degree of control over their respective national GHG emissions and (b) the policies and procedures that will best meet their national objectives in dealing with climate change.

The emissions of non-$CO_2$ GHG can be reduced without resource to heroic measures. Anthropogenic methane emissions can be readily reduced by preventing leakage of the gas from gas wells, pipes, tanks, tankers, and coal mines and land fills. However, nitrous oxide emission control is less certain because we do not fully understand the sources of emission of this gas. The manufacture of CFCs is being phased out worldwide as a consequence of an international treaty, the Montreal Convention of 1987. Unfortunately, because of their very long lifetimes in the atmosphere, CFCs will contribute to global warming for many decades to come.

The reduction of emissions of carbon dioxide is a problem of a different kind and magnitude than reducing emissions of the other GHG pollutants. Eighty-six percent of the world's primary energy

---

[4] See Chapter 9.

[5] See Chapter 12.

sources and 63% of electricity generation is supplied by fossil fuels. Reducing $CO_2$ emissions to the atmosphere simply means lowering the rate of consumption of fossil fuels. Burning less fossil fuels or replacing them by other energy sources would involve a radical change in our energy supply structure. Traditionally, population and economic growth has always been associated with an increase of fossil fuel usage, not the inverse. Providing a burgeoning world population with energy and bridging the energy consumption inequalities that now exist between developed and developing nations would normally require more fossil energy consumption, not less. There are no clear-cut solutions to this dilemma.

In this book we described some of the alternatives to increased fossil fuel consumption, such as replacing fossil-fueled electricity production by nuclear and renewable sources, energy efficiency improvements in fossil fuel usage, and sequestration methods for carbon dioxide. Most of these alternatives require much greater capital investments and/or higher prices for commodities, including electricity, that use primary energy for their production. These measures imply great changes in how energy is supplied and utilized and will undoubtedly require some level of government intervention in the energy marketplace.

### 13.3.1  Coping with Climate Change

As a consequence of international agreements, the industrialized nations of the world are considering programs that would reduce GHG emissions, of which the major component is carbon dioxide, with the goal of reducing them to only a fraction of their 1990 value by the year 2050. The purpose of the rollback is twofold: (a) to prevent the increase of GHG to a level that would produce irreversible climate effects that could not be undone by future emissions control and (b) to make possible further reductions to levels that would restore the climate to a sustainable level for the indefinite future, while still providing adequate global energy supplies.

The first of these goals is a formidable task, especially given the necessity to reduce radically emissions by 2050 so as to bring about a reversal in global warming rather than simply a reduction in the rate of increase of warming. The second is equally difficult and more uncertain of attainment, as it lies beyond the horizon of prediction for current economies and technology. Attention has properly focused on the first goal as a necessary but not sufficient step in achieving the goal of preventing damage to the global environment.

The technologies that can contribute to substantial reductions in GHG emissions are all very capital intensive. To completely replace the energy infrastructure of industrial economies by 2050 will require an annual capital investment over that period that is a significant component of the national economies, as measured by the GDP. This is in sharp contrast to the much smaller effects of limiting local or even regional environmental damage, where abatement costs are relatively small compared with national economic output. The planning of climate change control programs inevitably must be treated as a component of national and global economic growth.

In the United States, macroeconomic modeling of desirable GHG emission reduction scenarios has become the standard for examining possible strategies. These commonly involve a regulatory system in which economic incentives motivate energy consumers and producers to utilize more expensive technologies that meet energy needs while reducing GHG emissions per unit of energy consumed. The mechanism used is that national governments would impose a tax on emissions or a cap on emissions through emission allowances that must be purchased by emitters from a national pool of limited size (called *cap and trade*). Either scheme imposes a cost on emissions that impels emitters to adopt low-emission technology that will lower their annual cost of energy consumption by avoiding the GHG tax, which inevitably will increase over time as greater reductions are mandated.

Consider, for example, the production of electricity by a pulverized coal power plant. One possible way to reduce its GHG emissions (which are predominantly $CO_2$ emissions) is to replace the plant by a nuclear power plant or a wind farm. The capital costs of these alternatives are much higher than that of the coal plant so that their production cost of electricity is greater than that of the coal plant, even when accounting for the lower cost of fuel burned. But if a sufficiently large $CO_2$ emission tax is imposed on the coal plant, its cost of electricity production will exceed that of the alternatives, and the production cost of the latter will become an advantage to replace the old technology with the new. The average cost of electricity will increase as low- or zero-emission plants come on line. Consumers of electric energy will replace old consumption technology with more efficient new technology so as to offset their annual electricity bill increase, despite the increased cost of the newer, but more energy-efficient technology. Inevitably, as the GHG emission tax is increased, energy use will become a larger fraction of the GDP.

An example of a cap-and-trade scheme for reducing U.S. GHG emissions from all energy sectors is illustrated in Figure 13.1 for the years 2005–2050. The methodology incorporates economic models of all energy-producing or -consuming technologies (power plants, automobiles, material processing, etc.) now in use or expected to be available in the time frame 2005–2050, together with their economic costs, capital, and operating. The model determines the least costly path for reaching a goal of 50% GHG emissions reduction below the level of 1990 by 2050. Along this time path, the model determines the marginal cost of each unit of emission reduction ($/ton $CO_2$ equivalent). Of course, the model accounts for many more details than are shown in Figure 13.1,

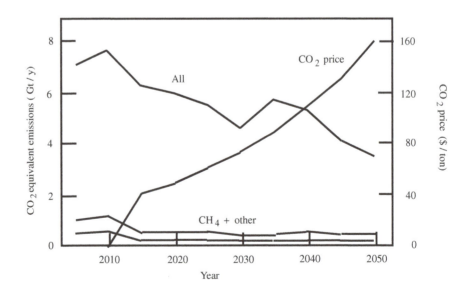

**FIGURE 13.1**    An illustration of a cap-and-trade greenhouse gas emission reduction plan for the United States for the period 2005–2050, showing the components of greenhouse gas emission rate on the left ordinate and $CO_2$ emission price on the right ordinate. (Data from Stavins, R. *A U.S. Cap-and-Trade System to Address Global Climate Change*. Washington, DC: The Brookings Institution, 2007.)

such as the consumption rate of various fossil fuels and the distribution of electric power production among all types of electric power plants.

In Figure 13.1, three curves of GHG equivalents are shown for combinations of $CO_2$, $CH_4$, and "other" GHG (CFC, $N_2O$, etc.). The uppermost curve, labeled "all," is the sum of all GHG; the middle curve, labeled "$CH_4$ + other," is the non-$CO_2$ contribution, so the difference between the upper and middle curves is the $CO_2$ contribution. The lowest curve, unlabeled, is the contribution from "other." Approximately, the proportional contributions of $CO_2$, $CH_4$, and "other" are 90, 5, and 5 %. This information, along with similar models for other nations, can be used to model the concentration of these GHG in the atmosphere in the years 2005–2050 and thereby the future changes in atmospheric temperature ("global warming") and climactic effects ("global climate change").

The curve "$CO_2$ price" in Figure 13.1 is the price (in units of \$/ton $CO_2$ equivalent) of each unit reduction in emissions along the path from 2005–2050.[6] It measures the least costly way of replacing old technology with new technology, more costly but lower emitting, which will bring about the required reduction in GHG emissions. Notice that this marginal cost increases approximately linearly with time as the annual emission rate declines. The annual cost of abatement is the product of the marginal cost and the emission rate, reaching a maximum in 2050 of about \$1 trillion per year in 2050. This would be about 1–2% of the expected U.S. GDP in 2050. GHG control will take a bigger bite of the GDP as emissions are reduced.

## 13.4  CONCLUSION

In conclusion, while urban and regional environmental pollution is still a major problem, especially in developing nations, the experience of industrialized nations has shown that it is technically and economically solvable and politically manageable on the time scale of several decades. It remains to be seen whether the much greater problem of global warming can be solved with comparable measures sustained over the next centuries.

---

[6]Prices are given in terms of 2005 dollars.

# APPENDIX A: MEASURING ENERGY

Energy in materials, such as fossil or nuclear fuels, and electrical energy delivered by power lines are commodities in industrialized economies that are traded in the marketplace. There must be standards for measurement of the energy content and other pertinent properties of these commodities for this market to operate efficiently. These standards of measurement are derived from the development of modern science and technology, where agreement among researchers about how to quantify the results of their experiments is essential to continued scientific progress.

By common agreement among scientists of all nations, a system of units of measurement has been selected: the International System of Units, or SI for short. The SI system defines seven base units of measurement that are mutually independent of each other, it being impossible to measure one unit in terms of any other. Furthermore, all other physical quantities may be measured in terms of one or more of these units. The magnitudes of the base units have been arbitrarily chosen, but are clearly defined by agreement. The defined base units are the meter (length), kilogram (mass), second (time), ampere (electric current), kelvin (thermodynamic temperature), mole (amount of substance), and candela (luminous intensity). The first six of these are used in this text. They are listed in the first section of Table A.1 of this Appendix, "Base units," together with their abbreviated symbols.[1]

There are many physical quantities that arise in scientific studies for which it is useful to define a unit of measurement that is derived from the base units by a well-known physical law. For example, the SI unit of force, the newton, is defined as the magnitude of force that, when applied to a 1 kg mass, will cause the mass to experience an acceleration of 1 m/s$^2$. By Newton's law of motion (force = mass × acceleration), 1 newton must equal 1 kg m/s$^2$. For the physical and chemical quantities of interest in this book, Table A.1 lists these derived units in its second section, "Derived units."

---

[1] Units named after a scientist (newton, ampere, kelvin, etc.) are not capitalized when spelled out but their abbreviations (N, A, K, etc.) are capitalized.

**TABLE A.1**   SI Units

| Measurement | Unit | Symbol | SI base unit value [a] |
|---|---|---|---|
| **Base units** | | | |
| Length | meter | m | m |
| Mass | kilogram | kg | kg |
| Time | second | s | s |
| Electric current | ampere | A | A |
| Thermodynamic temperature | kelvin | K | K |
| Amount of substance | mole | mol | mol |
| **Derived units** | | | |
| Plane angle | radian | rad | 1 |
| Solid angle | stearadian | st | 1 |
| Frequency | hertz | Hz | $1/s$ |
| Force | newton | N | $kg\,m/s^2$ |
| Pressure | pascal | Pa | $N/m^2 = kg/m\,s^2$ |
| Energy | joule | J | $Nm = kg\,m^2/s^2$ |
| Power | watt | W | $J/s = kg\,m^2/s^3$ |
| Electric charge | coulomb | C | $A\,s$ |
| Electric potential | volt | V | $W/A = kg\,m^2/A\,s^3$ |
| Electric capacitance | farad | F | $C/V = A^2\,s^4/kg\,m^2$ |
| Electric resistance | ohm | Ω | $V/A = m^2\,kg/A^2\,s^3$ |
| Magnetic flux | weber | Wb | $V\,s = kg\,m^2/A\,s^2$ |
| Magnetic flux density | tesla | T | $Wb/m^2 = kg/A\,s^2$ |
| Inductance | henry | H | $Wb/A = kg\,m^2/A^2\,s^2$ |
| Activity (radionuclide) | becquerel | Bq | $1/s$ |
| Absorbed dose | gray | Gy | $J/kg = m^2/s^2$ |
| Dose equivalent | sievert | Sv | $J/kg = m^2/s^2$ |
| **Defined units** [a] | | | |
| Plane angle | degree (angular) | ° | $(\pi/180)$ rad |
| Length | nautical mile | nmile | **1,852** m |
| Speed | knot (nmile/h) | kt | **0.51444** m/s |
| Area | hectare | ha | $10^4$ m$^2$ |
| Volume | liter | L | **1E(−3)** m$^2$ |
| Mass | ton (metric) | ton | **1E(3)** kg |
| Pressure | bar | bar | **1E(5)** Pa |
| Time | minute | min | **60** s |
| Time | hour | h | **60** min = **3,600** s |
| Time | day | d | **24** h = **86,400** s |
| Time | year (365 days) | y | **3.1536 E(7)** s |
| Temperature | centigrade | °C | K |
| Activity (radionuclide) | curie | Ci | **3.7E(10)** Bq |
| Absorbed dose | rad | rd | **1E(−2)** Gy |
| Dose equivalent | rem | rem | **1E(−2)** Sv |
| Energy | calorie (Int. table) | $cal_{IT}$ | **4.1868** J |

[a] Boldface values are exact.

Among the derived SI units, the unit of energy is the joule (J) and that of power, the time rate of energy use, is the watt (W), which equals 1 joule per second (J/s). In terms of mechanical units, a joule equals 1 newton meter (Nm), or unit force times unit distance, and a watt is 1 newton meter per second (Nm/s), or unit force times unit velocity. In terms of common electrical units, a joule equals 1 volt ampere second (VAs), or unit charge times unit electric potential, and a watt equals 1 volt ampere (VA), or unit current times unit electric potential.

For various practical reasons, including the needs of commerce and historical usage that preceded modern science, additional SI units have been defined. The ones pertinent to this text are listed in the third section of Table A.1, labeled "Defined units." Among these we note the nautical mile, which is the distance along the earth's surface corresponding to a minute of latitude, and the velocity of a knot, or 1 nautical mile per hour, the usual unit of wind speed. The unit for small geographic areas is the hectare (ha), and the laboratory scale measure of volume is the liter (L). Large masses are usually measured in metric tons (ton). The chemist's unit of energy is the calorie, which equals the amount of heat required to warm a gram of water by 1 degree centigrade.

Unlike all the other industrialized nations, the United States has not adopted the metric system of measurement for domestic and commercial purposes, but continues with the system of units it inherited from England. To facilitate conversion to SI units, the values of pertinent U.S. commercial units are listed in Table A.2.

Many of these units are commonly used in the United States. For example, the gallon (gal) is the unit volume for retail sales of vehicle fuel, whereas the barrel is the preferred unit volume of international petroleum suppliers and refiners. The unit of force is the pound force (lbf), while the unit of mass is the pound mass (lbm). For large amounts of mass, the short ton is used, to be distinguished from the metric ton. The energy unit is the British thermal unit (Btu), the amount of heat required to warm a pound mass of water by 1 degree Fahrenheit. The unit of power is the horsepower (hp). A common pressure unit is the pound force per square inch (psi).

**TABLE A.2**  U.S. Commercial Units

| Quantity | Unit | Symbol | SI unit value [a] |
|---|---|---|---|
| Length | inch | in | **2.54E(−2)** m |
| Length | foot | ft | **3.048E(−1)** m |
| Length | mile (statute) | mile | 5280 ft = 1.609E(3) m |
| Area | acre | | 4.0469E(3) $m^2$ |
| Volume | gallon (US) | gal | 3.7854E(−3) $m^3$ |
| Volume | barrel | bbl | 42 gal (US) = 1.5899E(−1) $m^3$ |
| Force | pound (force) | lbf | 4.448 N |
| Mass | pound (mass) | lbm | 4.5359E(−1) kg |
| Mass | ton (short) | | 2000 lbm = 9.07185E(2) kg |
| Energy | British thermal unit | Btu | 1.05506E(3) J |
| Energy | Quad (1E(15) Btu) | Q | 1.05506E(18) J |
| Energy | therm | therm | 1.05506E(8) J |
| Power | horsepower | hp | **7.46 E(2)** W |
| Pressure | pound (force)/square inch | psi | 6.895E(3) Pa |
| Temperature | Fahrenheit | °F | **(5/9)** K |

[a] Boldface values are exact.

The conversion of quantities from one system of units is straitforward. For example, to convert $x$ Btu/lbm to SI units of J/kg, multiply by the conversion factors from Table A.2:

$$x\,\text{Btu/lbm} = x\left(\frac{\text{Btu}}{\text{lbm}}\right)\left(\frac{1.05506\text{E}(3)\,\text{J}}{\text{Btu}}\right)\left(\frac{\text{lbm}}{4.5359\text{E}(-1)\,\text{kg}}\right) = 2.326\text{E}(3)\,x\,\text{J/kg}.$$

In explaining the technology of energy systems in this text we need to invoke the principles of physics, chemistry, and thermodynamics. The quantitative application of these sciences requires the measurement of certain properties of atoms and molecules that are used in defining important physical constants. Some of these that we use are listed in Table A.3. Among these are the electron

**TABLE A.3**  Measured Quantities

| Quantity | Symbol | SI unit value |
|---|---|---|
| Charge of electron | e | $1.6030\text{E}(-19)$ C |
| Electron volt | eV | $1.6030\text{E}(-19)$ J |
| Faraday | $\mathcal{F}$ | $9.6485\text{E}(4)$ C/mol |
| Planck's constant | h | $6.6256\text{E}(-34)$ Js |
| Boltzmann's constant | k | $1.3805\text{E}(-23)$ J/K |
| Stefan–Boltzmann constant | $\sigma$ | $5.6704\text{E}(-8)$ W/m$^2$ K$^4$ |
| Unified atomic mass unit | amu | $1.66054\text{E}(-27)$ kg |
| Avogadro's number | $N_0$ | $6.0221\text{E}(23)$/mol |
| Universal gas constant | R | $8.3143\text{E}(3)$ J/kg K |
| Standard gravitational acceleration | g | $9.80665$ m/s$^2$ |
| Standard atmospheric pressure | | $1.01325\text{E}(5)$ Pa |
| Melting point of ice ($0°\text{C} = 32°\text{F}$) | | $273.15$ K |

**TABLE A.4**  SI Unit Prefixes

| Factor | Prefix | Symbol | U.S. word modifier |
|---|---|---|---|
| 1E(18) | exa | E | |
| 1E(15) | peta | P | Quadrillion |
| 1E(12) | tera | T | Trillion |
| 1E(9) | giga | G | Billion |
| 1E(6) | mega | M | Million |
| 1E(3) | kilo | k | Thousand |
| 1E(2) | hecto | h | Hundred |
| 1E(−1) | deci | d | |
| 1E(−2) | centi | c | Percent |
| 1E(−3) | milli | m | |
| 1E(−6) | micro | $\mu$ | |
| 1E(−9) | nano | n | |
| 1E(−12) | pico | p | |

volt (eV), the Faraday ($\mathcal{F}$), Avogadro's number ($N_0$), and the universal gas constant (R), used in analyzing electrochemical processes.

The SI units of Table A.1 are often of inconvenient size. Just as paper currency comes in different denominations, physical quantities need to have different sizes to accommodate different uses. The SI system includes the use of prefixes to change the size of units by factors of 10, up or down—for example, kilometer (km), centimeter (cm), micrometer ($\mu$m). Table A.4 lists the SI unit prefixes that cover a range of 30 orders of magnitude, enough for most practical purposes. Occasionally we use practical, if not always logical, units such as the unit of electrical energy, the kilowatt hour (kWh), which equals 3.6 megajoules (MJ). The kilowatt hour, an amount of energy that will light a kilowatt bulb for 1 hour, is a better unit for commercial use than the megajoule.

# INDEX

Acid deposition, 121, 125
  acidity, 290
  modeling, 293
    source apportionment, 294
Acid mine drainage, 298
Adiabatic combustion
    temperature, 46, 58, 68
Adiabatic process, 40
Advanced cycles, 129
Aerodynamic drag, 244, 251, 252
Aerosols, 275
Afforestation, 293
Air pollutants, 6, 272f
  haze and visibility
      impairment, 295
  health and environmental
      effects, 277
  photochemical smog, 12
  primary and secondary, 277
Air pollution, 272
  meteorology, 279
  modeling, 281
    area source, 284
    Gaussian plume, 282
    line source, 284
    plume rise, 283
  standards, 275
    ambient, 275
    emission, 273
    U.S. ambient, 275
    U.S. vehicle, 274

Air quality modeling, 281
Air separation unit, 130
Albedo, 309
Alternating current. *See* Current,
    alternating
Alternator, 85
Ammonia slip, 127
Anode, 76
Armature, 89
Atmospheric deposition, 300
Automotive transportation, 157.
    *See also* Highway vehicles
  energy, 5, 8
Availability, 47
Avogadro's number, 77

Baghouse, 120
Base load, 104
Battery
  energy efficiency, 100
  properties, 100t
  storage, 96
    lead-acid, 97
Best available control
    technology, 117, 275
Biochemical synthesis, 73
  ethanol, 74
Biofuels, 169–172
Biomass
  energy, 162, 168, 345
  fuels, 69

environmental effects, 172
  photosynthesis, 169
Bitumen, 29
Black body radiation, 308
Boiler, 49, 108f
Bottom ash, 106
Brayton cycle, 52
Breeding ratio, 152
Briggs plume rise, 283
Burner, 106f

Capacitor, 94
Carbon cycle, 319
Carbon dioxide, 319f
  capture, 330f
  concentration, 313, 320
  control, 329
  hydrate, 342
  sequestration, 337f
  utilization, 344
Carbon emissions, 17, 329
  control, 329
Carnot cycle, 45
Catalytic converter, 263
Cathode, 76
Chain reaction, 138
Chemical absorption, 330
Chlorofluorocarbons, 309, 326
Claus process, 122, 131, 334
Clausius inequality, 40
Clean Air Act Amendments, 273
Climate change, 305, 318

Coal, 26f
  anthracite, 26
  ash, 90
  bituminous, 26
  characteristics, 26
  gasification, 104, 333
  heating values, 26
  lignite, 26
  reserves, 26
  resources, 27
  washing, 106, 121
Coefficient of performance, 56
Cogeneration, 86, 132
Collector. *See* Solar energy
Combined cycle, 52, 54, 129, 130
Combustion, 67–69
  adiabatic temperature, 46, 58, 68
  chamber, 52
  power plant, 90
  products, 68
  reactants, 68
  reciprocating engines, 193
  thermodynamic properties, 66t
Compression ratio, 51
Compressor, 52
Condenser
  refrigeration, 55
  steam, 48, 96
Control rods, 146
Coolant, reactor, 145
Cooling tower, 113
Coriolis force, 318
Corona discharge, 100
Criteria pollutants, 276
Critical point, 49
Current, electric
  alternating, 91, 92, 98
  direct, 92, 98
Current, thermohaline, 318

DeLaval nozzle, 110
Deutsch equation, 118
Dispersion modeling, 281
District heating, 116
Dry adiabatic lapse rate, 280
Dry deposition, 290
Dry scrubber, 124

Earth radiance, 307
Economizer, 108

Efficiency
  electrical, 91
  fuel, 203–208
  thermodynamic, 46
    fuel cell, 78
El Niño, 319
Electric
  capacitance, 94
  current. *See* Current, electric
  efficiency, 91
  field, 89
  generator, 88
    induction, 76
  inverter, 93
  motor, 88
  permittivity, 94
  potential, 90
  power, 9
    demand, 87
    transmission, 92
  rectifier, 93
  resistance, 91
Electrochemical cell, 76
  battery, 75
  fuel cell. *See* Fuel cell
Electrochemical energy storage, 81–83
Electrochemical reactions, 74
Electrode, 76
Electrolyte, 76, 81
Electrostatic precipitator, 117
Emissions, 273–274
  control, 116f
  standards, 273
    stationary sources, 274
    mobile sources, 274
Empirical kinetic modeling approach (EKMA), 288
Energy, 7, 35–37
  chemical, 36
  conservation, 7, 36, 283
  consumption. *See* Energy consumption
  electric, 37
  gravitational, 100
  internal, 36
  kinetic, 35
  magnetic, 37
  nuclear, 32, 138
  potential, 35
  renewable. *See* Renewable energy
  sources. *See* Energy supply
  storage, 94
    efficiency, 100t

  electrochemical, 81–83
  electrostatic, 94
  flywheel, 100
  hydropower, 99
  magnetic, 95
  mechanical, 99
  properties, 100, 100t
  supply. *See* Energy supply
  thermodynamic, 35
  transportation, 5, 10
  total, 37
Energy consumption, 15f
  electricity, 20
    U.S., 22
  global, 7, 12, 15
  U.S., 15, 16
  U.S. commercial, 24
  U.S. industrial, 22
  U.S. residential, 24
  U.S. transportation, 25
Energy supply
  coal, 26
  global, 26
  hydro-electric, 26–31
  natural gas, 29
  nuclear, 26
  petroleum, 27
  renewable, 26
Enhanced oil recovery, 237
Enthalpy, 41
Entropy, 40
Environment, 11
Evaporator, 55
Extinction coefficient, 296

Faraday constant, 77, 82
Feed water, 108
Fertile isotope, 140
First law of thermodynamics, 34, 39
Fissile isotope, 140
Fission, 128
Flame, 106
  temperature, 107
  turbulent, 107
Flue gas, 107
  denitrification, 107, 275
  desulfurization, 122, 275, 293
Fly ash, 106, 299
Forest burning, 17
Fossil fuel, 64
  combustion, 67
    stoichiometric ratio, 67

synthesis, 71
  synthesis gas, 72
Fossil fuel power plant, 104–136
  advanced cycles, 120–133
    coal gasification combined
      cycle, 130
    cogeneration, 86, 132
    combined cycle, 52, 54, 129
    fuel cell, 75–80, 132
  boiler, 48, 108
    superheater, 50, 109
  burner, 106
  condenser, 48, 112
  cooling tower, 113, 299
  emission control, 116f
    electrostatic precipitator,
      119
    nitrogen oxide, 125
    particles, 117
    sulfur, 121
  flue gas desulfurization, 123,
    275, 299
  fuel storage and preparation,
    106
  gas turbine, 30, 52, 111, 129
  generator, 114
  low-$NO_x$ burner, 125
  selective reduction, 127
  steam turbine, 48, 109
  waste disposal, 129
Free energy, 42, 59, 75
  Gibbs, 42
Fuel consumption, specific, 62
Fuel economy. *See* Highway
  vehicles
  vehicle, 248–253
Fuel cell, 75–80, 132
  anode, 76
  cathode, 76
  electrode, 76
  electrolyte, 76
Fuel heating value, 68, 66t
  higher, 26, 69
  lower, 68, 243
Fuel, synthetic, 70, 71t
  production efficiency, 71t
Fuel (thermal) efficiency, 60
Fuel rod, nuclear, 145
Fusion, 128, 156f

Gas turbine, 30, 52, 111, 129
  combined cycle, 20, 129
Gaseous diffusion, 155
Gaussian plume, 282

Generator, electric, 85, 88
Geopressurized methane, 30
Geothermal energy, 162,
  173–176
  environmental effects, 176
  heat pump, 175–176
  installed power, 174t
  power plant, 153
  resource, 175
Global warming, 300, 305f
  climate change, 305f
  effects, 317
  modeling, 311
  observed trend, 316
  potential, 312
  radiative forcing, 313
Greenhouse effect, 306f
Greenhouse gas, 5, 305, 313
  emissions, 319f
Gibbs free energy. *See* Free
  energy

Halocarbons, 326
Haze and visibility impairment,
  121, 278, 295
Heat, 37
  exchange, 43–44, 109
  interaction, 39
  pump, 56, 176
    geothermal, 176
  transfer, 39, 43–44
    coefficient, 44
Heat engine, 44
  combined cycle, 54
  ideal cycles, 45–56
    Brayton, 52
    Carnot, 45
    Otto, 50
    Rankine, 47
  thermodynamic efficiency, 45
Heating
  regenerative feed water, 50,
    108
  reheating, 50
  superheating, 50, 109
Highway vehicles, 234, 234t
  aerodynamic drag, 244, 252,
    257
  catalytic converters, 263
  characteristics, 242, 243t
  electric drive, 254–259
    battery-powered, 254–255,
      254t
    fuel cell, 256–259, 258t

  hybrid, 255
  emissions, 235, 259–267
    Federal test procedure, 259
    reduction, 261–267
    U.S. emission standards,
      259–261, 216t
  engine performance, 252
  evaporative emissions, 265
  fuel cell, 256–259
    characteristics, 258t
  fuel consumption, 235t
  fuel economy, 251–253
    corporate average fuel
      economy,
      248–251, 249f
    driving cycles, 250, 250f
  fuel efficiency, 248–253
  hybrid, 255
    characteristics, 243t
  internal combustion engine.
    *See* Internal combustion
    engine
  mass, 251
  passenger vehicle
    characteristics, 243t
  power and performance,
    244–248
  rolling resistance, 245,
    251–252
  transmission, 246
  vehicle use, 235, 235t
Hurricane, 318
Hydrocarbons, 65
Hydrogen economy, 80
  energy storage, 81
  synthetic hydrogen, 81
  vehicle propulsion, 81
Hydropower, 86, 164
  development, 146t
  environmental effects, 167
  hydroturbines, 166
  pumped storage, 88, 89
    efficiency, 100t

Inductance, 96
Induction, 95
Industrial pollution, 13
Integrated gasification combined
  cycle, 130
  with carbon capture, 333
Internal combustion engine, 45,
  233, 236–238
  brake mean effective pressure,
    241

Internal combustion
    engine (*Cont.*)
    brake power, 240
    brake specific fuel
        consumption, 242
    combustion, 230
    compression ignition, 236
    efficiency, 242–244, 244f
    engine displacement, 241
    four-stroke cycle, 237, 237f
    nitric oxide formation, 239
    power and performance,
        240–244
    spark ignition, 236
    thermal efficiency, 242
    two-stroke cycle, 237
Inversion, 279
Ionizing radiation, 141
Ionosphere, 309
Irradiance, 308. *See also* Solar
    energy, irradiance
Isotope, 138

Land pollution, 301
Laser enrichment, 155
Lead pollution, 300
Lighting
    fluorescent, 24
    incandescent, 24
Liquefied natural gas, 106
Linear-no-threshold hypothesis,
    145

Magnetic
    energy, 95
    field, 89
Magnetic (*Cont.*)
    inductance, 96
    induction, 95
    inductor, 95
    permeability, 95
Marketable permits, 275
Mass deficit, 138–139
Maximum achievable control
    technology, 275
Mechanical energy storage, 99
Membrane separation, 335
Methane, 29, 320, 327
    hydrate, 30
Mercury, 128
    emissions, 128
    deposition, 300
Mesosphere, 311

Mixing layer, 279
Moderator, 145
    beryllium, 146
    graphite, 146
    heavy water, 146
    light water, 146
Motor, electric, 88

National Ambient Air Quality
    Standard,
    277
Natural gas, 29f
    composition, 29
    compressed, 90
    heating value, 30
    life time, 31
    liquefied, 106
    reserves, 30
    resources, 30
    unconventional, 30
Neutron, 138f
    absorption, 146
    economy, 146
    scattering, 146
    thermal, 146
Nitrogen dioxide, 286
Nitric oxide, 261–263
Nitrogen oxides, 286
    control, 125f
    fuel $NO_x$, 125, 286
    thermal $NO_x$, 125, 286
Nitrous oxide, 314, 322, 327
Nuclear energy, 137–159
    chain reaction, 138
    fertile isotope, 140
    fissile isotope, 140
    fission, 140
    fusion, 138, 156
        laser fusion, 158
        magnetic confinement, 157
Nuclear fuel cycle, 153
    enrichment, 154
    spent fuel reprocessing, 155
    waste disposal, 156
    waste storage, 156
Nuclear reactor, 145f
    boiling water, 147
    breeder, 151
        breeding ratio, 152
    CANDU, 150
    control rods, 145
    coolant, 145
    critical, 146
    fuel rods, 145

    gas cooled, 151
    moderator, 146
    pressurized water, 149
Nucleons, 138

Ocean sequestration, 341
Ocean thermal power, 163,
    225–226
Ocean wave power, 163,
    216–224
    environmental impacts, 224
    farms, 224
    probability distribution, 219
    systems, 220–224
Oil desulfurization, 122
Oil sands, 29
Oil shale, 29
Oil spills, 298
Open cycle, 50
Otto cycle, 50
Oxyfuel combustion, 332
Ozone, 109, 285f, 305, 314
    hole, 286
    stratosphere, 286
    troposphere, 286

Permeability, magnetic, 95
Photochemical smog, 10
Photo-oxidants, 285f
    modeling, 287
        Empirical Kinetic Modeling
            Approach, 288
        Regional Oxidant Model,
            289
Photovoltaic cell, 162, 185–190
    efficiency, 190
    photovoltaic farms, 190, 191t
Particulate matter, 117f
    control, 117
    emission standards, 274
Pasquill-Gifford stability, 280
Petroleum, 27f
    composition, 28
    crude, 28
    heating value, 28
    life time, 28
    reserves, 27
    resources, 27
    unconventional, 29
Physical absorbtion, 334
Plutonium, 140
Pollutant, 272f
    criteria, 276

toxic, 276
Polycyclic aromatic
    hydrocarbons, 300
Power plant
    efficiency, 105
    fossil fuel *See* Fossil fuel
        power plant
    nuclear, 3. *See also* Nuclear
        reactor
    steam, 3. *See also* Fossil fuel
        power plant
Pressure swing absorption, 335
Pumped storage, 72
Pyritic sulfur, 106

Quenching, 131

Radiance, 307
Radiation, 141f
    alpha, 141
    beta, 141
    biological effects, 144
    dosage, 143
        gray, 144
        rad, 144
    dose equivalent, 144
        rem, 144
        sievert, 144
    gamma, 142
    infrared, 307
    ionizing, 141, 144
    protection, 144
Radiative forcing, 313
Radiative height, 312
Radiative temperature, 310
Radioactivity, 141f
    decay rate, 142
    half life, 142
    isotope, 143
    units, 143
        becquerel, 143
        curie, 143
Radon, 153
Rankine cycle, 47, 104
Receptor modeling, 281
Refrigeration, 24, 55
Regional haze, 295
Regulating energy use, 350, 354t
Reheating, 49
Renewable energy, 5, 8, 162
    biomass. *See* Biomass energy
    capacity factor, 164
    capital cost, 226–228, 227t

effectiveness, 164
energy flux, 164t
geothermal. *See* Geothermal
    energy
hydropower. *See* Hydropower
ocean thermal. *See* Ocean
    thermal power
ocean tidal. *See* Tidal power
ocean wave. *See* Ocean wave
    power
photovoltaic. *See* Photovoltaic
    cell
production, 163t
solar thermal. *See* Solar
    energy
wind. *See* Wind power
Rolling resistance, 200

Sea level, 317
Second law of thermodynamics,
    34, 40
Selective catalytic reduction,
    127
Selective non-catalytic
    reduction, 127
Sequestration
    carbon dioxide, 337f
    coal seams, 338
    deep ocean, 341
    deep sedimentary basins, 329
    oil and gas reservoirs, 337
SI units, 357t
    prefixes, 359t
Solar energy, 176
    clear sky irradiance, 179t
    flat plate collector, 181
        efficiency, 182
        heat transfer coefficient, 184
    focusing collector, 183
    irradiance, 176, 177f, 179t,
        307
        beam, diffuse, 178
    spectral distribution, 156f
Solar spectrum, 177f, 307
    irradiance, 307
Solar thermal farms, 185, 186t
Solid waste, 129, 299
Soot, 273, 301
Sorbent injection, 123
Source-receptor modeling, 281
Specific fuel consumption, 61
Specific heat
    constant-pressure, 42
    constant-volume, 42

Stability categories, 280
Stack height, 281
Stack plume, 281
Staged combustion, 125
State implementation plan, 276
Steady flow, 43
Steam power plant, 108f
Stoichiometric ratio, 41, 107
Stratosphere, 286, 311
Sulfur oxides, 121f
    emission control, 116, 274
    emission rate, 274
    emission standard, 274
Superheating, 50, 109
Synthetic fuel, 70
    efficiency, 71t

Temperature
    absolute, 40
    adiabatic combustion, 46, 58,
        59
    radiative, 310
    surface, 310
Temperature swing absorption,
    335
Terrestrial radiation, 308
Thermal pollution, 113, 299
Thermodynamics, 34
    efficiency, 46
        fuel cell, 75
        heat engine, 46
    laws, 34
        first law, 34, 39
        second law, 34, 40
    properties, 41
        extensive, 41
        fuel combustion, 44t
        intensive, 41
        specific extensive, 41
    state variables, 36
Thermohaline current, 318
Thermosphere, 311
Thorium, 137
Tidal current power, 214, 215f
Tidal power, 162, 210, 211t
    capacity factor, 213
    effectiveness, 217
    environmental effects, 216
    ideal power, 212
    plant characteristics, 212t
    tidal period, 211
Toxic pollutants, 275, 300
Transformer, 92
Troposphere, 286, 311

Turbine
    gas, 52, 111
    steam, 48, 109
    wind. *See* Wind turbine

Uranium, 153f
    hexafluoride, 153
    oxide, 146, 153
    yellow cake, 153
Urban airshed model, 289
U.S. commercial units, 305t

Vapor compression cycle, 55
Visibility impairment, 121, 295
Volatile organic compounds, 287

Water gas shift reaction, 333
Water pollution, 297

acid mine drainage, 298
atmospheric deposition to
    surface
    waters, 300
coal washing, 298
thermal pollution, 299
Wet deposition, 290
Wet scrubber, 123
Wind advection, 279
Wind statistics, 234
    variability, 202
Wind power, 162, 192
    capacity factor, 200
    effectiveness, 202
    energy flux, 198
    environmental effects, 210
farm, 205
    array, 209
    footprint, 206
    network integration, 208

probability distribution, 189,
    200f
resources, 198, 199f
turbine, 193
    aerodynamics, 193
    power, 193
    power coefficient, 195
Wind turbine, 193. *Also see*
    Wind power
    capital cost, 205
    components, 196
    economical design, 204
    properties, 198t
Work, 33
    interaction, 37

X-rays, 141

Zircalloy, 146